Praise for Larry Gonick and His Work

"MY GLANCE THROUGH GONICK'S *CARTOON GUIDE TO STATISTICS* BEGAN WITH PROFESSIONAL SKEPTICISM, AND ENDED UP WITH ITS ADOPTION AS THE ONLY REFERENCE TEXT FOR MY GENERAL EDUCATION COURSE 'REAL-LIFE STATISTICS: YOUR CHANCE FOR HAPPINESS (OR MISERY)'."

—XIAO-LI MENG, CHAIRMAN, STATISTICS DEPARTMENT, HARVARD UNIVERSITY

"SO CONSISTENTLY WITTY AND CLEVER THAT THE READER IS BARELY AWARE OF BEING GIVEN A THOROUGH GROUNDING IN THE SUBJECT."

—*OMNI* MAGAZINE

"GONICK IS ONE OF A KIND."

—*DISCOVER* MAGAZINE

"*THE CARTOON HISTORY OF THE UNIVERSE* IS OBVIOUSLY ONE OF THE GREAT BOOKS OF ALL TIME."

—TERRY JONES, *MONTY PYTHON*

"A MASTERPIECE!"

—STEVE MARTIN

"SUPERB ARTISTRY AND STAND-UP WIT! INSIDIOUSLY DISGUISED AS CARTOON BOOKS, GONICK'S WELL-RESEARCHED AND HILARIOUSLY ILLUSTRATED GRAPHIC TEXTS SHOULD BE IN EVERY LIBRARY. THEY ARE CAPABLE OF MAKING THE DENSEST AND MOST RESISTANT CEREBELLUM ABSORB AND RETAIN REAL INFORMATION! GONICK'S BOOKS ARE FOOD FOR THOUGHT, RICH WITH HUMOR—AND THEY LEAVE YOU WAITING FOR THE NEXT COURSE."

—LYNN JOHNSTON, CREATOR OF *FOR BETTER OR FOR WORSE*

"LARRY GONICK SHOULD GET AN OSCAR FOR HUMOR AND A PULITZER FOR HISTORY."

—RICHARD SAUL WURMAN, CREATOR OF THE TED CONFERENCES

"LARRY GONICK HAS CREATED A GENRE ALL HIS OWN. THE USE OF COMIC ART TO TELL SERIOUS HISTORY IS A BRILLIANT APPLICATION OF THE MEDIUM. THE UNDERLYING SCHOLARSHIP IN THIS WORK REINFORCES AND DEMONSTRATES THE CAPABILITY OF CARTOONS AS A VALID TEACHING FORM. . . . BEST OF ALL, HE IS WEDDING LEARNING WITH FUN. BRAVO!"

—WILL EISNER

"LIKE ANY GOOD HISTORIAN, LARRY GONICK SEASONS HIS FACTS WITH A GOOD DOSE OF PERSPECTIVE, AND LIKE ANY GOOD CARTOONIST, HE MIXES HIS DRAMA WITH A GOOD DOSE OF HUMOR."

—JEFFREY BROWN, AUTHOR OF *CLUMSY* AND *FUNNY MISSHAPEN BODY*

"ON WINTRY EVENINGS WHEN MY MOOD NEEDS CHEERING, I CURL UP WITH LARRY GONICK'S *CARTOON HISTORY OF THE UNIVERSE*. GONICK'S DRAWINGS AND TEXTS ARE SO IRREVERENT, SO UNABASHEDLY CYNICAL (YET MORE INFORMATIVE THAN MANY A 'SERIOUS' HISTORY BOOK) THAT I FIND MYSELF WIPING AWAY TEARS OF LAUGHTER, ALL THE WHILE MARVELING AT WHAT KINDS OF TANGLED COMEDIES AND HORRORS IT TOOK FOR HUMANKIND TO EVOLVE INTO GREAT CIVILIZATIONS—YET WITH SO LITTLE 'HUMANIST' PROGRESS! THERE'S NOTHING BETTER FOR RESTORING ONE'S PERSPECTIVE THAN TO BE BOUNCED THROUGH A FEW THOUSAND YEARS OF WAR, LECHERY, AND CUNNING."

—RITA DOVE, FORMER U.S. POET LAUREATE, IN THE *WASHINGTON POST*

"WELLS, TOYNBEE, MCNEILL, DURANT—MOVE OVER! MAKE ROOM FOR A COLLEAGUE—A WITTY CHRONICLER-COMMENTATOR WORKING IN A MEDIUM MORE SUITED THAN PROSE TO EFFICIENTLY WEDDING FACT AND INTERPRETATION. LARRY GONICK'S *CARTOON HISTORY* . . . IS ONE OF THE MOST AMUSING, PROVOCATIVE SURVEYS OF THE PLANET'S PROGRESS EVER MADE. IT'S HARD TO IMAGINE HOW GONICK'S ACHIEVEMENT COULD BE EQUALED, LET ALONE BETTERED." —*BOOKLIST*

"EACH NEW VOLUME OF LARRY GONICK'S EPIC AND IRREVERENT RENDITION OF HISTORY IS INFORMATIVE, FUNNY, AND A TRIUMPH OF THE CARTOONIST'S CRAFT—IN OTHER WORDS, CAUSE FOR CELEBRATION." —CHARLES JOHNSON, WINNER OF THE NATIONAL BOOK AWARD

"GONICK'S TAKE ON HISTORY IS WHIP-SMART, SKEPTICAL ABOUT FAMILIAR BUT QUESTIONABLE STORIES, AND ABSOLUTELY IN COMMAND OF DOZENS OF SIMULTANEOUS HISTORICAL THREADS. HE'S ALSO VERY FUNNY." —*PUBLISHERS WEEKLY*

"[*THE CARTOON HISTORY OF THE MODERN WORLD* IS] INCISIVELY IRREVERENT. . . . WHEN IT COMES TO THE SHEER AUDACITY OF THE LENGTHS TO WHICH HE SEEKS TO TRANSPORT HIS READERS, GEOGRAPHICALLY AND TEMPORALLY, GONICK IS IN A CLASS BY HIMSELF. WHO ELSE COULD CRAM INTO A SINGLE BOOK DETAILS ABOUT PETER THE GREAT'S ADMIRATION FOR THE FRENCH, THE OTTOMAN EMPIRE'S FAILURE TO CAPTURE VIENNA (AND THE SUBSEQUENT INVENTION OF THE CROISSANT), JESUIT ACTIVITIES IN CHINA AND JAPAN, AND THE DUTCH ATTEMPTS TO ALTER AFRICAN CURRENCY SYSTEMS? AND THAT'S JUST ON PAGE 233."

—JEFFREY WASSERSTROM, PROFESSOR OF HISTORY, UNIVERSITY OF CALIFORNIA AT IRVINE, IN *NEWSWEEK*

"NEVER BEFORE HAS SUCH A WEALTH OF PALAEONTOLOGICAL & ANTHROPOLOGICAL INFORMATION BEEN PUBLISHED THAT IS BOTH ACCURATE AND HUMOROUS. I WOULD VERY MUCH LIKE TO BUY A LARGE NUMBER." —R. G. KRUSZYNSKI, SCIENTIFIC OFFICER, BRITISH MUSEUM OF NATURAL HISTORY

GONICK'S *CARTOON HISTORIES* AND *CARTOON GUIDES* HAVE BEEN REQUIRED READING IN COURSES AT BISMARK HIGH SCHOOL, BISMARK, NORTH DAKOTA; BLOOMSBURG UNIVERSITY; BOSTON COLLEGE; BUCKINGHAM BROWN & NICHOLS SCHOOL, CAMBRIDGE, MASSACHUSETTS; CALIFORNIA INSTITUTE OF THE ARTS; CALIFORNIA STATE UNIVERSITY AT CHICO; CARNEGIE-MELLON UNIVERSITY; COLUMBIA UNIVERSITY; CORNELL UNIVERSITY; DARTMOUTH COLLEGE; DUKE UNIVERSITY; GIRVAN ACADEMY, SCOTLAND; HARVARD UNIVERSITY; HUMBOLDT STATE UNIVERSITY; HUNTINGDON COLLEGE; ILLINOIS STATE UNIVERSITY; JOHN JAY COLLEGE; THE JOHNS HOPKINS UNIVERSITY; KENT SCHOOL DISTRICT, KENT, WASHINGTON; KENYON COLLEGE; LANCASTER UNIVERSITY, ENGLAND; LICK-WILMERDING HIGH SCHOOL, SAN FRANCISCO, CALIFORNIA; LIVERPOOL UNIVERSITY, ENGLAND; LOGAN HIGH SCHOOL, LOGAN, UTAH; LONDON SCHOOL OF ECONOMICS; LOUISIANA STATE UNIVERSITY; LOWELL HIGH SCHOOL, SAN FRANCISCO, CALIFORNIA; THE MARIN ACADEMY; MARQUETTE HIGH SCHOOL, CHESTERFIELD, MISSOURI; MIT; NEW YORK UNIVERSITY; NORTH CAROLINA STATE UNIVERSITY; NORTHWESTERN UNIVERSITY; NUEVA SCHOOL, HILLSBOROUGH, CALIFORNIA; OHIO STATE UNIVERSITY; PENNSYLVANIA STATE UNIVERSITY; PHILIPPINE HIGH SCHOOL, DILMAN, PHILIPPINES; REDBUD ACADEMY, AMARILLO, TEXAS; ROCHESTER INSTITUTE OF TECHNOLOGY; RUTGERS UNIVERSITY; SAINT IGNATIUS HIGH SCHOOL, SAN FRANCISCO, CALIFORNIA; SAN DIEGO STATE UNIVERSITY; SAN DIEGO SUPERCOMPUTER CENTER; SOUTHEAST MISSOURI STATE UNIVERSITY; SOUTHWOOD HIGH SCHOOL, SHREVEPORT, LOUISIANA; STANFORD UNIVERSITY; SWARTHMORE COLLEGE; TEMPLE UNIVERSITY; UNIVERSITEIT UTRECHT, NETHERLANDS; UNIVERSITY OF ALABAMA; THE UNIVERSITY OF CALIFORNIA AT BERKELEY, LOS ANGELES, SANTA BARBARA, SANTA CRUZ, AND SAN DIEGO; THE UNIVERSITY OF CHICAGO; THE UNIVERSITY OF EDINBURGH, SCOTLAND; THE UNIVERSITY OF FLORIDA; THE UNIVERSITY OF IDAHO; THE UNIVERSITY OF ILLINOIS; THE UNIVERSITY OF LEICESTER, ENGLAND; THE UNIVERSITY OF MARYLAND; THE UNIVERSITY OF MIAMI, FLORIDA; THE UNIVERSITIES OF MICHIGAN, MISSOURI, NEBRASKA, NEW BRUNSWICK, SCRANTON, SOUTH FLORIDA, TEXAS, TORONTO, WASHINGTON, AND WISCONSIN; YALE UNIVERSITY; AND MANY MORE.

THE CARTOON GUIDE TO
CALCULUS

ALSO BY LARRY GONICK

THE CARTOON HISTORY OF THE UNIVERSE, VOLUMES 1-7
THE CARTOON HISTORY OF THE UNIVERSE, VOLUMES 8-13
THE CARTOON HISTORY OF THE UNIVERSE, VOLUMES 14-19
THE CARTOON HISTORY OF THE MODERN WORLD, PART 1
THE CARTOON HISTORY OF THE MODERN WORLD, PART 2
THE CARTOON HISTORY OF THE UNITED STATES
THE CARTOON GUIDE TO CHEMISTRY (WITH CRAIG CRIDDLE)
THE CARTOON GUIDE TO THE COMPUTER
THE CARTOON GUIDE TO THE ENVIRONMENT (WITH ALICE OUTWATER)
THE CARTOON GUIDE TO GENETICS (WITH MARK WHEELIS)
THE CARTOON GUIDE TO (NON)COMMUNICATION
THE CARTOON GUIDE TO PHYSICS (WITH ART HUFFMAN)
THE CARTOON GUIDE TO SEX (WITH CHRISTINE DEVAULT)
THE CARTOON GUIDE TO STATISTICS (WITH WOOLLCOTT SMITH)
THE ATTACK OF THE SMART PIES

THE CARTOON GUIDE TO
CALCULUS
LARRY GONICK
wm
WILLIAM MORROW
An Imprint of HarperCollinsPublishers

HARPERCOLLINS BOOKS MAY BE PURCHASED FOR EDUCATIONAL, BUSINESS, OR SALES PROMOTIONAL USE. FOR INFORMATION, PLEASE E-MAIL THE SPECIAL MARKETS DEPARTMENT AT SPSALES@HARPERCOLLINS.COM.

INCORPORATING CORRECTIONS TO THE FIRST EDITION

LIBRARY OF CONGRESS CATALOGING-IN-PUBLICATION DATA IS AVAILABLE UPON REQUEST.

ISBN 978-0-06-168909-3

17 18 19 20 RRD 10 9

CONTENTS

Acknowledgments

THE HARVARD MATH DEPARTMENT OF ANOTHER ERA FILLED THE AUTHOR'S HEAD WITH THIS STUFF: JOHN TATE, MY FIRST CALCULUS TEACHER, LYNN LOOMIS, SHLOMO STERNBERG, RAOUL BOTT, DAVID MUMFORD, BARRY MAZUR, ANDREW GLEASON, LARS AHLFORS, AND GEORGE MACKEY, WHOSE SON FOUNDED THE WHOLE FOODS GROCERY CHAIN, SOURCE OF MUCH OF THE CHOCOLATE THAT FUELED THE WRITING OF THIS BOOK. DOWN THE WAY AT MIT, VICTOR GUILLEMIN ADVISED MY NEVER-FINISHED THESIS, AND NAGISETTY RAO FROM THE TATA INSTITUTE IN BOMBAY TAUGHT ME TO APPRECIATE THE "NUTS AND BOLTS" OF ANALYSIS WITHOUT SO MUCH ALGEBRA. MORE RECENTLY, A NUMBER OF PEOPLE HAVE HELPED ME THINK ABOUT CALCULUS AGAIN: JAMES MAGEE VETTED THE FIRST FEW CHAPTERS AND URGED ME TO KEEP CLOSELY TO THE CURRICULUM; SEVERAL VIGOROUS DISCUSSIONS WITH DAVID MUMFORD CLARIFIED QUESTIONS ABOUT RIGOR AND INTUITION; CRAIG BENHAM, ANDREW MOSS, AND MARK WHEELIS ENDURED MY RANTS ABOUT SPEEDOMETERS, PARALLEL AXES, AND VARIOUS RELATED ISSUES. THANKS TO ALL, AND SPECIAL THANKS TO THE PEOPLE WHO CREATED FONTOGRAPHER, THE MARVELOUS PIECE OF SOFTWARE THAT MADE IT POSSIBLE TO DO "HANDWRITTEN" MATHEMATICAL TYPESETTING!

NOTE: THIS EDITION CORRECTS SOME MINOR ERRORS (AND A COUPLE OF LARGER ONES) FOUND IN EARLIER PRINTINGS. IN PARTICULAR, A FEW OF THE PROBLEM SETS HAVE BEEN CHANGED. A LINK TO SOLUTIONS TO SELECTED PROBLEMS (IN PDF) MAY BE FOUND ON THE AUTHOR'S WEB PAGE HTTP://WWW.LARRYGONICK.COM/TITLES/SCIENCE/CARTOON-GUIDE-TO-CALCULUS-2/.

TO DAVID MUMFORD,
MENTOR, BENEFACTOR, AND FRIEND

Initial Conditions

OPEN A FEW CALCULUS BOOKS, AND YOU'LL SEE SOMETHING STRANGE: THEY ALL LOOK ALIKE!!
THE SAME PAGEFULS OF FORMULAS, THE SAME SET OF TOPICS, MOSTLY, AND EVEN THE SAME FONT! THEY ALL WEIGH A TON, TOO...
SIGH... ALL THOSE FORMULAS... CALCULUS IS BASED ON SOME BEAUTIFUL IDEAS, BUT THE PAYOFF IS IN THE FORMULAS! THIS BOOK WILL BE FULL OF 'EM, TOO... SORRY!
ON THE OTHER HAND, I WANT TO USE MY PEN TO EXPLAIN THE BRILLIANT AND ELEGANT THINKING BEHIND ALL THOSE EQUATIONS...
SO HERE AND NOW I OFFER AN IRONCLAD GUARANTEE: THIS CALCULUS BOOK WILL BE DIFFERENT!!!
FOR ONE THING, IT DOESN'T WEIGH VERY MUCH... PLUS, JUST LOOK AT THE FONT!

Chapter -1
Speed, Velocity, Change

BASIC IDEA #1

CALCULUS IS THE MATHEMATICS OF CHANGE, AND CHANGE IS MYSTERIOUS. SOME THINGS GROW IMPERCEPTIBLY... OTHERS ZOOM... HAIR GROWS SLOWLY AND IS SUDDENLY CUT... TEMPERATURES RISE AND FALL... SMOKE CURLS THROUGH THE AIR... PLANETS WHEEL THROUGH SPACE... AND TIME, TIME NEVER STOPS...

THINK HARD ABOUT CHANGE, AND YOU MAY REACH SOME PRETTY STRANGE CONCLUSIONS. IN ANCIENT GREECE, FOR EXAMPLE, **ZENO OF ELEA** THOUGHT ABOUT CHANGE AND CONVINCED HIMSELF THAT **MOTION IS IMPOSSIBLE.** HE REASONED LIKE SO:

MOTION IS A CHANGE OF POSITION OVER TIME.
AT ANY INSTANT, NO CHANGE OF POSITION TAKES PLACE.
THEREFORE, THERE CAN BE NO MOTION AT ANY INSTANT.
BUT TIME IS A SUCCESSION OF INSTANTS.
THEREFORE, MOTION NEVER TAKES PLACE!
HEY! HOW DID I GET OVER HERE?

EVEN TIME MOVES... IT'S SO WEIRD...

IN THE LATE 1600s, ROUGHLY 2,000 YEARS AFTER ZENO, TWO OTHER GUYS HAD A DIFFERENT IDEA.
ACTUALLY, I HAD THE IDEA AND YOU STOLE IT!
YOU TOOK THE WORDS RIGHT OUT OF MY MOUTH...

ISAAC NEWTON AND **GOTTFRIED LEIBNIZ** LOOKED AT THE PROBLEM THIS WAY: EVEN THOUGH A MOVING CANNONBALL GOES NOWHERE IN AN INSTANT, STILL IT HAS **SOMETHING** THAT INDICATES MOTION.

WHAT IT HAS IS VELOCITY, A NUMBER. YOU MIGHT SAY THAT EVERY OBJECT CARRIES AROUND AN INVISIBLE METER THAT READS OUT THE OBJECT'S SPEED AND DIRECTION AT ALL TIMES.
OH, NOW I'M BEGINNING TO SEE...
IN OTHER WORDS, WE CAN IMAGINE THAT EVERYTHING HAS A SORT OF SPEEDOMETER, JUST LIKE THE ONE IN A CAR (EXCEPT THAT THIS SPEEDOMETER INDICATES DIRECTION TOO).

A PRETTY SHARP IDEA FOR NEWTON AND LEIBNIZ TO HAVE HAD, CONSIDERING THAT SPEEDOMETERS WOULDN'T BE INVENTED FOR ANOTHER 200 YEARS YET...
WHAT'S A SPEEDOMETER?

WHAT'S A CAR?
HOW DID OUR TWO GENIUSES GET THE IDEA? TO ANSWER THIS, LET'S EXPLORE A CAR'S SPEEDOMETER READING.

ACTUALLY, WE WANT A **VELOCIMETER,** NOT A SPEEDOMETER. A VELOCIMETER LOOKS JUST LIKE A SPEEDOMETER, EXCEPT THAT IT ATTACHES A **MINUS SIGN** TO THE SPEED WHEN THE CAR IS BACKING UP. VELOCITY IS THE NEGATIVE OF THE SPEED WHEN YOU GO IN REVERSE.

0
-20
20
-40
40
-60
60

FORSOOTH!

TO APPRECIATE THE DIFFERENCE BETWEEN SPEED AND VELOCITY, IMAGINE A CAR MOVING FORWARD FOR ONE HOUR AT A STEADY RATE OF 50 KM/HR, THEN TURNING AROUND AND COMING BACK (IN A "NEGATIVE DIRECTION") FOR ANOTHER HOUR AT THE SAME SPEED.

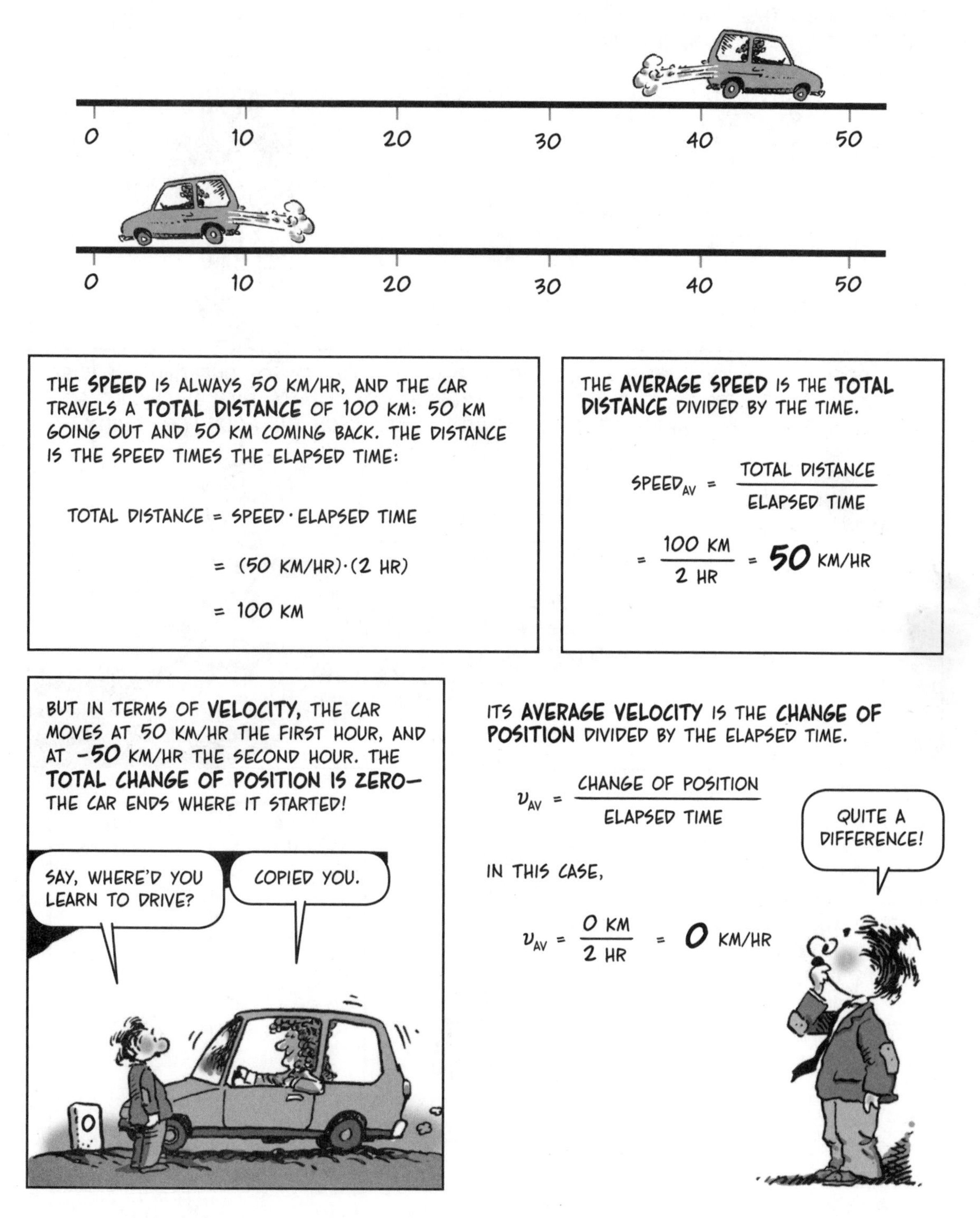

0 10 20 30 40 50
0 10 20 30 40 50
THE **SPEED** IS ALWAYS 50 KM/HR, AND THE CAR TRAVELS A **TOTAL DISTANCE** OF 100 KM: 50 KM GOING OUT AND 50 KM COMING BACK. THE DISTANCE IS THE SPEED TIMES THE ELAPSED TIME:
$$\text{TOTAL DISTANCE} = \text{SPEED} \cdot \text{ELAPSED TIME} = (50\ \text{KM/HR}) \cdot (2\ \text{HR}) = 100\ \text{KM}$$
THE **AVERAGE SPEED** IS THE **TOTAL DISTANCE** DIVIDED BY THE TIME.
$$\text{SPEED}_{AV} = \frac{\text{TOTAL DISTANCE}}{\text{ELAPSED TIME}} = \frac{100\ \text{KM}}{2\ \text{HR}} = \mathbf{50}\ \text{KM/HR}$$
BUT IN TERMS OF **VELOCITY**, THE CAR MOVES AT 50 KM/HR THE FIRST HOUR, AND AT **−50** KM/HR THE SECOND HOUR. THE **TOTAL CHANGE OF POSITION IS ZERO**— THE CAR ENDS WHERE IT STARTED!
SAY, WHERE'D YOU LEARN TO DRIVE?
COPIED YOU.
0
ITS **AVERAGE VELOCITY** IS THE **CHANGE OF POSITION** DIVIDED BY THE ELAPSED TIME.
$$v_{AV} = \frac{\text{CHANGE OF POSITION}}{\text{ELAPSED TIME}}$$
IN THIS CASE,
$$v_{AV} = \frac{0\ \text{KM}}{2\ \text{HR}} = \mathbf{0}\ \text{KM/HR}$$
QUITE A DIFFERENCE!

IN SYMBOLS: IF t_1 AND t_2 ARE ANY TWO TIMES, AND AN OBJECT IS AT POSITION s_1 AT TIME t_1 AND AT POSITION s_2 AT TIME t_2, THEN THE OBJECT'S **AVERAGE VELOCITY** OVER THE TIME INTERVAL BETWEEN t_1 AND t_2 IS

$$v_{AV} = \frac{s_2 - s_1}{t_2 - t_1}$$

OR

$$s_2 - s_1 = v_{AV}(t_2 - t_1)$$

NOW WE NEED A BETTER DRIVER—SOMEONE WITH A STEADIER FOOT—SO LET'S PUT MY FRIEND **DELTA WYE** BEHIND THE WHEEL...

YO!

WHAT DOES IT MEAN WHEN DELTA'S VELOCIMETER READS 100 KM/HR? FOR ONE THING, IT MUST MEAN THAT IF SHE WERE TO HOLD HER VELOCITY **PERFECTLY STEADY,** THEN SHE WOULD GO 100 KM IN ONE HOUR, RIGHT? (DELTA HAS MOUNTED A CLOCK ON THE ROOF FOR CLARITY.)

IF I START HERE AT NOON...
I ARRIVE HERE AT ONE O'CLOCK!
100

AND WE'D GO 200 KM IN 2 HOURS, 50 KM IN HALF AN HOUR, $100t$ KILOMETERS IN t HOURS... A FORMULA THAT SHOULD WORK EVEN FOR **SHORT TIME INTERVALS.** AT A PERFECTLY STEADY 100 KM/HR, DELTA GOES 1 KM IN $\frac{1}{100}$ HOUR (36 SECONDS), 0.1 KM IN 0.001 HOUR (3.6 SECONDS), AND 0.001 KM, ONE METER, IN 0.00001 HR, OR 0.036 SECONDS.

ER... LOGICAL, I GUESS...

$t_2 - t_1$ (HOURS)	$s_2 - s_1$ (KILOMETERS)
10	1000
9	900
5	500
1	100
0.5	50
0.1	10
0.01	1
0.001	0.1
0.0001	0.01
0.0000001	0.00001

THAT'S **IF** THE VELOCITY REMAINS PERFECTLY STEADY... BUT IN THE REAL WORLD, VELOCITY CHANGES AS A CAR SLOWS DOWN AND SPEEDS UP. WHAT DOES THE READING MEAN THEN? (NOW SHE'S ADDED A VELOCIMETER UP TOP AS WELL.)

VELOCITY = 0
SPEEDING UP
VELOCITY HIGH
SLOWING DOWN
VELOCITY LOWER

THE ANSWER IS A LITTLE SUBTLE: YOU'VE SURELY NOTICED THAT OVER A **VERY SHORT TIME PERIOD,** A SPEEDOMETER **DOESN'T CHANGE MUCH.** EVEN IF YOU FLOOR IT, v IS NEARLY CONSTANT OVER A TIME SPAN OF, SAY, 1/500 SEC. A PHOTO TAKEN WITH A SHORT EXPOSURE WOULD SHOW A VELOCIMETER IMAGE WITH VIRTUALLY NO BLUR.

WHAT'S A PHOTO?

THIS WAS NEWTON'S AND LEIBNIZ'S

Basic Idea:

CALCULATE THE RATIO $(s_2 - s_1)/(t_2 - t_1)$ OVER A VERY SHORT TIME INTERVAL. FOR ALL INTENTS AND PURPOSES, THIS RATIO IS THE VELOCITY AT TIME t_1 (AND ALSO AT t_2, THEY'RE SO CLOSE!).

TO PUT IT ANOTHER WAY, A BODY'S **INSTANTANEOUS VELOCITY** IS **CLOSELY APPROXIMATED** BY $(s_2 - s_1)/(t_2 - t_1)$ **WHEN $t_2 - t_1$ IS SMALL.** (YOU MIGHT WONDER HOW NEWTON AND LEIBNIZ THOUGHT THEY MIGHT ACTUALLY MEASURE A CHANGE OF POSITION OVER A TIME INTERVAL OF, SAY, 0.00001 SEC., BUT NEVER MIND THAT!)
ARRHEFFF! IT'S THE PRINCIPLE OF THE THING...

BUT NEWTON AND LEIBNIZ WANTED MORE THAN AN APPROXIMATION: THEY WANTED THE VELOCITY'S **EXACT VALUE...** AND WHAT'S MORE, THEY SHOWED HOW TO **GET IT!** FORGET MEASUREMENT: THEY USED **MATH,** A NEW KIND OF MATH THEY INVENTED ESPECIALLY FOR THE PURPOSE.

AND WE'LL CALL IT **FLUXIONS!**
NO. WE WON'T.

WE CALL IT **CALCULUS.**

IF A BODY'S **POSITION** DEPENDS ON TIME ACCORDING TO SOME FORMULA, THEN CALCULUS POPS OUT A NEW, EXACT FORMULA FOR THE **VELOCITY** AT ANY TIME.

THIS SEEMED SO MAGICAL THAT MORE THAN A FEW PEOPLE FOUND IT SUSPICIOUS... WEIRD... BASED ON STRANGE, UNFOUNDED ASSUMPTIONS... SOMEHOW... WRONG...

YOU'RE **ALMOST** DIVIDING BY ZERO!
(LEIBNIZ'S APPROACH SEEMED ESPECIALLY FISHY: HE WAS HAPPY TO DIVIDE ONE THING BY ANOTHER NOT ONLY WHEN THE QUANTITIES WERE SMALL, BUT ALSO WHEN THEY WERE "INFINITELY SMALL" BUT NOT ZERO, WHATEVER THAT MEANT.)
FREAK!

FISHY FOUNDATIONS OR NOT, CALCULUS WORKED, AND IT WORKED BEAUTIFULLY. IT WAS AMAZINGLY EFFECTIVE. IT PRODUCED RESULTS!

MANY, MANY, MANY RESULTS...
WHOA!
ASTRONOMY, COMMUNICATIONS, ELECTRICITY, BIOLOGY, CHEMISTRY, MECHANICS, STATISTICS, COMPUTER SCIENCE, PSYCHOLOGY, ECONOMICS...
POPULATION DYNAMICS...

SO PEOPLE PUT CALCULUS TO WORK... NOT ONLY FINDING VELOCITIES, BUT ALSO THE RATE OF CHANGE OF ALL KINDS OF FLUCTUATING QUANTITIES. CALCULUS IS USED EVERYWHERE!

EVENTUALLY, THEY EVEN FIXED THE FOUNDATIONS, MORE OR LESS... UNFORTUNATELY, WE LACK THE SPACE TO EXPLAIN FULLY HOW THIS WAS DONE, OR TO DESCRIBE THE TROUBLESOME ISSUES RAISED BY CALCULUS... LET'S JUST SAY THAT SOME OF ZENO'S SUBTLETIES REMAIN A CHALLENGE TO THIS DAY...

HEY, MAN, YOU WORRY TOO MUCH!
YEAH, C'MON! WHATEVER WORKS...

Chapter 0
Meet the Functions

IN WHICH WE LEARN SOMETHING ABOUT RELATIONSHIPS

WE BEGIN WITH ONE OF THE MOST BEAUTIFUL AND FRUITFUL IDEAS OF MODERN MATHEMATICS: THE **FUNCTION.** EVERYTHING IN THIS BOOK WILL BE ABOUT FUNCTIONS. SO... WHAT'S A FUNCTION?

UM... I GUESS **I** AM?
WHOOF! I THOUGHT YOU SAID FRUITFUL **AND** BEAUTIFUL!

A FUNCTION IS A SORT OF **INPUT-OUTPUT DEVICE** OR **NUMBER-PROCESSOR.** A FUNCTION (CALL IT f) EATS AND SPEWS NUMBERS IN A SPECIFIC WAY. FOR EACH NUMBER EATEN (CALL IT x), f OUTPUTS A SINGLE, UNIQUE NUMBER, $f(x)$, PRONOUNCED "EFF OF ECKS." f IS LIKE A RULE THAT TRANSFORMS x INTO $f(x)$. IN GOES x, OUT COMES $f(x)$.

f
2
f
f(2)
DON'T WORRY... IT'S CLEAN, ABSTRACT, NUMERICAL OUTPUT...
f
3
f
f(3)

IF YOU DON'T LIKE YOUR OUTPUT FLOATING AROUND IN THE AIR LIKE SWAMP GAS, THEN THINK OF NUMBERS AS LYING ALONG A LINE. IN THAT CASE, YOU CAN IMAGINE A FUNCTION f EATING NUMBERS FROM ONE LINE AND MERELY **POINTING** TO THE CORRESPONDING OUTPUT VALUES ON THE OTHER LINE.

f
x
f(x)
MARGINALLY LESS GROSS...

FOR EXAMPLE, A CAR'S POSITION s IS A FUNCTION OF TIME t. YOU CAN THINK OF s AS READING TIME (OR EATING IT AS INPUT!) FROM A TIMELINE AND POINTING TO THE CAR'S POSITION $s(t)$ ON THE TRACK.

t
s(t)

More Examples:

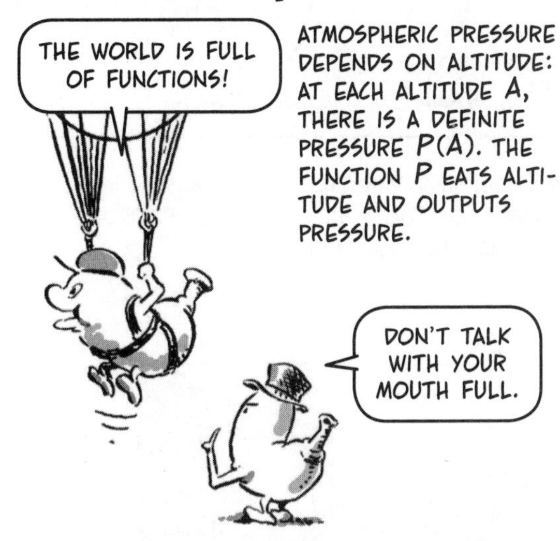
THE WORLD IS FULL OF FUNCTIONS!
DON'T TALK WITH YOUR MOUTH FULL.

ATMOSPHERIC PRESSURE DEPENDS ON ALTITUDE: AT EACH ALTITUDE A, THERE IS A DEFINITE PRESSURE $P(A)$. THE FUNCTION P EATS ALTITUDE AND OUTPUTS PRESSURE.

AS A SPHERICAL BALLOON INFLATES, ITS VOLUME IS A FUNCTION OF THE RADIUS. EACH RADIUS r DETERMINES A UNIQUE VOLUME $V(r)$.

$\frac{4}{3}\pi r^3$
r
SOAP

ON A STRAIGHT MOUNTAIN TRAIL, ALTITUDE IS A FUNCTION OF POSITION ALONG THE TRAIL. EACH POSITION x HAS A UNIQUE ALTITUDE $A(x)$.

A(x)
x

IN THE EXAMPLE OF THE SPHERICAL BALLOON, THE VOLUME FUNCTION V WAS CALCULATED FROM THE RADIUS r BY MEANS OF A **FORMULA:**

$$V(r) = \frac{4\pi r^3}{3}$$

TO FIND THE VOLUME ASSOCIATED WITH A PARTICULAR RADIUS, SAY $r = 10$, WE INPUT, OR **PLUG IN,** THAT NUMBER IN PLACE OF r:

$$V(10) = \frac{4\pi(10)^3}{3} = \frac{4000}{3}\pi$$

$$\approx 4{,}188.79\ldots$$

(THE SIGN "≈" MEANS "IS APPROXIMATELY EQUAL TO.")

CUBE THIS AND MULTIPLY THE RESULT BY (4π/3)!
CAN I USE A CALCULATOR?
4

IMPORTANT: THE LETTERS WE ASSIGN TO THE FUNCTION AND VARIABLE DON'T MATTER! HERE ARE THREE FORMULAS THAT ALL DEFINE THE SAME FUNCTION BECAUSE THEY PRODUCE THE SAME OUTPUT FOR ANY GIVEN INPUT. THEY ALL DESCRIBE THE SAME RULE.

$$V(r) = \frac{4\pi r^3}{3}$$

$$f(t) = \frac{4\pi t^3}{3}$$

$$g(u) = \frac{4\pi u^3}{3}$$

4π/3 (INPUT)³
V
f
WHAT'S IN A NAME?
NOTHING! SO YOU WON'T MIND IF I CALL YOU "MELONHEAD"?...

HERE IS A SLIGHTLY MORE COMPLICATED EXAMPLE. SUPPOSE h IS GIVEN BY THIS FORMULA:
$h(x) = \sqrt{x^2 - 1}$
WE COMPUTE A FEW VALUES...
$h(1) = \sqrt{1^2 - 1} = 0$
$h(2) = \sqrt{2^2 - 1} = \sqrt{3}$
$h(\sqrt{5}) = \sqrt{5 - 1} = 2$
ETC...
AND COMPILE A LITTLE TABLE. IT'S FULL OF GAPS, BUT YOU CAN FILL IN MANY MISSING VALUES... EXCEPT...
EVEN IN HERE?
x | $h(x)$
-3 | $\sqrt{8}$
-2.9 | $\sqrt{7.41}$
-2.8 | $\sqrt{6.84}$
-2 | $\sqrt{3}$
-1 | 0
1 | 0
2 | $\sqrt{3}$
$\sqrt{5}$ | 2
3 | $\sqrt{8}$
ETC...

WHEN x IS BETWEEN −1 AND 1, THE EXPRESSION INSIDE THE SQUARE ROOT SIGN IS NEGATIVE: $x^2 - 1 < 0$. IN THAT CASE, $h(x)$ IS **UNDEFINED,** BECAUSE NEGATIVE NUMBERS HAVE NO (REAL) SQUARE ROOT. EVERY INPUT ACCEPTED BY h MUST HAVE A VALUE EITHER ≥ 1 OR ≤ -1. NOTHING ELSE IS ALLOWED!

AK! IT'S A DEAD ZONE!
-4
-3
-2
-1
0
1
2
3
4

GIVEN ANY FUNCTION, ITS **DOMAIN** IS THE SET OF ALL NUMBERS WHERE THE FUNCTION IS DEFINED. A FUNCTION f WILL ACCEPT INPUTS **ONLY** FROM WITHIN ITS DOMAIN.

ANYTHING ELSE IS INDIGESTIBLE!

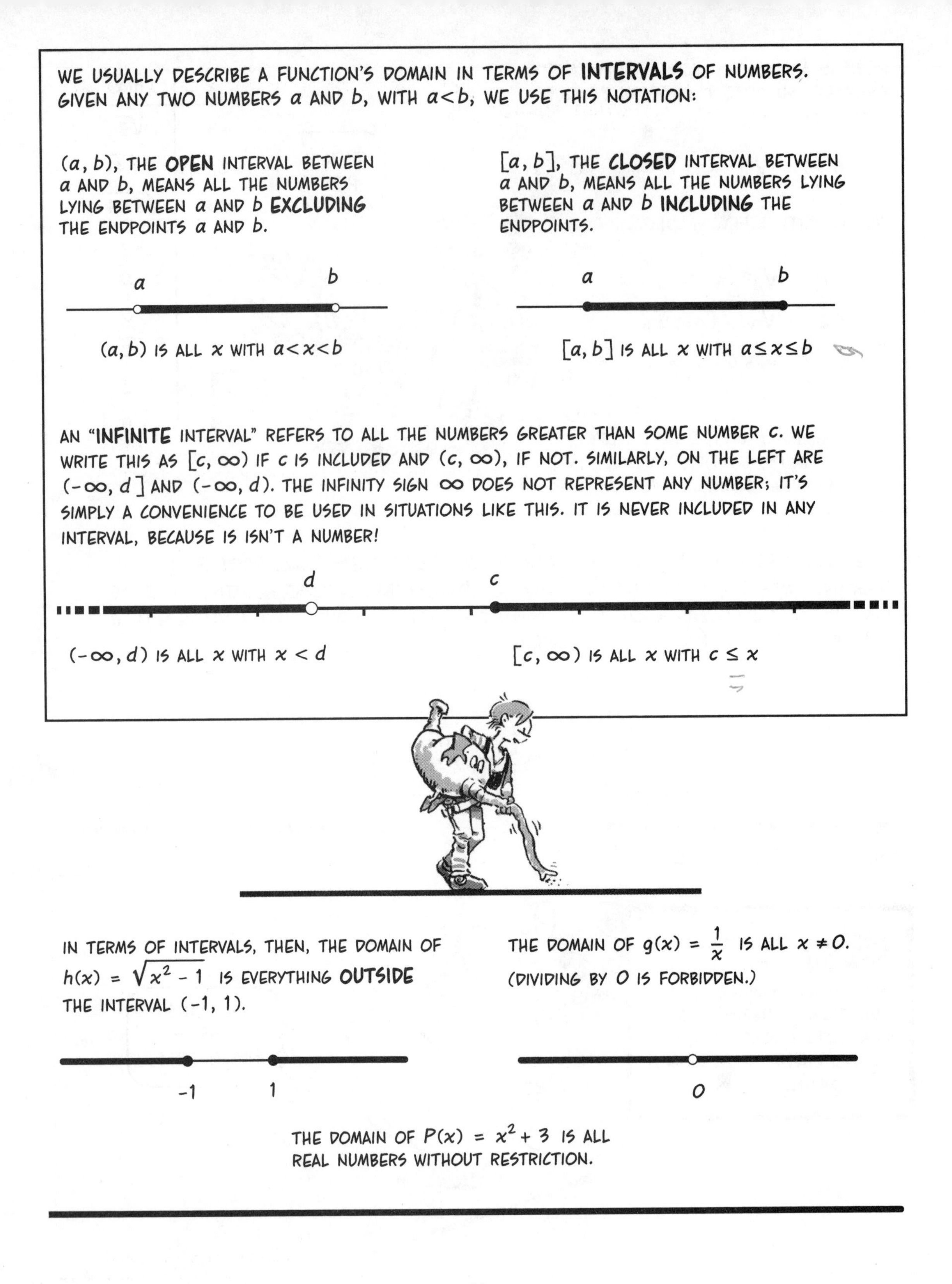
WE USUALLY DESCRIBE A FUNCTION'S DOMAIN IN TERMS OF **INTERVALS** OF NUMBERS. GIVEN ANY TWO NUMBERS a AND b, WITH $a<b$, WE USE THIS NOTATION:
(a, b), THE **OPEN** INTERVAL BETWEEN a AND b, MEANS ALL THE NUMBERS LYING BETWEEN a AND b **EXCLUDING** THE ENDPOINTS a AND b.
a
b
(a, b) IS ALL x WITH $a<x<b$
$[a, b]$, THE **CLOSED** INTERVAL BETWEEN a AND b, MEANS ALL THE NUMBERS LYING BETWEEN a AND b **INCLUDING** THE ENDPOINTS.
a
b
$[a, b]$ IS ALL x WITH $a \leq x \leq b$
AN "**INFINITE** INTERVAL" REFERS TO ALL THE NUMBERS GREATER THAN SOME NUMBER c. WE WRITE THIS AS $[c, \infty)$ IF c IS INCLUDED AND (c, ∞), IF NOT. SIMILARLY, ON THE LEFT ARE $(-\infty, d]$ AND $(-\infty, d)$. THE INFINITY SIGN ∞ DOES NOT REPRESENT ANY NUMBER; IT'S SIMPLY A CONVENIENCE TO BE USED IN SITUATIONS LIKE THIS. IT IS NEVER INCLUDED IN ANY INTERVAL, BECAUSE IS ISN'T A NUMBER!
d
c
$(-\infty, d)$ IS ALL x WITH $x < d$
$[c, \infty)$ IS ALL x WITH $c \leq x$
IN TERMS OF INTERVALS, THEN, THE DOMAIN OF $h(x) = \sqrt{x^2 - 1}$ IS EVERYTHING **OUTSIDE** THE INTERVAL $(-1, 1)$.
-1
1
THE DOMAIN OF $g(x) = \frac{1}{x}$ IS ALL $x \neq 0$. (DIVIDING BY 0 IS FORBIDDEN.)
0
THE DOMAIN OF $P(x) = x^2 + 3$ IS ALL REAL NUMBERS WITHOUT RESTRICTION.

NOW RETURN TO OUR IMAGE OF A FUNCTION PICKING UP INPUTS FROM ONE NUMBER LINE AND POINTING TO OUTPUTS ON ANOTHER NUMBER LINE.

a
$f(a)$

IF WE LIKE, WE CAN LET THE FUNCTION'S CARTOON BODY FADE AWAY AND CONCENTRATE ON THE ACT OF **POINTING.**

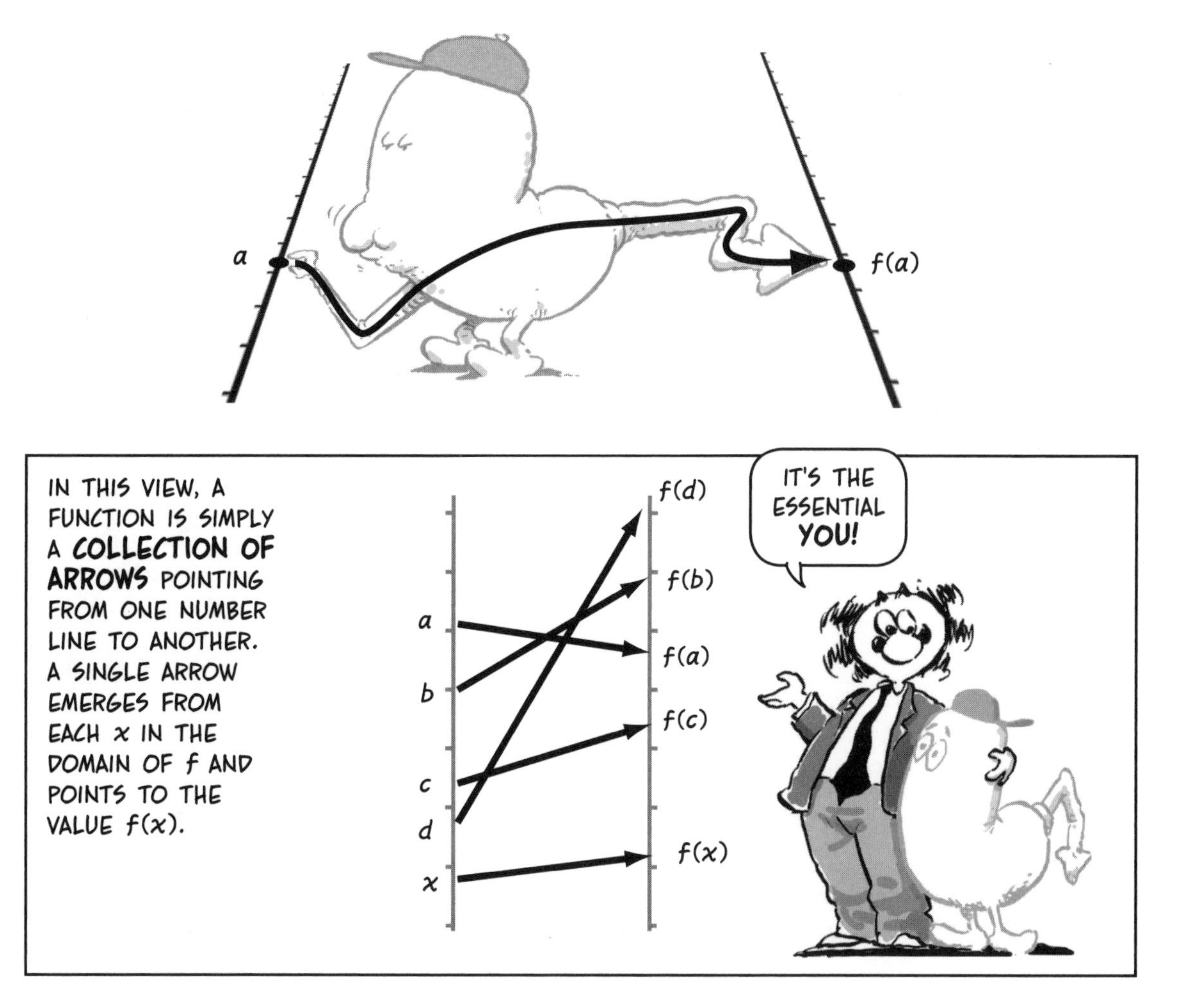

a
$f(a)$
$f(d)$
$f(b)$
$f(a)$
$f(c)$
$f(x)$
a
b
c
d
x
IT'S THE ESSENTIAL **YOU!**

IN THIS VIEW, A FUNCTION IS SIMPLY A **COLLECTION OF ARROWS** POINTING FROM ONE NUMBER LINE TO ANOTHER. A SINGLE ARROW EMERGES FROM EACH x IN THE DOMAIN OF f AND POINTS TO THE VALUE $f(x)$.

NOW LET'S PLAY WITH THOSE ARROWS.

HM... WHAT IF I TURN ONE OF THESE LINES SIDEWAYS?

WHEN THE FIRST LINE, OR **AXIS,** IS TURNED SIDEWAYS, WE CAN VIEW A FUNCTION AS A **GRAPH.** THE INPUTS x ARE ON THE HORIZONTAL AXIS, THE OUTPUTS y ARE ON THE VERTICAL AXIS, AND ABOVE (OR BELOW) ANY POINT a ON THE x-AXIS WE PLOT A POINT $(a, f(a))$, WITH y-COORDINATE EQUAL TO THE VALUE OF THE FUNCTION f AT a.

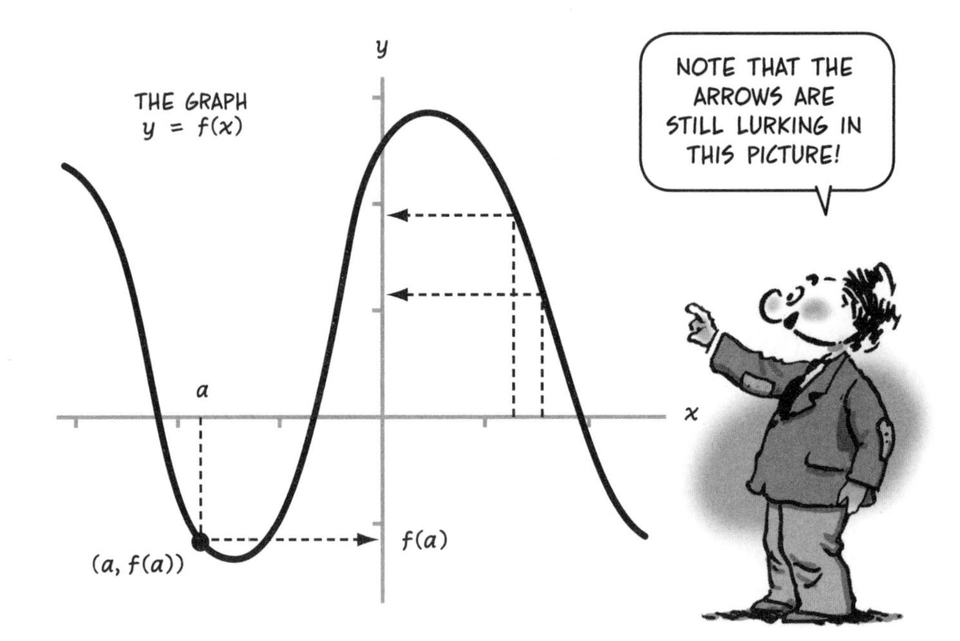
y
THE GRAPH
y = f(x)
a
x
f(a)
(a, f(a))
NOTE THAT THE ARROWS ARE STILL LURKING IN THIS PICTURE!

THE CURVE CONSISTS OF ALL POINTS (x, y) WITH $y = f(x)$, A PHRASE WE ABBREVIATE BY SAYING "THE **GRAPH** $y = f(x)$."

HERE ARE SOME SIMPLE EXAMPLES.

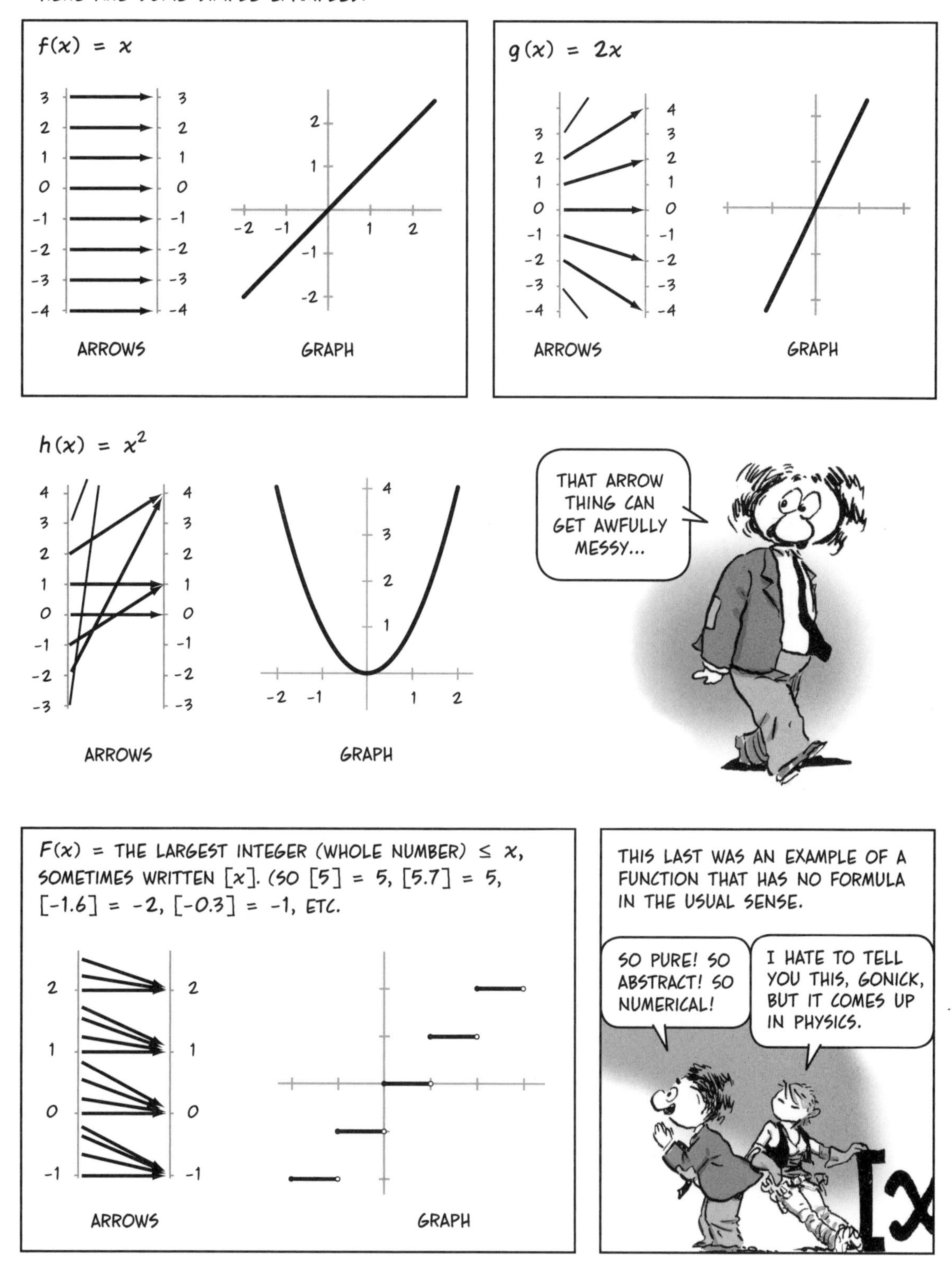

$f(x) = x$
ARROWS
GRAPH
$g(x) = 2x$
ARROWS
GRAPH
$h(x) = x^2$
ARROWS
GRAPH
THAT ARROW THING CAN GET AWFULLY MESSY...
$F(x)$ = THE LARGEST INTEGER (WHOLE NUMBER) $\leq x$, SOMETIMES WRITTEN $[x]$. (SO $[5] = 5$, $[5.7] = 5$, $[-1.6] = -2$, $[-0.3] = -1$, ETC.
ARROWS
GRAPH
THIS LAST WAS AN EXAMPLE OF A FUNCTION THAT HAS NO FORMULA IN THE USUAL SENSE.
SO PURE! SO ABSTRACT! SO NUMERICAL!
I HATE TO TELL YOU THIS, GONICK, BUT IT COMES UP IN PHYSICS.

Add, Multiply, Divide

FUNCTIONS CAN BE COMBINED IN VARIOUS WAYS, JUST AS NUMBERS CAN. IF f AND g HAVE OVERLAPPING DOMAINS, WE CAN ADD, MULTIPLY, AND DIVIDE THE FUNCTIONS WHEREVER THEY ARE BOTH DEFINED. THIS PRODUCES NEW FUNCTIONS $f + g$, fg, AND f/g (AS LONG AS WE'RE CAREFUL NEVER TO DIVIDE BY ZERO).

$(f + g)(x) = f(x) + g(x)$

$(fg)(x) = f(x)g(x)$

$(f/g)(x) = f(x)/g(x)$ EXCEPT WHERE $g(x) = 0$.

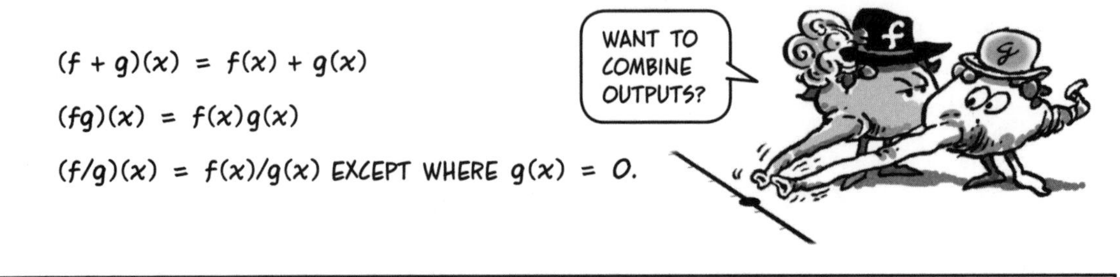
WANT TO COMBINE OUTPUTS?
f
g

THE GRAPH OF $f + g$ CAN BE BUILT FROM THE GRAPHS OF f AND g BY ADDING THE y-COORDINATES AT EACH POINT x IN THEIR COMMON DOMAIN.

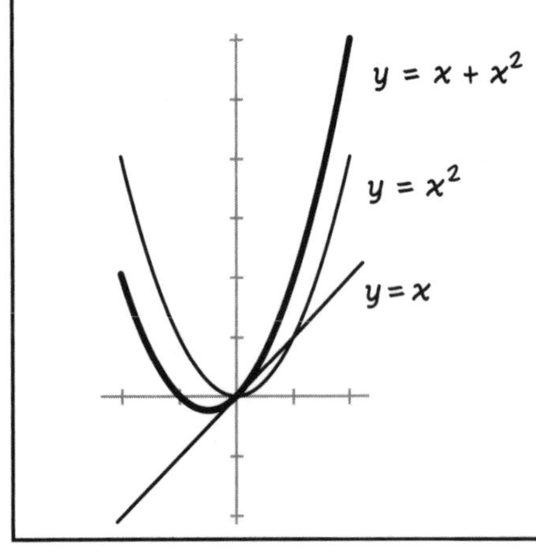
$y = x + x^2$
$y = x^2$
$y = x$

THE DIFFERENCE BETWEEN TWO FUNCTIONS CAN BE VISUALIZED AS THE DIRECTED DISTANCE BETWEEN THEIR GRAPHS.

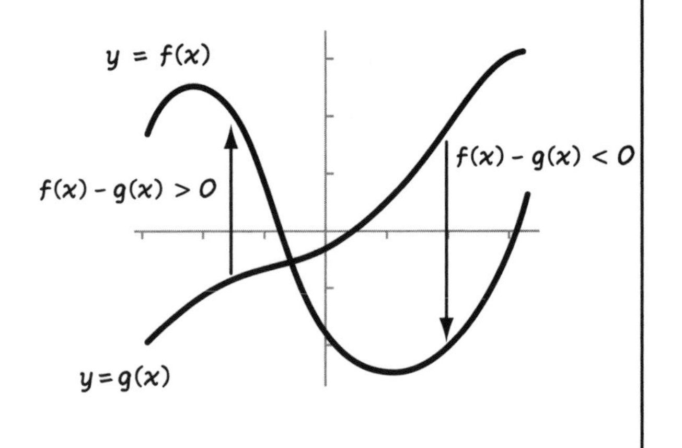
$y = f(x)$
$f(x) - g(x) > 0$
$f(x) - g(x) < 0$
$y = g(x)$

IN GENERAL, THE GRAPHS OF PRODUCTS fg AND QUOTIENTS f/g ARE NOT SO EASILY SEEN IN TERMS OF f AND g. USUALLY THEY MUST BE CALCULATED POINT BY POINT.

SOMETIMES IT'S NOT TOO BAD, THOUGH...
HM?
$y = x$
$y = \sin x$
$y = x \sin x$

The Elementary Functions

NOW THAT WE'VE COVERED SOME BASIC IDEAS ABOUT FUNCTIONS, LET'S REVIEW A FEW COMMON EXAMPLES, FUNCTIONS TO WHICH WE WILL REFER THROUGHOUT THE REMAINDER OF THIS BOOK.

YOU LOOKED SO... DIFFERENT IN PRE-CALC!

THESE FUNCTIONS ARE CALLED **ELEMENTARY** FUNCTIONS, BECAUSE, LIKE CHEMICAL ELEMENTS, THEY CAN BE COMBINED IN AN INEXHAUSTIBLE VARIETY OF WAYS...

Absolute Value

CALCULUS IS ABOUT APPROXIMATIONS, AND THE **ABSOLUTE VALUE FUNCTION** MEASURES HOW CLOSELY ONE NUMBER APPROXIMATES ANOTHER.

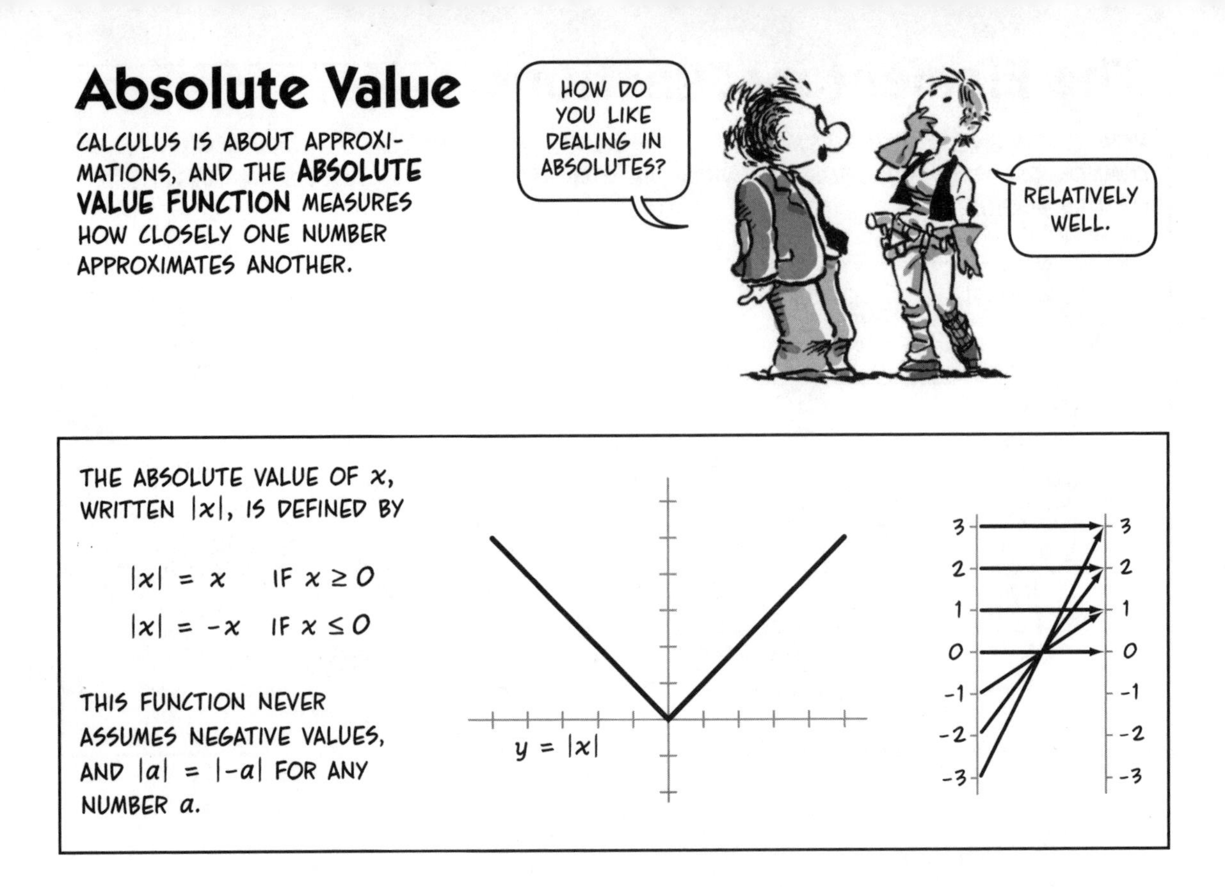

HOW DO YOU LIKE DEALING IN ABSOLUTES?
RELATIVELY WELL.
$y = |x|$
3
2
1
0
−1
−2
−3

THE ABSOLUTE VALUE OF x, WRITTEN $|x|$, IS DEFINED BY

$$|x| = x \quad \text{IF } x \geq 0$$

$$|x| = -x \quad \text{IF } x \leq 0$$

THIS FUNCTION NEVER ASSUMES NEGATIVE VALUES, AND $|a| = |-a|$ FOR ANY NUMBER a.

YOU CAN THINK OF $|a|$ AS THE (POSITIVE, ABSOLUTE) **DISTANCE** OF a FROM 0 ON THE NUMBER LINE, AND $|a - b| = |b - a|$ AS THE **DISTANCE BETWEEN a AND b.**

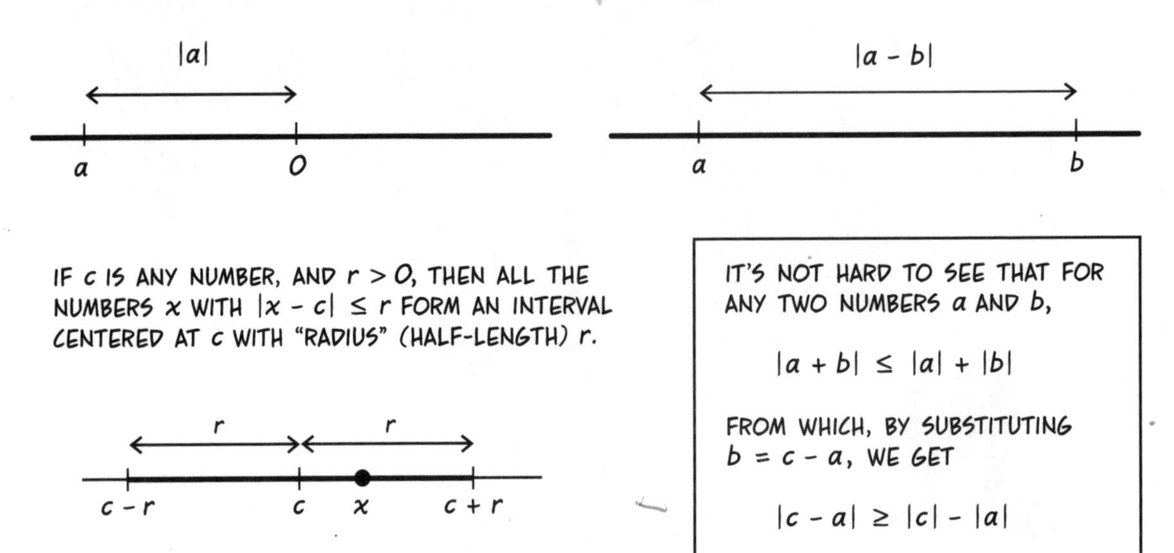

$|a|$
a
0
$|a - b|$
a
b
r
r
$c - r$
c
x
$c + r$
$|x - c| \leq r$

IF c IS ANY NUMBER, AND $r > 0$, THEN ALL THE NUMBERS x WITH $|x - c| \leq r$ FORM AN INTERVAL CENTERED AT c WITH "RADIUS" (HALF-LENGTH) r.

IT'S NOT HARD TO SEE THAT FOR ANY TWO NUMBERS a AND b,

$$|a + b| \leq |a| + |b|$$

FROM WHICH, BY SUBSTITUTING $b = c - a$, WE GET

$$|c - a| \geq |c| - |a|$$

FOR ANY TWO NUMBERS a AND c.

Constants

IF C IS ANY FIXED NUMBER, THEN THERE IS A VERY SIMPLE-MINDED FUNCTION f DEFINED BY $f(x) = C$ FOR ALL x. NOT MUCH OF A FUNCTION, YOU MIGHT SAY, BUT IT IS A FUNCTION! ITS GRAPH IS THE HORIZONTAL LINE $y = C$. ALL ARROWS POINT TO THE SAME NUMBER.

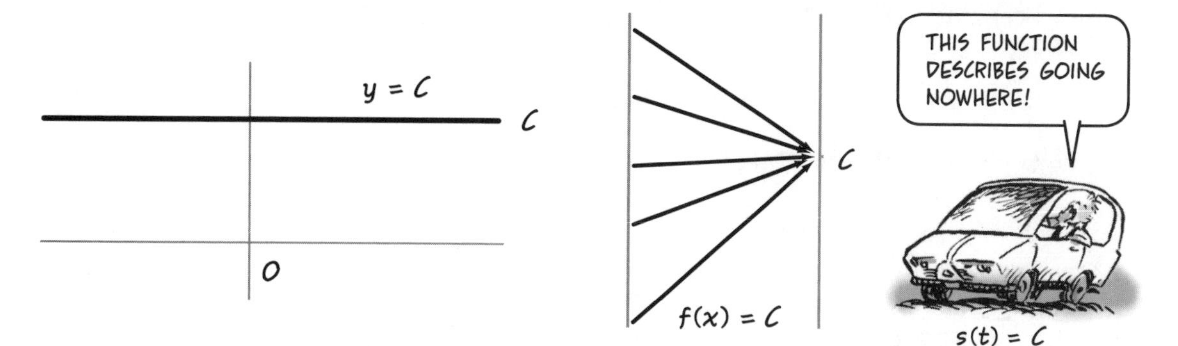

y = C
C
O
C
f(x) = C
THIS FUNCTION DESCRIBES GOING NOWHERE!
s(t) = C

Power Functions

THESE ARE THE FUNCTIONS WITH FORMULA x, x^2, x^3, ..., x^{17}, ... x^n... WHERE n IS A POSITIVE INTEGER. WHEN n IS **EVEN,** THESE FUNCTIONS ALL HAVE BOWL-SHAPED GRAPHS, BECAUSE $(-x)^n = x^n$. POSITIVE AND NEGATIVE INPUTS "LAND" IN THE SAME PLACE. IF n IS **ODD,** THEN $(-x)^n = -(x^n)$, AND THE GRAPHS BEND DOWNWARD ON THE LEFT.

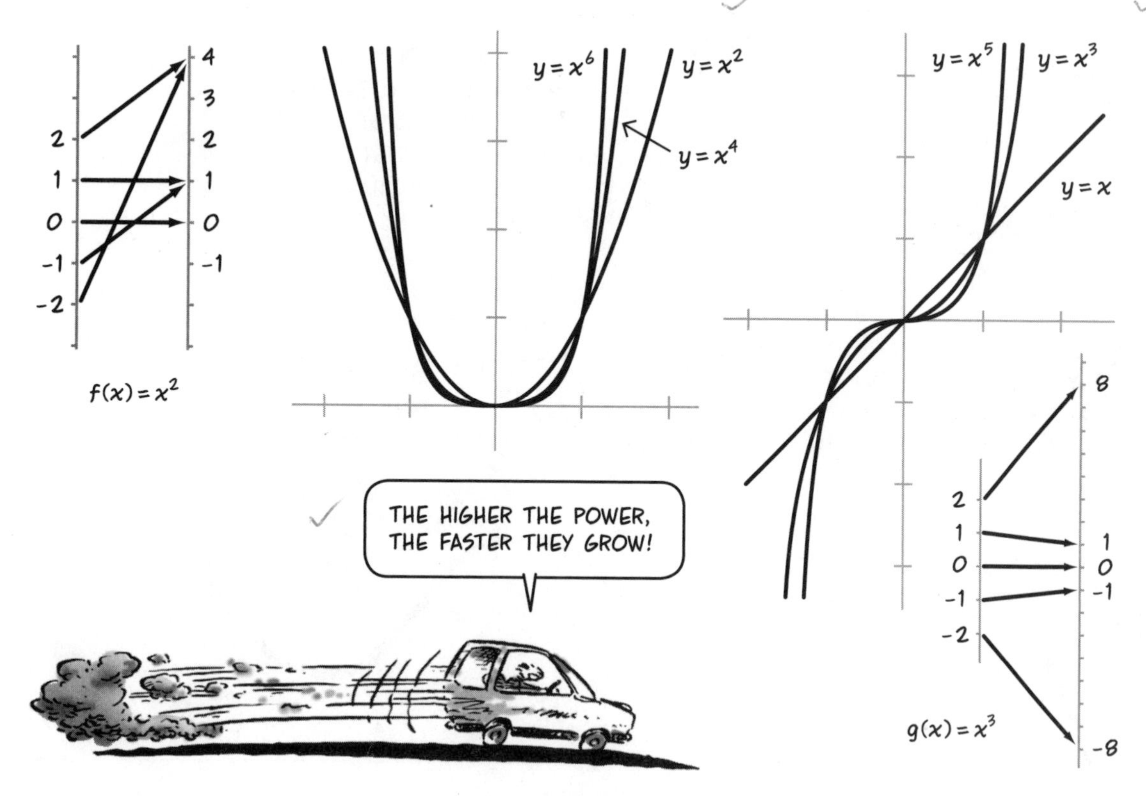

2
1
0
-1
-2
4
3
2
1
0
-1
f(x)=x²
y=x⁶
y=x²
y=x⁴
y=x⁵
y=x³
y=x
2
1
0
-1
-2
8
1
0
-1
-8
g(x)=x³
THE HIGHER THE POWER, THE FASTER THEY GROW!

Polynomials

WE ADD CONSTANTS AND MULTIPLES OF POWER FUNCTIONS TO MAKE **POLYNOMIALS,** WHICH HAVE FORMULAS LIKE $2x^2 + x + 41$ OR $x^{15} - x^{14} - 9x$. THE CONSTANT FACTORS ARE CALLED THE POLYNOMIAL'S **COEFFICIENTS,** AND THE LARGEST POWER OF x WITH A NON-ZERO COEFFICIENT IS CALLED THE POLYNOMIAL'S **DEGREE.**

$P(x) = 7x^{10} + 395x^4 + x^3 + 11$ HAS DEGREE 10.

$Q(x) = -x + 9$ HAS DEGREE 1

ALGEBRA TEACHES US THAT A POLYNOMIAL P OF DEGREE n HAS NO MORE THAN n **ROOTS,** MEANING NUMBERS $x_1, x_2, \ldots x_m$, WHERE $P(x_i) = 0$.

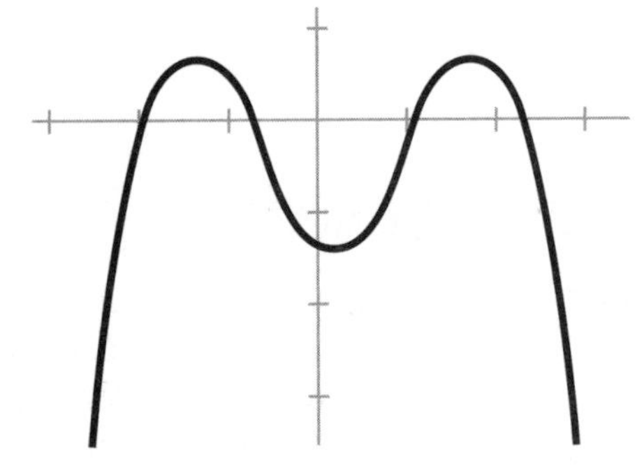

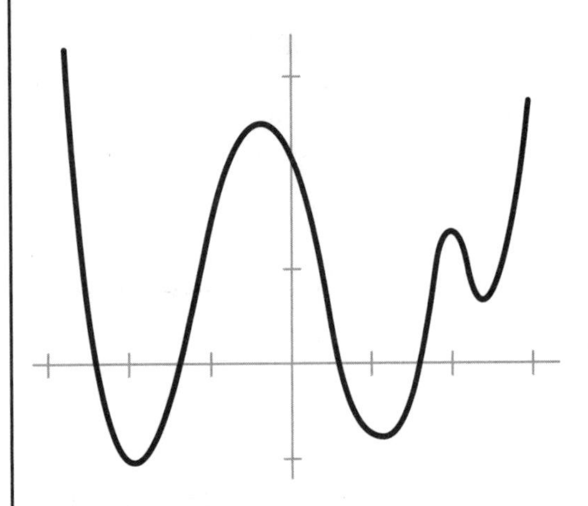

THIS MEANS THAT THE GRAPH OF AN nTH DEGREE POLYNOMIAL CROSSES THE x-AXIS NO MORE THAN n TIMES. IN FACT, WE WILL SEE THAT THE GRAPH HAS AT MOST $n - 1$ "TURNINGS" WHERE IT CHANGES FROM RISING TO FALLING OR VICE VERSA.

WE'LL ALSO SEE THAT THE GRAPH OF ANY POLYNOMIAL ZOOMS OFF TO INFINITY (EITHER POSITIVE OR NEGATIVE) AS x GOES OFF TO THE LEFT AND RIGHT WITHOUT BOUNDS.

"TO" INFINITY?
WELL, AWAY FROM EVERYTHING ELSE, ANYWAY.

Negative Powers

THESE ARE THE FUNCTIONS

$$f(x) = \frac{1}{x^n}, \quad n = 1, 2, 3, \ldots$$

THEY ARE ALSO WRITTEN

$$f(x) = x^{-n}$$

NEGATIVE POWER FUNCTIONS ARE DEFINED FOR ALL $x \neq 0$, AND, LIKE THE POSITIVE POWERS, THEIR GRAPHS DIFFER DEPENDING ON WHETHER n IS ODD OR EVEN.

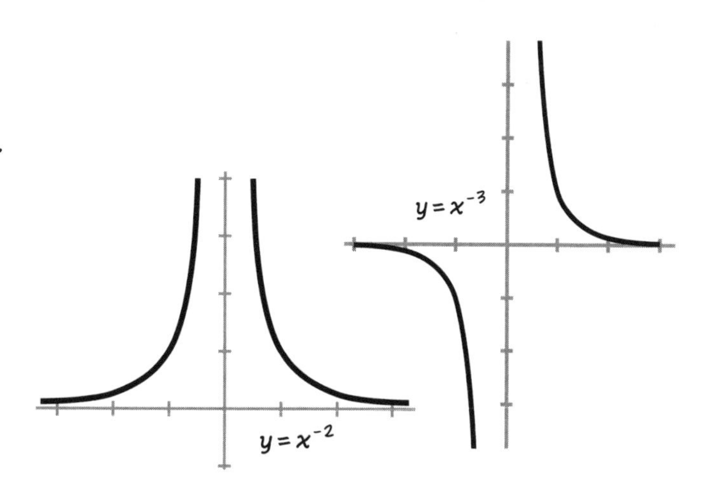

$y = x^{-3}$
$y = x^{-2}$

Fractional Powers

IF n IS A POSITIVE INTEGER, $x^{\frac{1}{n}}$ MEANS THE nTH ROOT OF x, $\sqrt[n]{x}$. THE FRACTIONAL NOTATION IS USED TO MAKE THIS FORMULA WORK:

$$(x^{\frac{1}{n}})^n = x^{\frac{1}{n} \cdot n} = x$$

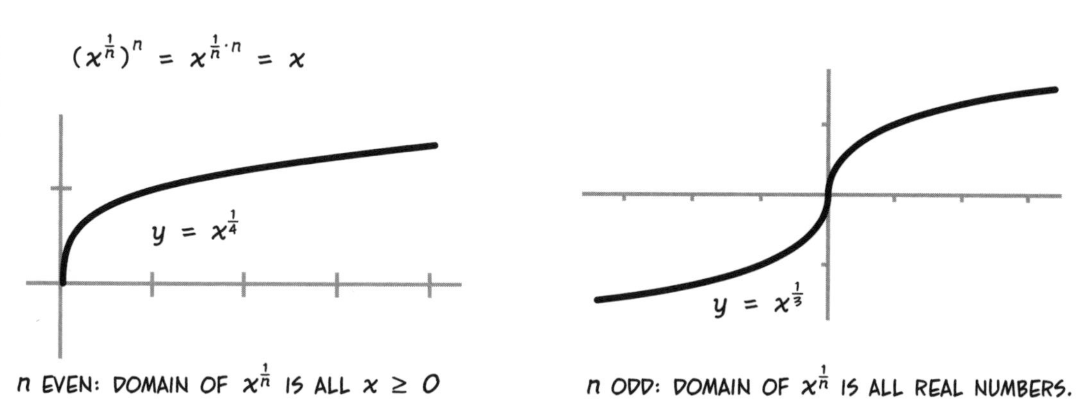

$y = x^{\frac{1}{4}}$
$y = x^{\frac{1}{3}}$

n EVEN: DOMAIN OF $x^{\frac{1}{n}}$ IS ALL $x \geq 0$

n ODD: DOMAIN OF $x^{\frac{1}{n}}$ IS ALL REAL NUMBERS.

THERE CAN BE NEGATIVE FRACTIONAL POWERS, TOO.

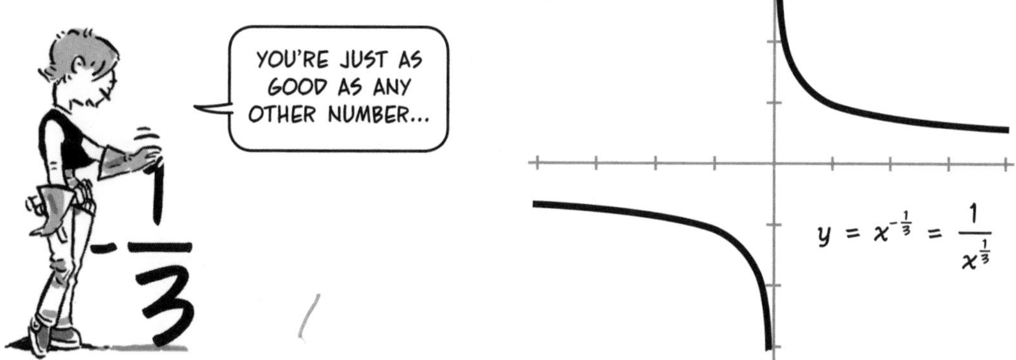

YOU'RE JUST AS GOOD AS ANY OTHER NUMBER...
$-\frac{1}{3}$
$y = x^{-\frac{1}{3}} = \frac{1}{x^{\frac{1}{3}}}$

Rational Functions

THESE ARE FUNCTIONS GIVEN BY **RATIOS** OF POLYNOMIALS

$$R(x) = \frac{P(x)}{Q(x)}$$

THEY ARE DEFINED WHEREVER $Q(x) \neq 0$. FOR EXAMPLE,

$$R(x) = \frac{3x^2 + 9x + 1}{x^3 + 16}, \quad x \neq \sqrt[3]{-16}$$

$$T(x) = \frac{x}{x^2 - 1}, \quad x \neq \pm 1$$

CRAZY!
NO, RATIONAL.
$y = \frac{x}{x^2-1}$

WE HAVE THREE THINGS TO SAY ABOUT RATIONAL FUNCTIONS. FIRST IS THAT YOU CAN SKIP THIS SECTION AND HEAD FOR PAGE 29 IF YOU WANT TO...
ER... 'BYE...

SECOND, WE CAN ASSUME THAT P HAS LOWER DEGREE THAN Q. IF IT DOESN'T, YOU CAN DO **LONG DIVISION OF POLYNOMIALS*** TO MAKE P/Q LOOK LIKE

$$P_1(x) + \frac{R(x)}{Q(x)}$$

WHERE P_1 IS A POLYNOMIAL, AND R, THE REMAINDER, IS A POLYNOMIAL WITH DEGREE LOWER THAN THAT OF Q.

HA! THOSE PAGE-SKIPPERS ARE GOING TO MISS THE FIRST TURNS OF THE **ALGEBRA CRANK!**

*IF YOU'VE NEVER DONE LONG DIVISION OF POLYNOMIALS, IT'S JUST LIKE LONG DIVISION OF NUMBERS, ONLY EASIER. LOOK IT UP SOMEWHERE; YOU'LL LIKE IT!

THIRD, ANY RATIONAL FUNCTION CAN BE WRITTEN AS A **SUM** OF SIMPLER "PARTIAL FRACTIONS" OF THESE TWO KINDS:

$$\frac{a}{(x+p)^n} \quad \text{OR} \quad \frac{bx+c}{(x^2+qx+r)^m},$$

WHERE a, b, c, p, q, AND r ARE CONSTANTS, AND n AND m ARE POSITIVE INTEGERS. IN OTHER WORDS, THE DENOMINATORS ARE POWERS OF FIRST- OR SECOND-DEGREE POLYNOMIALS.

THIS BECOMES USEFUL LATER, WHEN WE DO INTEGRATION.
P/Q

FINDING THESE CONSTANTS CAN BE MESSY IN PRACTICE—FOR STARTERS, YOU HAVE TO FACTOR $Q(x)$—BUT HERE ARE TWO EXAMPLES TO SHOW HOW IT WORKS.

Example: SUPPOSE

$$F(x) = \frac{x}{(x-1)^2}$$

NOW LET'S PUT THIS CRANK IN MOTION...

FIRST WRITE IT AS

$$\left(\frac{x}{x-1}\right)\left(\frac{1}{x-1}\right)$$

THE FIRST FACTOR CAN BE REDUCED BY LONG DIVISION:

$$\left(\frac{x}{x-1}\right) = \frac{1}{x-1} + 1$$

PLUGGING THAT IN AND EXPANDING GIVES

$$\left(\frac{1}{x-1} + 1\right)\left(\frac{1}{x-1}\right) = \frac{1}{(x-1)^2} + \frac{1}{x-1}$$

JUST AS PROMISED—NOTHING BUT CONSTANTS IN THE NUMERATORS OF FRACTIONS WITH DENOMINATORS OF THE FORM $(x+p)^n$.

OH, **YEAH!!!**

Example:

$$R(x) = \frac{-2x^2 + 7x - 3}{x^3 + 1}$$

THE FIRST STEP IS ALWAYS TO FACTOR THE DENOMINATOR. RECALL FROM ALGEBRA:

$$x^3 + 1 = (x + 1)(x^2 - x + 1).$$

NOW, ASSUME THERE IS AN ANSWER.

ALWAYS A GOOD IDEA IN ALGEBRA!

IT WOULD LOOK LIKE THIS:

$$\frac{-2x^2+7x-3}{x^3+1} = \frac{Ax+B}{(x^2-x+1)} + \frac{C}{x+1}$$

WE WISH TO SOLVE FOR A, B, AND C. COMBINING THE FRACTIONS ON THE RIGHT PRODUCES THIS NUMERATOR:

$$(A + C)x^2 + (A + B - C)x + (B + C)$$

THIS BEING THE SAME AS THE NUMERATOR OF THE ORIGINAL FRACTION, WE MUST HAVE

$$A + C = -2$$
$$A + B - C = 7$$
$$B + C = -3$$

THESE ARE THREE EQUATIONS IN THREE UNKNOWNS. WE DO SOME ALGEBRA AND FIND...

$A = 2$, $B = 1$, AND $C = -4$, SO:

$$R(x) = \frac{2x + 1}{x^2 - x + 1} + \frac{-4}{x + 1}$$

YOU CAN CHECK THE ANSWER BY ADDING TOGETHER THESE FRACTIONS, WHICH SHOULD COMBINE TO GIVE THE ORIGINAL FUNCTION.

WHEW!

AND NOW FOR SOMETHING YOU WON'T WANT TO MISS... THIS NEXT FUNCTION WILL REALLY GROW ON YOU...

Exponential Functions

NOT IF I CAN HELP IT! WE'RE HAVING RABBIT STEW TONIGHT...

EXPONENTIAL FUNCTIONS ARE GIVEN BY FORMULAS LIKE THIS:

$$f(x) = a^x$$

HERE THE "BASE" a IS FIXED, AND THE EXPONENT x VARIES. BY CONVENTION, WE ASSUME $a > 1$. THESE FUNCTIONS DESCRIBE CERTAIN KINDS OF GROWTH (POPULATION INCREASE, FOR EXAMPLE).

AMONG ALL POSSIBLE BASES a, MATHEMATICIANS SINGLE OUT ONE AS ESPECIALLY "NATURAL." THIS NUMBER, KNOWN AS e, HAS A DECIMAL EXPANSION THAT BEGINS LIKE THIS:

2.718281828459045235360287471352662497757247093699959574966967627724076630
353547594571382178525166427427466391932003059921817413596629043572900334
295260595630738132328627943490763233829880753195251019011573834187930702 1
54089149934884167509244761460668082264800168477411853742345442437107539077
744992069551702761838606261331384583000752044933826560297606737113200709 3
287091274437470472306969772093101416928368190255151086574637721112523897844
250569536967707854499699679468644549059879316368892300987931277361782154 2
499922957635148220826989519366803318252886939849646510582093923982948879
332036250944311730123819706841614039701983767932068328237646480429531180 23
287825098194558153017567173613320698112509961818815930416903515988885193458
072738667385894228792284998920868058257492796104841984443634632449684875
602336248270419786232090021609902353043699418491463140934317381436405462 5
315209618369088870701676839642437814059271456354906130 7505101
157477041718986106873969655212671546889570350354021234 10681701
210056278802351930332247450158539047304199577770935036604169973297250886
876966403555707162268447162560798826517871341951246652010305921236677194 32
527867539855894489697096409754591856956380236370162112047 7427228
364896134225164450781824423529486363721417402388934412479 6357437
026375529444833799801612549227850925778256209262264832 62779333
865664816277251640191059004916449982893150566047258027 7863186415519
565324425869829469593080191529872117255634754639644791 01459040905
862984967912874068705048958586717479854667757573205681 28845920541
334053922000113786300945560688166740016984205580403 3637...

MORE OR LESS...

WE CAN SEE WHY e IS NATURAL BY THINKING ABOUT **COMPOUND INTEREST.** IMAGINE A GENEROUS BANK (!) IS PAYING ANNUAL INTEREST OF **100%** ON YOUR SAVINGS ACCOUNT.

FIRST GENEROUS BANK

IF YOU START WITH \$1, AT THE END OF THE YEAR YOUR ACCOUNT WOULD HAVE DOUBLED TO \$2. PRETTY GOOD!

$$\$1 + 100\% \cdot (\$1) = \$2$$

BUT NOT GOOD ENOUGH, YOU COMPLAIN: YOU WANT YOUR INTEREST COMPOUNDED **MORE OFTEN.** YOU ASK THE BANK TO ADD ON 50% EVERY SIX MONTHS (100% PER YEAR TIMES HALF A YEAR), FOR THIS YEAR-END DOLLAR TOTAL:

$$(1 + \tfrac{1}{2}) + \tfrac{1}{2}(1 + \tfrac{1}{2}) = 2.25$$

BETTER!
50
50
25

NOW YOU DO A LITTLE ARITHMETIC: YOU NOTICE THAT

$$(1 + \tfrac{1}{2}) + \tfrac{1}{2}(1 + \tfrac{1}{2}) = (1 + \tfrac{1}{2})^2$$

AND THE NEXT TIME INTEREST IS ADDED, YOUR DOLLAR TOTAL WILL BE $(1 + \frac{1}{2})^3$, NEXT TIME $(1 + \frac{1}{2})^4$, NEXT TIME $(1 + \frac{1}{2})^5$...

AH, MATH!

SIMILARLY, IF YOU COMPOUND AT 100% **THREE** TIMES A YEAR, YOUR TOTAL AFTER ONE YEAR (THREE PAYMENTS) IS

$$\$(1 + \tfrac{1}{3})^3$$

IF COMPOUNDED n TIMES PER YEAR, YOUR YEAR-END TOTAL WOULD BE

$$\$(1 + \tfrac{1}{n})^n$$

AND YOU DECIDE TO FIND OUT JUST HOW MUCH MONEY THIS WOULD BE! USING YOUR CALCULATOR, YOU FIND:

PAYMENTS PER YEAR	TOTAL AFTER 1 YEAR	
1	$(1+1)^1$	$= \$2$
2	$(1+\frac{1}{2})^2$	$= \$2.25$
3	$(1+\frac{1}{3})^3$	$\approx \$2.37$
4	$(1+\frac{1}{4})^4$	$\approx \$2.44$
5	$(1+\frac{1}{5})^5$	$\approx \$2.49$
	...	
100	$(1+\frac{1}{100})^{100}$	$\approx \$2.705...$
1000	$(1+\frac{1}{1000})^{1000}$	$\approx \$2.718...$
	...	

THE TOTAL APPEARS TO BE APPROACHING e DOLLARS.

IF n IS VERY, VERY LARGE, YOU CAN THINK OF YOUR MONEY AS BEING COMPOUNDED **CONTINUOUSLY, ALL THE TIME.** IN THAT CASE, YOUR TOTAL BALANCE AT THE END OF ONE YEAR WOULD BE **EXACTLY e DOLLARS.**

ONE, TWO, TWO SEVENTY, TWO SEVENTY-ONE, TWO SEVENTY-ONE AND EIGHT TENTHS...

THE NUMBER e IS NATURAL BECAUSE CONTINUOUS COMPOUNDING IS NATURAL: IT DOESN'T DEPEND ON ANY PARTICULAR UNIT OF TIME.

WHAT'S SO SPECIAL ABOUT A YEAR?

THIS ALSO SHOWS THAT e IS THE MOST YOU CAN POSSIBLY MAKE IN A YEAR FROM ONE DOLLAR AT 100% INTEREST!

HEY! THERE'S ONLY $2.7182818284590452 HERE! I'VE BEEN SHORTCHANGED!!!

WE CAN USE THE FORMULA $(1+\frac{1}{n})^n$ TO CALCULATE e.
ALGEBRA TELLS US WE CAN EXPAND THAT BINOMIAL AS

$$1 + n\left(\frac{1}{n}\right) + \frac{n(n-1)}{2}\cdot\frac{1}{n^2} + \frac{n(n-1)(n-2)}{1\cdot2\cdot3}\cdot\frac{1}{n^3} + \frac{n(n-1)(n-2)(n-3)}{1\cdot2\cdot3\cdot4}\cdot\frac{1}{n^4} + \dots + \frac{1}{n^n}$$

WHEN n IS VERY LARGE, THE FRACTIONS $(n-1)/n$, $(n-2)/n$, ETC. ARE VERY NEARLY EQUAL TO 1, SO THE EARLY TERMS ARE VERY NEARLY

$$1 + 1 + \frac{1}{2} + \frac{1}{3!} + \frac{1}{4!} + \frac{1}{5!} + \dots$$

WHERE, IF m IS ANY INTEGER, $m!$ MEANS THE PRODUCT $1\cdot2\cdot3\cdot\ \dots\ \cdot m$.

NOW IF WE IMAGINE n GROWING "TO ∞," WE CAN CONCLUDE THAT e IS GIVEN BY A SUM WITH AN **INFINITE** NUMBER OF TERMS:

$$e = 1 + 1 + \frac{1}{2} + \frac{1}{3!} + \frac{1}{4!} + \frac{1}{5!} + \dots + \frac{1}{n!} + \dots$$

AND SO, IN FACT, IT IS.

SIGH... WHAT A BEAUTIFUL, BEAUTIFUL FORMULA.
HOW COME YOU NEVER SAY THAT TO ME...?
1/473!
1/952!
1/63!
1/1000!

BECAUSE OF THIS NUMBER'S SPECIAL, NATURAL STATUS, FROM NOW ON WE WILL REFER TO THE FUNCTION ***exp***, DEFINED BY

$$\mathbf{exp}(x) = e^x$$

AS **THE** EXPONENTIAL FUNCTION. e^x IS THE SUM YOU WOULD HAVE AFTER x YEARS IF ONE DOLLAR WERE COMPOUNDED CONTINUOUSLY AT 100% PER YEAR.

EXPONENTIAL FUNCTIONS GROW RAPIDLY WITH x. $f(x) = 2^x$, FOR EXAMPLE, DOUBLES EVERY TIME x INCREASES BY 1:

$$f(x+1) = 2^{x+1} = 2^x 2^1 = 2(2^x) = 2f(x)$$

e^x GROWS EVEN FASTER, AS YOU CAN EASILY CALCULATE. A POWER FUNCTION LIKE $g(x) = x^2$, BY COMPARISON, FALLS FAR BEHIND.

x	e^x	x^2
0	1.0	0
1	2.7183...	1
2	7.389...	4
3	20.085...	9
4	54.60...	16
5	148.41...	25
6	403.43...	36
7	1096.63...	49
8	2980.94...	64

IF a IS THE NUMBER WITH $e^a = 2$ ($a \approx 0.693$, AS YOU CAN CHECK ON YOUR CALCULATOR), THEN e^x DOUBLES WHENEVER x INCREASES BY a:

$$e^{(x+a)} = e^x e^a = 2e^x$$

AND IN PARTICULAR,

$$e^{na} = (e^a)^n = 2^n$$

$y = e^x$

$y = x^2$

IF r IS ANY POSITIVE NUMBER, THEN THE FUNCTION $h(x) = e^{rx}$ IS AN EXPONENTIAL FUNCTION, BECAUSE

$$e^{rx} = (e^r)^x$$

THE EXPONENTIAL WITH BASE e^r (NOTE THAT $e^r > 1$). IT INCREASES FASTER THAN $exp(x)$ IF $r > 1$ AND SLOWER IF $r < 1$.

$y = e^{rx}$
$r > 1$
$r = 1$
$r < 1$
EITHER WAY, IT GROWS!

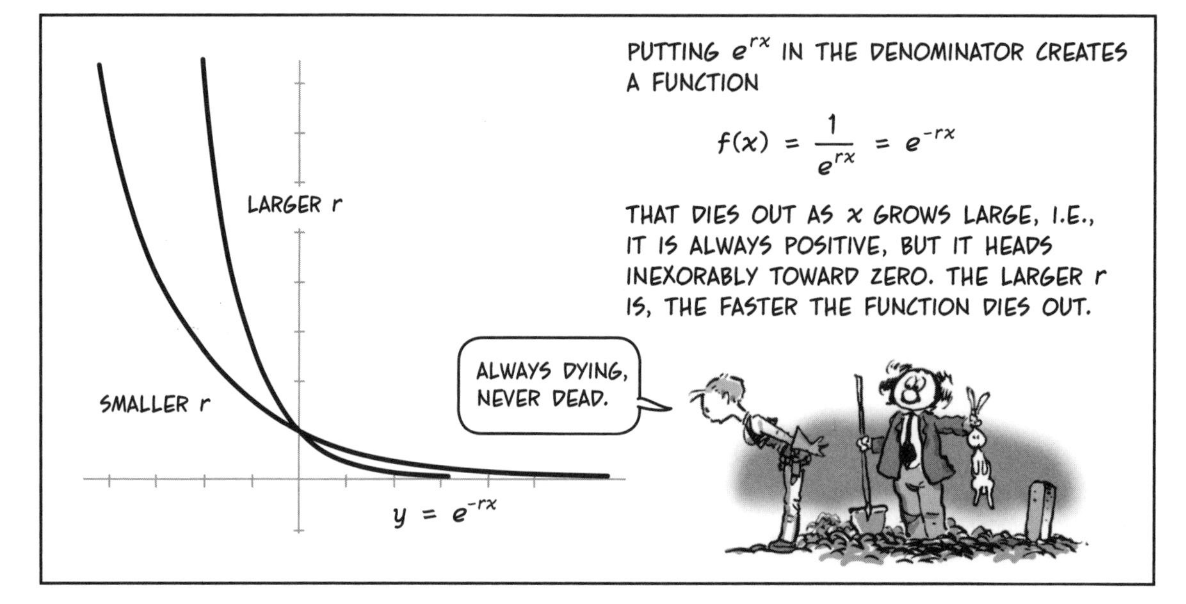

LARGER r
SMALLER r
$y = e^{-rx}$
ALWAYS DYING, NEVER DEAD.

PUTTING e^{rx} IN THE DENOMINATOR CREATES A FUNCTION

$$f(x) = \frac{1}{e^{rx}} = e^{-rx}$$

THAT DIES OUT AS x GROWS LARGE, I.E., IT IS ALWAYS POSITIVE, BUT IT HEADS INEXORABLY TOWARD ZERO. THE LARGER r IS, THE FASTER THE FUNCTION DIES OUT.

e^{-rx} DESCRIBES SUCH PHENOMENA AS RADIOACTIVE DECAY, WHERE THE **DECREASE** IN RADIATION IS PROPORTIONAL TO THE AMOUNT OF RADIOACTIVE MATERIAL PRESENT, RATHER LIKE COMPOUND INTEREST IN REVERSE.

IT'S AS IF THE BANK TOOK AWAY HALF YOUR MONEY EVERY SIX MONTHS...
YEH, ALMOST AS IF.

Circular Functions

OUR FINAL ELEMENTARY FUNCTIONS ARE THE **CIRCULAR,** OR **TRIG** FUNCTIONS: THE SINE, COSINE, TANGENT, AND SECANT. THESE DESCRIBE PROCESSES THAT GO BACK AND FORTH, UP AND DOWN, IN AND OUT, LIKE TIDES AND YO-YOS.

THESE FUNCTIONS ARISE EITHER IN CIRCLES OR RIGHT TRIANGLES. HERE IS A CIRCLE OF RADIUS 1, CENTERED AT THE ORIGIN. BEGINNING ON THE x-AXIS AT (1, 0), A POINT $P = (x_P, y_P)$ ORBITS COUNTERCLOCKWISE ALONG THE RIM. YOU CAN SEE A RIGHT TRIANGLE WITH HYPOTENEUSE OP.

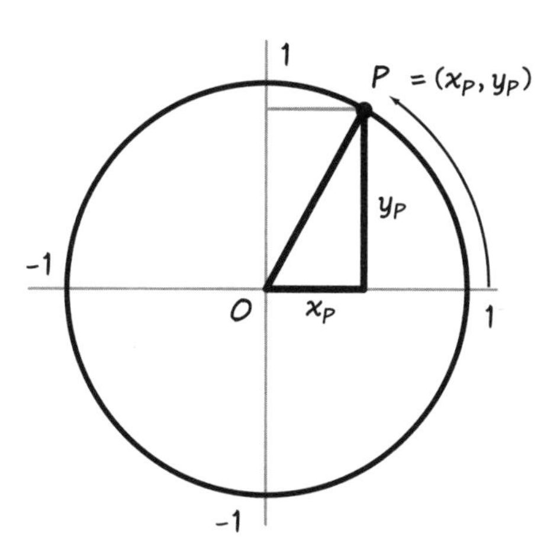

1
$P = (x_P, y_P)$
y_P
-1
O
x_P
1
-1

THE ANGLE θ (GREEK LETTER "THETA") BETWEEN OP AND THE x-AXIS IS MEASURED IN "NATURAL" UNITS, NAMELY THE **LENGTH OF THE ARC** TRAVELED BY P. THESE UNITS ARE CALLED **RADIANS.** SINCE THE CIRCLE'S CIRCUMFERENCE IS 2π, P TRAVELS 2π RADIANS IN ONE COMPLETE CIRCUIT. SMALLER ANGLES ARE PROPORTIONAL, AND MOVING CLOCKWISE GIVES NEGATIVE ANGLES. WHEN P DESCRIBES MORE THAN ONE CIRCUIT, THE ANGLE θ IS $>2\pi$.

θ
$-\theta$
$\theta + 2\pi$
WHY THE GREEK LETTER?
TO HONOR SOME DEAD WHITE GUYS. GOT A PROBLEM WITH THAT?
Pythagoras
Euclid
Archimedes

THE **SINE** AND **COSINE** OF θ ARE THE y AND x COORDINATES, RESPECTIVELY, OF THE POINT $P = (x_P, y_P)$. THE **TANGENT** OF θ IS THE RATIO y_P/x_P, WHEN $x_P \neq 0$.

$$\cos\theta = x_P$$

$$\sin\theta = y_P$$

$$\tan\theta = \frac{\sin\theta}{\cos\theta}$$

(YOU MAY HAVE LEARNED FROM THE ANCIENT GREEKS THAT $\sin\theta = y/r$, BUT HERE $r = 1$.)

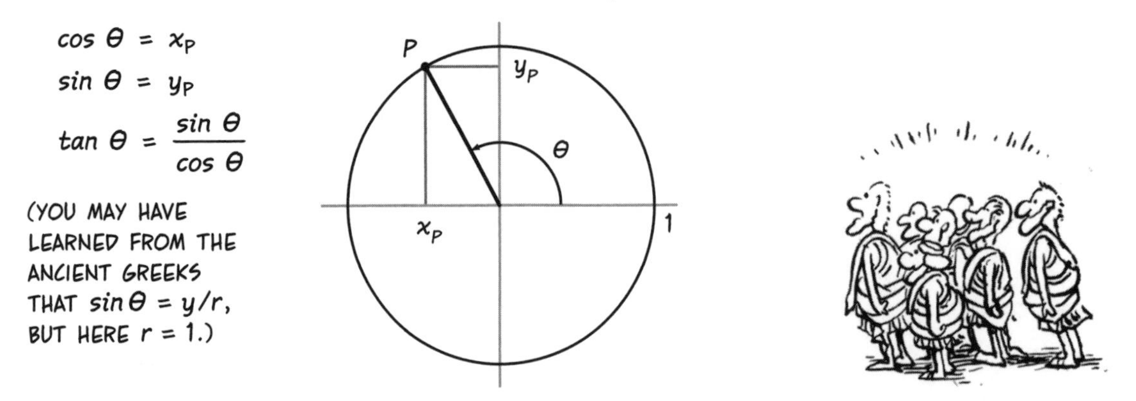
P
y_P
θ
x_P
1

THE SINE AND COSINE OSCILLATE BETWEEN −1 AND 1, REPEATING THEMSELVES EVERY 2π RADIANS. THE TANGENT REPEATS AFTER EVERY π RADIANS. THE TANGENT ZOOMS OFF TO INFINITY AT THE ODD HALVES OF π, WHERE THE COSINE IS ZERO.

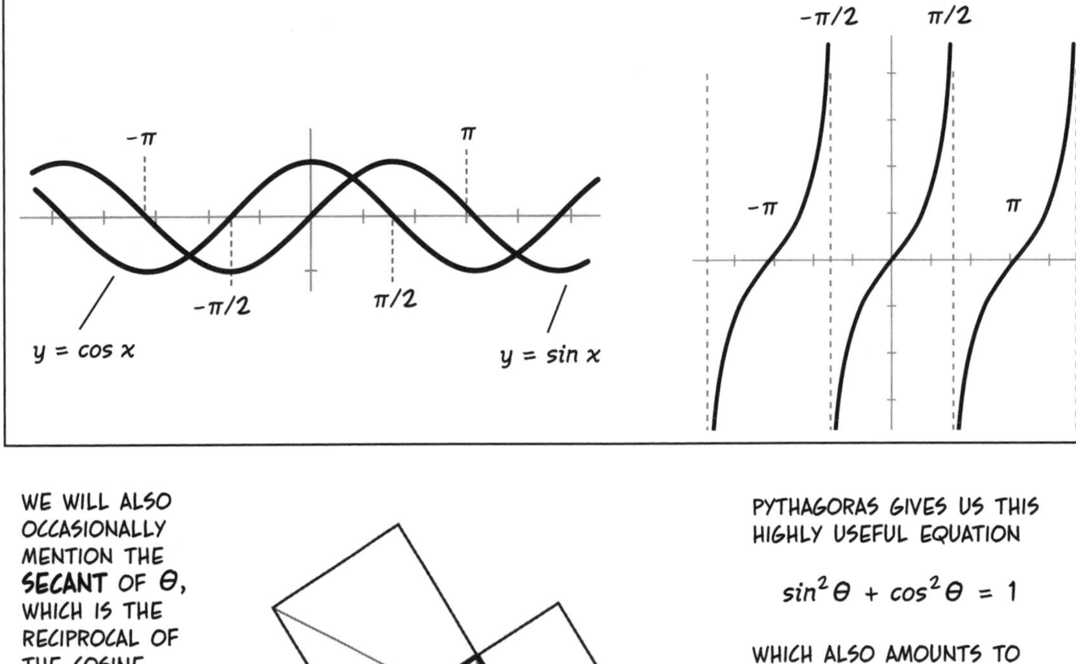
$-\pi$
π
$-\pi/2$
$\pi/2$
$y = \cos x$
$y = \sin x$
$-\pi/2$
$\pi/2$
$-\pi$
π

WE WILL ALSO OCCASIONALLY MENTION THE **SECANT** OF θ, WHICH IS THE RECIPROCAL OF THE COSINE, DEFINED WHEN $\cos\theta \neq 0$.

$$\sec\theta = \frac{1}{\cos\theta}$$

AND—

PYTHAGORAS GIVES US THIS HIGHLY USEFUL EQUATION

$$\sin^2\theta + \cos^2\theta = 1$$

WHICH ALSO AMOUNTS TO

$$\sec^2\theta = \tan^2\theta + 1$$

BECAUSE

$$\sec^2\theta = \frac{\sin^2\theta + \cos^2\theta}{\cos^2\theta}$$

ONE WAY TO VISUALIZE THE SINE AND COSINE IS TO IMAGINE THE POINT P IS A WEIGHT BEING SPUN AROUND AT THE END OF A 1-METER ROPE.

IMAGINE TWO OBSERVERS VIEWING THE CIRCLE EDGE-ON. ONE LOOKS ALONG THE x-AXIS, AND THE OTHER LOOKS DOWN THE y-AXIS.

x-GUY SEES THE WEIGHT START AT EYE LEVEL, THEN BOB UP AND DOWN, UP AND DOWN, UP AND DOWN. HE SEES THE y-VALUES, OR SINE.

y-GIRL, LOOKING DOWN, SEES **EXACTLY** THE SAME BACK-AND-FORTH MOTION, EXCEPT THAT THE WEIGHT START AT THE TOP OF ITS CYCLE. SHE SEES THE COSINE.

THIS CLEARLY SHOWS WHY THE SINE AND COSINE HAVE IDENTICAL GRAPHS, EXCEPT THAT ONE IS DISPLACED SIDEWAYS BY $\frac{\pi}{2}$.

$$\cos\theta = \sin(\theta + \frac{\pi}{2})$$

ALSO, SINCE $\cos(-\theta) = \cos\theta$,

$$\cos\theta = \sin(\frac{\pi}{2} - \theta)$$

AND

$$\sin\theta = \cos(\frac{\pi}{2} - \theta)$$

AND, AS I HOPE YOU'VE ALREADY LEARNED SOMEWHERE, THERE ARE COUNTLESS OTHER TRIGONOMETRIC IDENTITIES:

$$\sin(A+B) = \sin A\cos B + \sin B\cos A$$

$$\cos(A+B) = \cos A\cos B - \sin A\sin B$$

$$\sin^2\theta = \frac{1 - \cos 2\theta}{2}$$

$$\cos^2\theta = \frac{1 + \cos 2\theta}{2}$$

ETC.!

ANOTHER BASIC IDEA:

Composing Functions

SOMETIMES ONE FUNCTION IS "PLUGGED INTO" ANOTHER FUNCTION. FOR EXAMPLE, ON P. 15,

$$h(x) = \sqrt{x^2 - 1}$$

IS THE RESULT OF PLUGGING THE VALUE OF $f(x) = x^2 - 1$ INTO THE SQUARE ROOT FUNCTION $g(u) = \sqrt{u}$. FIRST WE EVALUATE $x^2 - 1$ AND THEN TAKE THE SQUARE ROOT. f IS CALLED THE **INSIDE** FUNCTION, AND g IS THE **OUTSIDE** FUNCTION.

f IS INSIDE THE RADICAL SIGN.
DOWN WITH BORING MATH BOOKS!
RADICAL SIGN OF A DIFFERENT KIND

Example 1:

$$F(x) = \tan^2 x + \tan x + 1$$

FIRST FIND $\tan x$, THEN PLUG IT INTO $g(y) = y^2 + y + 1$. THE INSIDE FUNCTION IS $f(x) = \tan x$ AND THE OUTSIDE FUNCTION IS g. WE WRITE

$$F(x) = g(f(x))$$

Example 2:

$$G(x) = e^{x^2}$$

INSIDE FUNCTION:

$$u(x) = x^2$$

OUTSIDE FUNCTION:

$$v(t) = e^t$$

$$G(x) = v(u(x))$$

Example 3:

$$H(x) = \tan(x^2 + x + 1)$$

INSIDE FUNCTION:

$$g(x) = x^2 + x + 1$$

OUTSIDE FUNCTION:

$$f(\theta) = \tan\theta$$

$$H(x) = f(g(x))$$

WHAT'S HAPPENING HERE IS THAT ONE FUNCTION'S OUTPUT BECOMES ANOTHER FUNCTION'S INPUT. THE FUNCTION g "EATS" THE OUTPUT OF THE FUNCTION f.

EEYEW! GROSS!
RELAX... THESE AREN'T BODILY FUNCTIONS...
f
g
x
f(x)
g(f(x))

IN EFFECT, THE ARROW OF f IS FOLLOWED BY THE ARROW OF g:

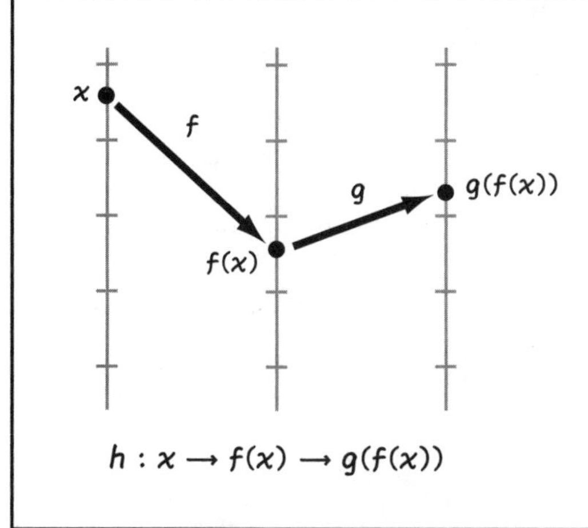

x
f
g
g(f(x))
f(x)

$$h : x \rightarrow f(x) \rightarrow g(f(x))$$

WE CALL THE FUNCTION h THE **COMPOSITION** OF g AND f, SOMETIMES WRITTEN $g \circ f$. NOTE THAT THE **INSIDE FUNCTION IS EVALUATED FIRST.** ITS ARROW IS ON THE **LEFT.** ALSO NOTE THAT **THE ORDER MATTERS.** IN GENERAL, $g \circ f \neq f \circ g$. IN EXAMPLES 1 AND 3 ON THE PREVIOUS PAGE, FOR INSTANCE,

$$f(g(x)) = \tan(x^2 + x + 1)$$
$$\neq \tan^2 x + \tan x + 1 = g(f(x))$$

YOU CAN EVEN HAVE A CHAIN COMPOSED OF MANY FUNCTIONS. WHY NOT!?

x
x²+1
√(x²+1)
cos(√(x²+1))
e^cos(√(x²+1))

COMPOSITION LEADS STRAIGHT TO

Fractional Powers

BY COMPOSING $f(x) = x^{\frac{1}{n}}$ WITH $g(y) = y^m$, WE CAN DEFINE FRACTIONAL POWERS OF x:

$$h(x) = x^{\frac{m}{n}} = (x^{\frac{1}{n}})^m = (x^m)^{\frac{1}{n}}.$$

FIRST TAKE THE nTH ROOT AND THEN THE mTH POWER, OR VICE VERSA. (HERE THE ORDER OF COMPOSITION DOESN'T MATTER.)

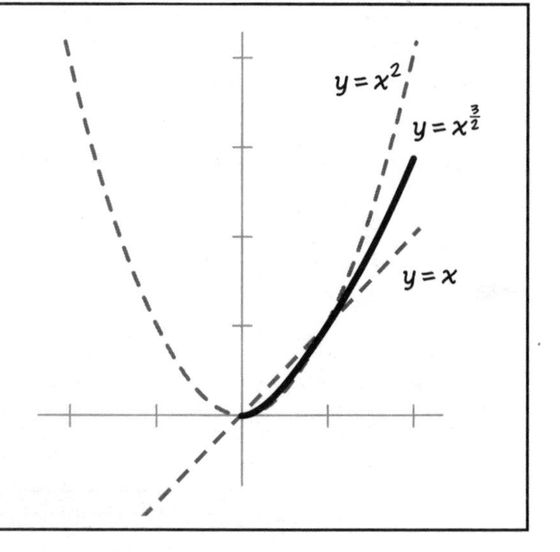

y = x²
y = x^(3/2)
y = x

NEXT BIG IDEA:

Inverting Functions

SOMETIMES WHEN WE COMPOSE TWO FUNCTIONS, A STRANGE THING HAPPENS: **NOTHING!**

WHAT'S WRONG WITH DOING NOTHING?*
x
x

Example: IF

$$f(x) = x^{\frac{1}{3}} \text{ AND } g(y) = y^3 \quad \text{THEN} \quad h(x) = g(f(x)) = (x^{\frac{1}{3}})^3 = x$$

PLUG x INTO $g \circ f$, AND OUT COMES x AGAIN. h CUBES THE CUBE ROOT, SO IN THE END THIS COMPOSITION DOESN'T DO ANYTHING! g "UNDOES" THE EFFECT OF f.

ALL THAT WORK... AND FOR WHAT?
f
216
6
g
216

IN WORDS, $g(x)$ IS "THE NUMBER WHOSE CUBE IS x." WE OFTEN WANT TO KNOW THIS KIND OF INFORMATION... SUCH THINGS AS:

THE NUMBER WHOSE SQUARE IS 4
THE NUMBER WHOSE SINE IS $\frac{1}{2}\sqrt{2}$
THE NUMBER WHOSE EXPONENTIAL IS 2

OR, IN SYMBOLS, WHAT NUMBER x, θ, OR t SOLVES THE EQUATIONS:

$$x^2 = 4$$
$$\sin\theta = \tfrac{1}{2}\sqrt{2}$$
$$e^t = 2$$

*WITH A TIP OF THE HAT TO THE CHINESE PHILOSOPHER ZHUANGZI!

BUT THERE'S A COMPLICATION... IT UNFORTUNATELY MAKES NO SENSE TO ASK FOR "THE" NUMBER WHOSE SQUARE IS 4, BECAUSE THERE ARE TWO OF THEM, 2 AND −2.

YOU'RE NOT VERY WELL-DEFINED, ARE YOU?
x^2
$\sqrt{y}$
2
4
sine
3π/4
π/4
−5π/4
9π/4

THE SINE IS EVEN WORSE. THE ANGLE $\pi/4$ SOLVES THE EQUATION:

$$\sin\theta = \tfrac{1}{2}\sqrt{2}$$

BUT SO DO A LOT OF OTHER ANGLES: $3\pi/4$, $-5\pi/4$, $9\pi/4$, $11\pi/4$, ETC.

$$\sin\left(\frac{\pi}{4} \pm 2\pi n\right) = \tfrac{1}{2}\sqrt{2}\,,\ n = 0, 1, 2, 3, \ldots$$

$$\sin\left(\frac{3\pi}{4} \pm 2\pi n\right) = \tfrac{1}{2}\sqrt{2}\,,\ n = 0, 1, 2, 3, \ldots$$

IN OTHER WORDS, THESE FUNCTIONS HAVE **MANY ARROWS** LANDING ON THE GIVEN NUMBER. A VALUE OF THE FUNCTION GENERALLY COMES FROM MANY DIFFERENT VALUES OF x.

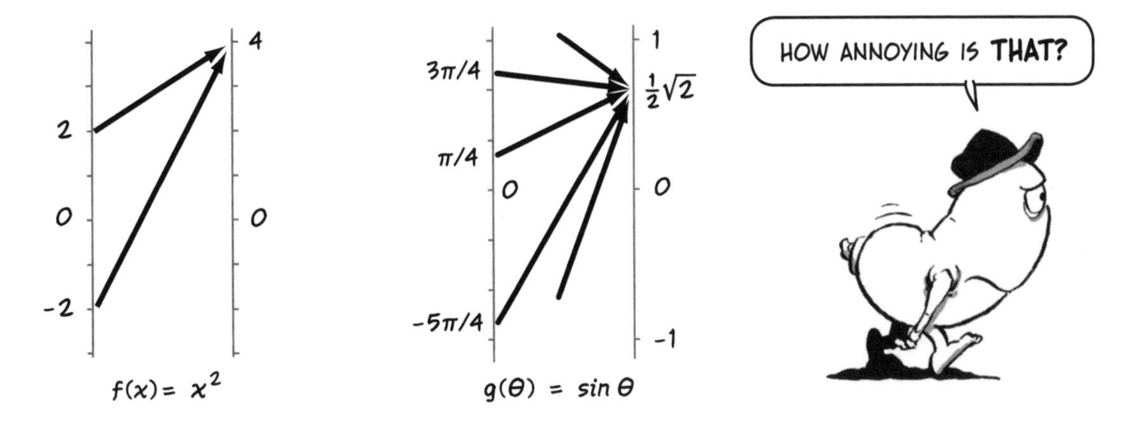

2
0
−2
4
0
$f(x) = x^2$
3π/4
π/4
0
−5π/4
1
$\frac{1}{2}\sqrt{2}$
0
−1
$g(\theta) = \sin\theta$
HOW ANNOYING IS **THAT?**

BUT NOT ALL FUNCTIONS ARE LIKE THAT: A FUNCTION IS CALLED **ONE-TO-ONE** IF NO TWO OF ITS ARROWS LAND IN THE SAME PLACE. IN SYMBOLS, IF $a \neq b$, THEN $f(a) \neq f(b)$. EACH VALUE OF f IS THE HEAD OF ONLY **ONE ARROW.**

I NEVER REPEAT MYSELF!
a
f(a)
b
f(b)

IF f IS ANY ONE-TO-ONE FUNCTION, WE CAN MAKE A NEW FUNCTION, f^{-1}, "**f-INVERSE,**" THAT UNAMBIGUOUSLY UNDOES THE ACTION OF f BY **REVERSING ITS ARROWS.** THE DOMAIN OF THE INVERSE FUNCTION f^{-1} IS ALL THE VALUES ASSUMED BY f, AND FOR ANY NUMBER $f(x)$ IN ITS DOMAIN, f^{-1} IS DEFINED BY

$$f^{-1}(f(x)) = x$$

YOU'RE A BACKWARD SORT OF FUNCTION, AREN'T YOU?

BECAUSE f^{-1} REVERSES THE ARROWS OF f, f OBVIOUSLY REVERSES THE ARROWS OF f^{-1} TOO—IT'S MUTUAL! SO IT FOLLOWS THAT

$$f(f^{-1}(y)) = y$$

THE TWO FUNCTIONS ARE INVERSES OF EACH OTHER! ORDER DOESN'T MATTER.

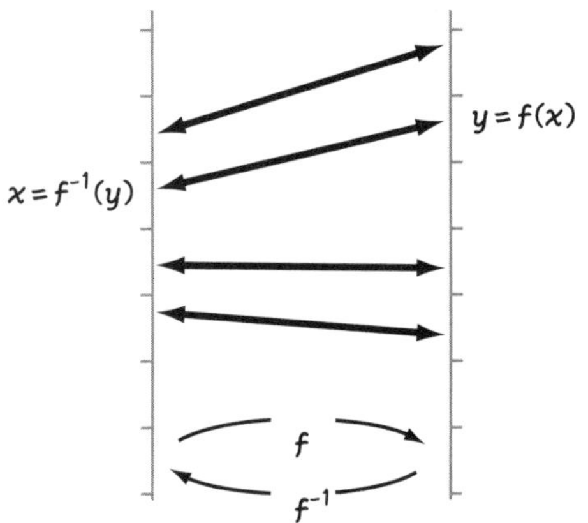
y = f(x)
x = f^{-1}(y)
f
f^{-1}

f^{-1}
EITHER WAY, WE CAN ACCOMPLISH NOTHING!

WHAT FUNCTIONS ARE ONE-TO-ONE? FOR OUR PURPOSES, IT WILL BE FUNCTIONS THAT ARE

increasing or decreasing.

WE DEFINE A FUNCTION TO BE **INCREASING,** OR **STRICTLY INCREASING,** IF THE VALUES $f(x)$ RISE AS x DOES. THAT IS, GIVEN ANY TWO POINTS a AND b IN THE DOMAIN OF f,

IF $a < b$, THEN $f(a) < f(b)$.

f IS **STRICTLY DECREASING** IF $a < b$ IMPLIES THAT $f(a) > f(b)$.* BECAUSE OF THE INEQUALITY, **EVERY INCREASING FUNCTION IS ONE-TO-ONE,** AND SO IS EVERY DECREASING FUNCTION.

THE VOLUME OF A SPHERE IS AN INCREASING FUNCTION OF RADIUS

AN INCREASING FUNCTION HAS A GRAPH THAT GOES UPHILL AS THE VARIABLE MOVES TO THE RIGHT. A DECREASING FUNCTION GOES DOWNHILL.

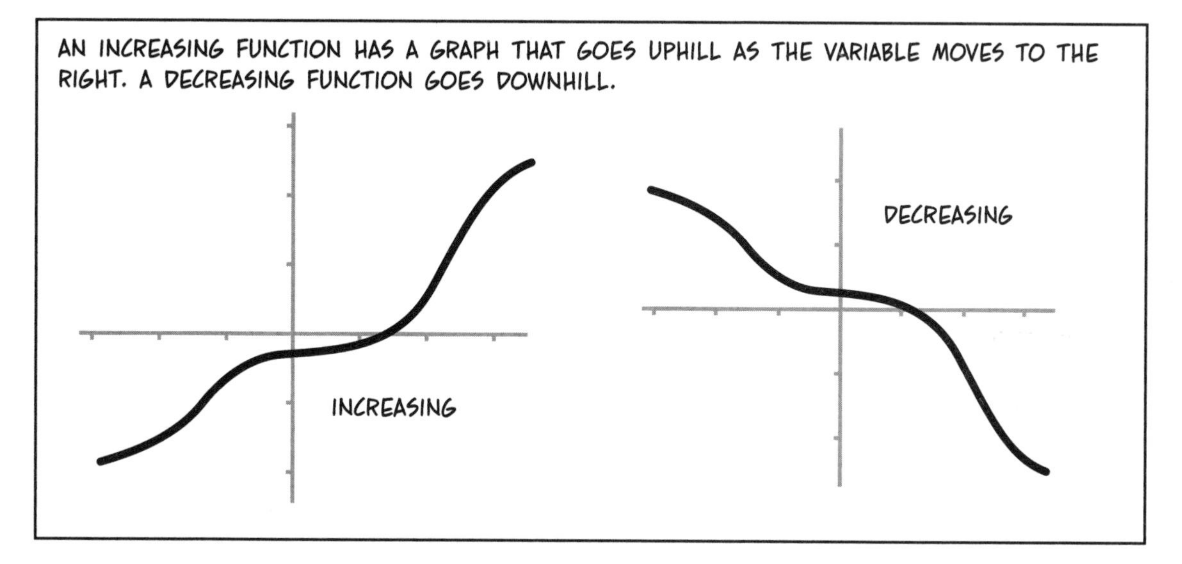
INCREASING
DECREASING

IN TERMS OF ARROWS, AN INCREASING FUNCTION'S ARROWS NEVER CROSS, BECAUSE THE VALUES $f(x)$ KEEP GOING UP THE LINE. **ALL** A DECREASING FUNCTION'S ARROWS CROSS EACH OTHER!

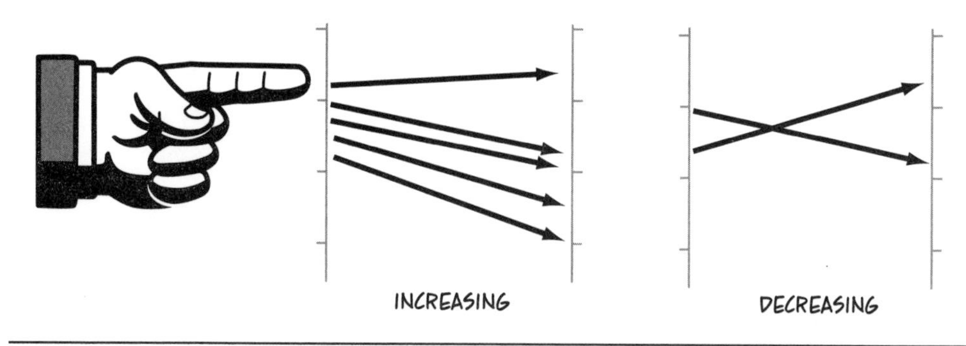
INCREASING
DECREASING

---

*NOTE THAT A FUNCTION f IS INCREASING IF AND ONLY IF $-f$ IS DECREASING.

SINCE AN INCREASING (OR DECREASING) FUNCTION IS ONE-TO-ONE, IT HAS AN INVERSE!

Little Example:

$f(x) = x^3$ IS INCREASING. ITS INVERSE IS

$$f^{-1}(x) = x^{\frac{1}{3}}$$

IN GENERAL, $g(x) = x^n$ IS INCREASING FOR ANY **ODD** INTEGER n, AND THE INVERSE IS

$$g^{-1}(x) = x^{\frac{1}{n}}$$

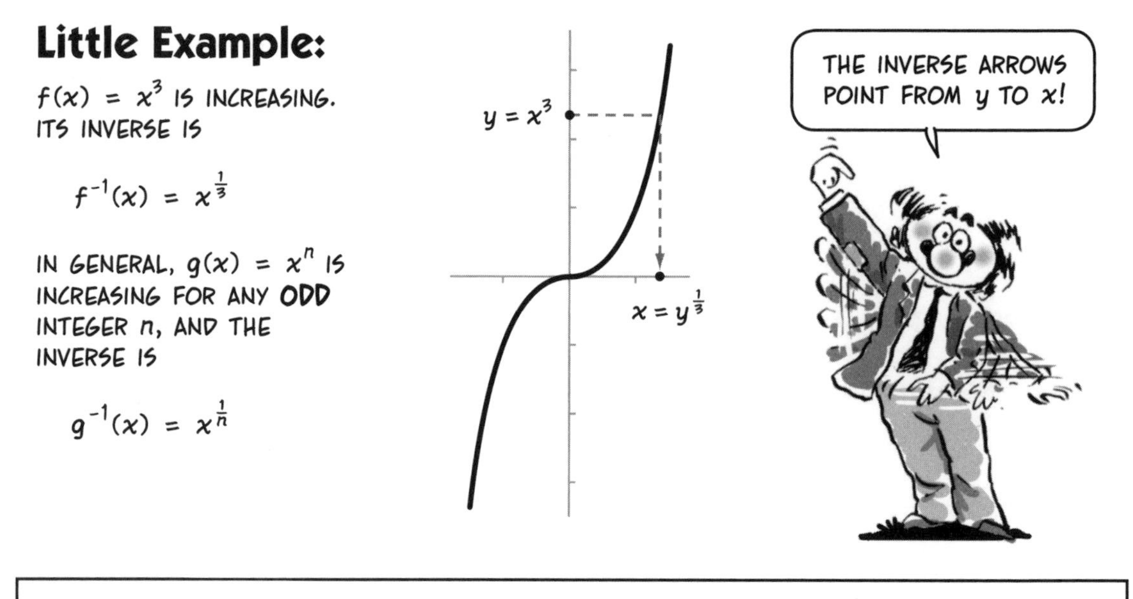

$y = x^3$
$x = y^{\frac{1}{3}}$
THE INVERSE ARROWS POINT FROM y TO x!

Big, Important Example: Natural Logarithm, Inverse of the Exponential

THE **EXPONENTIAL FUNCTION** $Exp(x) = e^x$ IS INCREASING.

PROOF: IF $a < b$, THEN

$$\frac{e^b}{e^a} = e^{(b-a)} > 1 \quad \text{BECAUSE } b - a > 0, \text{ SO}$$

$$e^b > e^a$$

ITS INVERSE FUNCTION IS CALLED THE **NATURAL LOGARITHM,** WRITTEN ln ("ELL-EN").

THE DOMAIN OF ln IS $(0, \infty)$ OR **ALL POSITIVE NUMBERS** BECAUSE e^x ASSUMES ALL VALUES GREATER THAN ZERO,* AND

$$e^{\ln y} = y \quad \text{AND} \quad \ln(e^x) = x$$

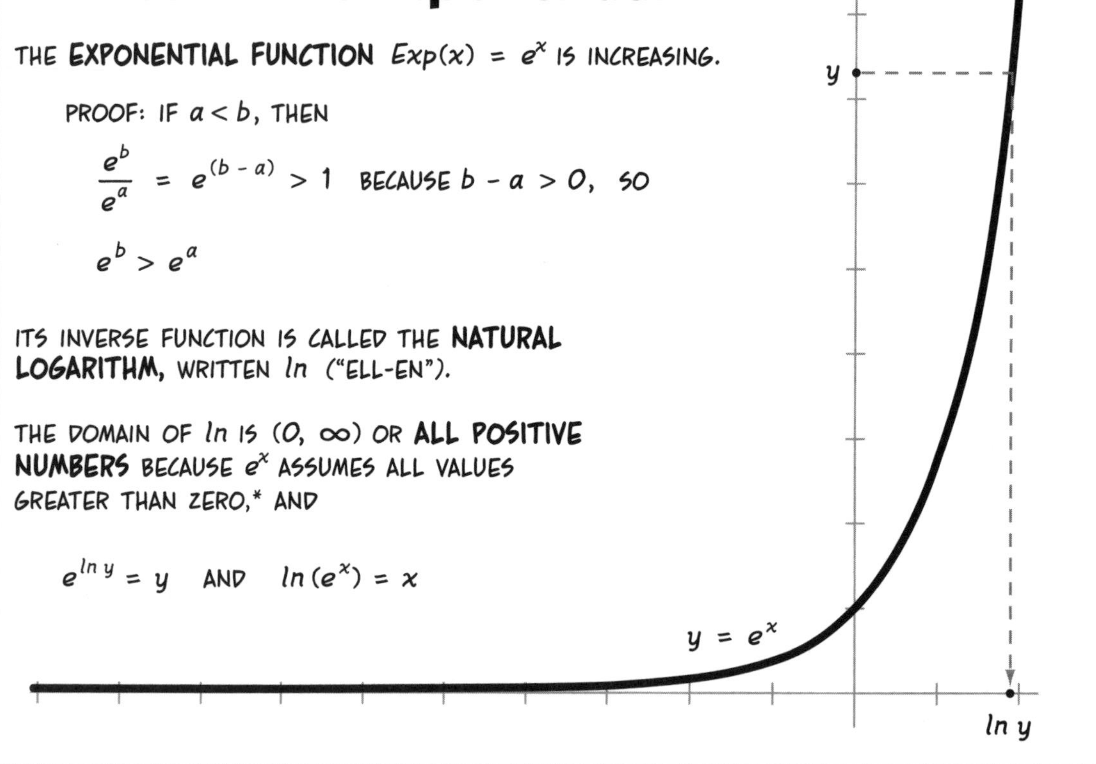

y
$y = e^x$
$ln\ y$

*SORRY, BUT YOU'RE ASKED TO TAKE THIS ON FAITH IN THIS BOOK.

EXPONENTS, YOU SHOULD RECALL, BEHAVE THIS WAY:

$$(e^x)(e^y) = e^{x+y} \qquad (e^x)^y = e^{xy}$$

THESE IMPLY THE FAMOUS LOG FORMULAS THAT USED TO BE SO IMPORTANT FOR MANAGING BIG CALCULATIONS BACK IN THE DAY BEFORE MECHANICAL AND ELECTRONIC COMPUTERS, WHEN EVERYTHING WAS DONE BY HAND.

$$\ln(xy) = \ln x + \ln y$$

$$\ln x^p = p \ln x$$

AND IN PARTICULAR, WHEN $p = -1$,

$$\ln \frac{1}{x} = \ln x^{-1} = -\ln x$$

THE LOGARITHM ENABLES US TO EXPRESS OTHER EXPONENTIALS IN TERMS OF "THE" EXPONENTIAL WITH BASE e. TAKE 2^x, FOR EXAMPLE. USING A CALCULATOR, YOU CAN FIND AN APPROXIMATE VALUE FOR $\ln 2$:

$\ln 2 \approx 0.693...$* FROM WHICH:

$$2^x = (e^{\ln 2})^x = e^{(\ln 2)x} = e^{0.693...x}$$

REPLACE 2 BY **ANY** NUMBER $a > 1$ AND THE EXPONENTIAL $A(x) = a^x$ CAN BE EXPRESSED SIMILARLY:

$$a^x = e^{rx}, \text{ WHERE } r = \ln a.$$

BY HAND? WHA—?
LOOK UP "LOGARITHM" ONLINE TO FIND OUT WHAT I'M TALKING ABOUT...
Table of Logarithms 1732 Edition

CONCLUSION: **EVERY EXPONENTIAL FUNCTION CAN BE EXPRESSED AS e^{rx} FOR SOME NUMBER r.**

*IT MAKES SENSE THAT $\ln 2$ IS BETWEEN 0 AND 1, BECAUSE 2 IS BETWEEN 1 ($= e^0$) AND e ($= e^1$).

Graphing Inverses

WE'VE SEEN HOW INVERSES LOOK IN TERMS OF ARROWS: f^{-1} SIMPLY TURNS ALL THE f ARROWS AROUND. HOW DOES THIS LOOK ON A GRAPH?

JUST FLIP SOME OF THESE AROUND... AND... UM...

ON THE GRAPH $y = f(x)$, FOLLOW AN ARROW FROM A POINT x TO $f(x) = y$. THE INVERSE FUNCTION f^{-1} REVERSES THAT ARROW, SO $f^{-1}(y) = x$.

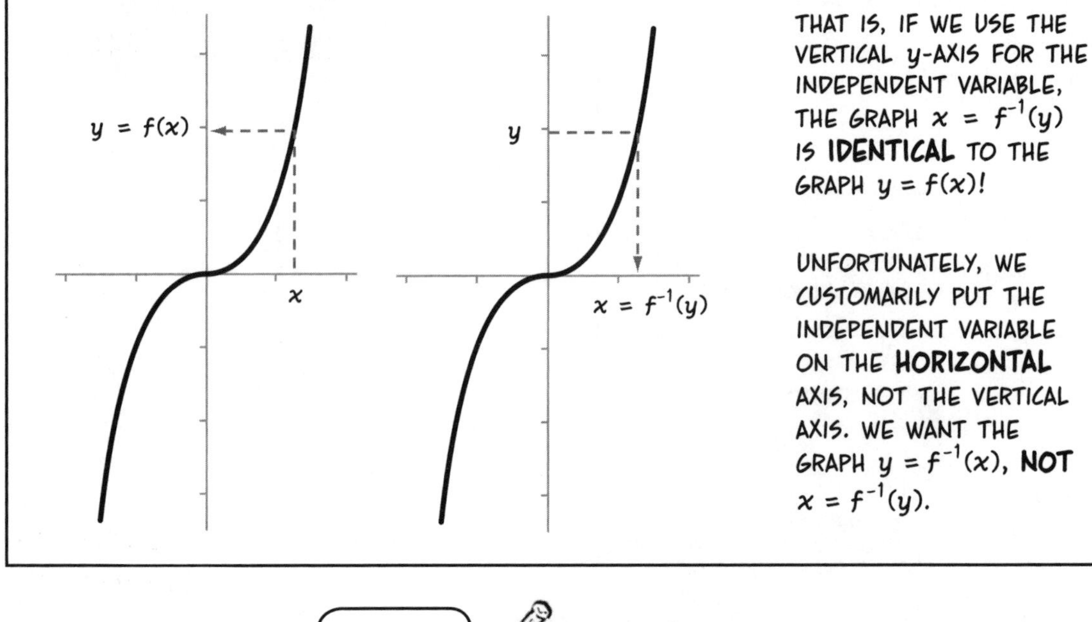

$y = f(x)$
x
y
$x = f^{-1}(y)$

THAT IS, IF WE USE THE VERTICAL y-AXIS FOR THE INDEPENDENT VARIABLE, THE GRAPH $x = f^{-1}(y)$ IS **IDENTICAL** TO THE GRAPH $y = f(x)$!

UNFORTUNATELY, WE CUSTOMARILY PUT THE INDEPENDENT VARIABLE ON THE **HORIZONTAL** AXIS, NOT THE VERTICAL AXIS. WE WANT THE GRAPH $y = f^{-1}(x)$, **NOT** $x = f^{-1}(y)$.

WHAT HAPPENS IF WE EXCHANGE x AND y?

NOTHING **TOO** CHAOTIC, I HOPE...

IF A POINT (a, b) IS ON THE GRAPH $y = f(x)$, THEN (b, a) IS ON THE GRAPH $y = f^{-1}(x)$. THE POINT (a, b) IS THE REFLECTION OF THE POINT (b, a) ACROSS THE LINE $y = x$, SO THE GRAPH $y = f^{-1}(x)$ IS THE **MIRROR IMAGE** OF THE GRAPH $y = f(x)$ REFLECTED ACROSS THE LINE $y = x$.

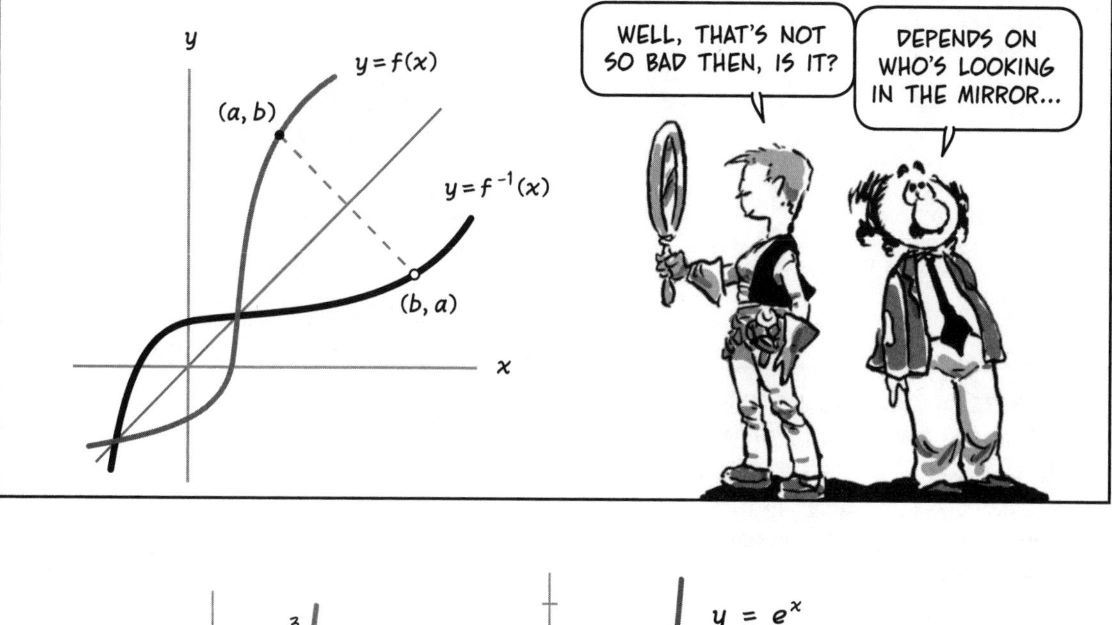

y
y = f(x)
(a, b)
y = f⁻¹(x)
(b, a)
x
WELL, THAT'S NOT SO BAD THEN, IS IT?
DEPENDS ON WHO'S LOOKING IN THE MIRROR...

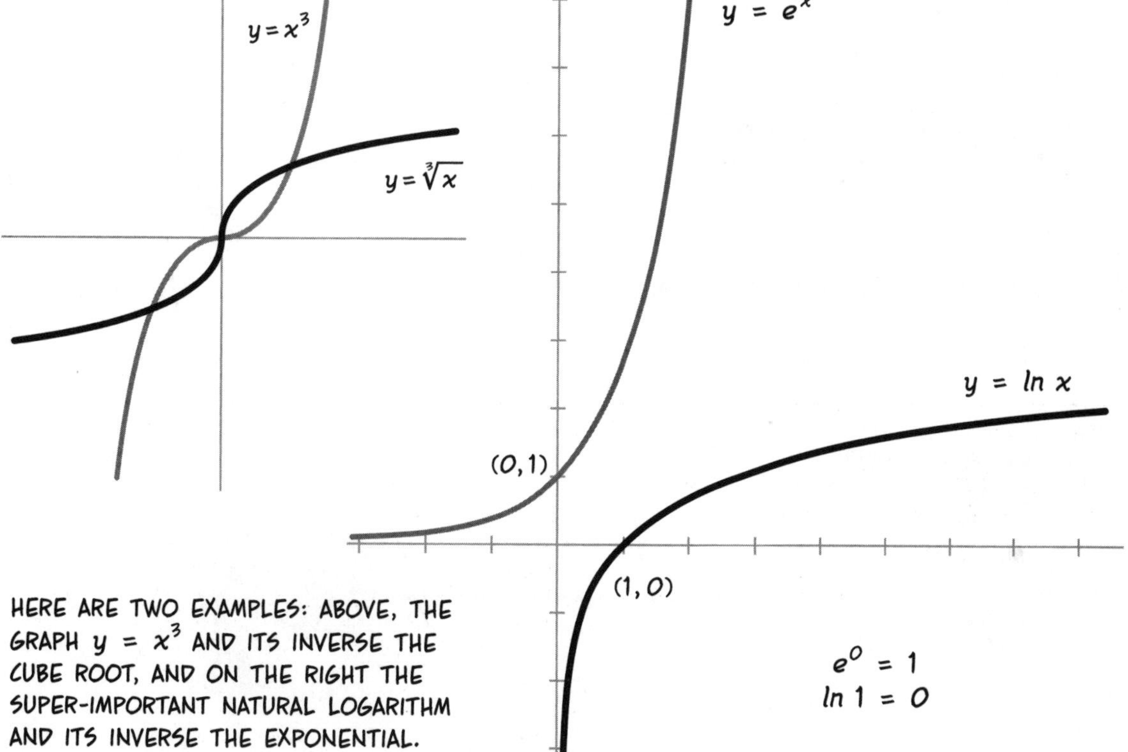

y = x³
y = ∛x
y = eˣ
y = ln x
(0, 1)
(1, 0)
e⁰ = 1
ln 1 = 0

HERE ARE TWO EXAMPLES: ABOVE, THE GRAPH $y = x^3$ AND ITS INVERSE THE CUBE ROOT, AND ON THE RIGHT THE SUPER-IMPORTANT NATURAL LOGARITHM AND ITS INVERSE THE EXPONENTIAL.

CAN WE INVERT A FUNCTION THAT IS NOT ONE-TO-ONE, THAT GOES UP AND DOWN? IF MANY ARROWS LAND AT A POINT y, WHICH ONE DO WE REVERSE? THE ANSWER IS: PICK WHICHEVER ONE YOU LIKE AND IGNORE THE REST!

I KNOW WHICH ONE I LIKE...

ONE SYSTEMATIC WAY TO DO THIS IS TO FLIP ONLY ARROWS ORIGINATING ON AN **INTERVAL WHERE THE FUNCTION IS ONE-TO-ONE.** FOR EXAMPLE, $f(x) = x^2$ IS INCREASING (AND SO ONE-TO-ONE) ON THE INTERVAL $[0, \infty)$. REVERSING ONLY THE ARROWS THAT START THERE MAKES AN INVERSE

$$f^{-1}(x) = \sqrt{x}$$

THAT ALWAYS GIVES THE **NON-NEGATIVE SQUARE ROOT.** THEN FOR ALL $x \geq 0$,

$$f(f^{-1}(x)) = x$$

$$f^{-1}(f(x)) = x \quad \text{(NO NEGATIVE } x \text{ ALLOWED!)}$$

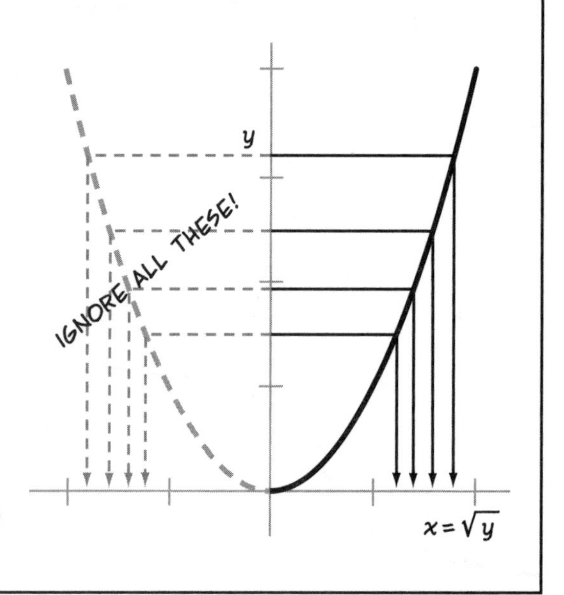
y
IGNORE ALL THESE!
$x = \sqrt{y}$

THIS WORKS FOR ANY FUNCTION f: **RESTRICT ITS DOMAIN** TO AN INTERVAL WHERE f IS INCREASING (OR DECREASING), AND ON THIS INTERVAL, f HAS AN INVERSE.

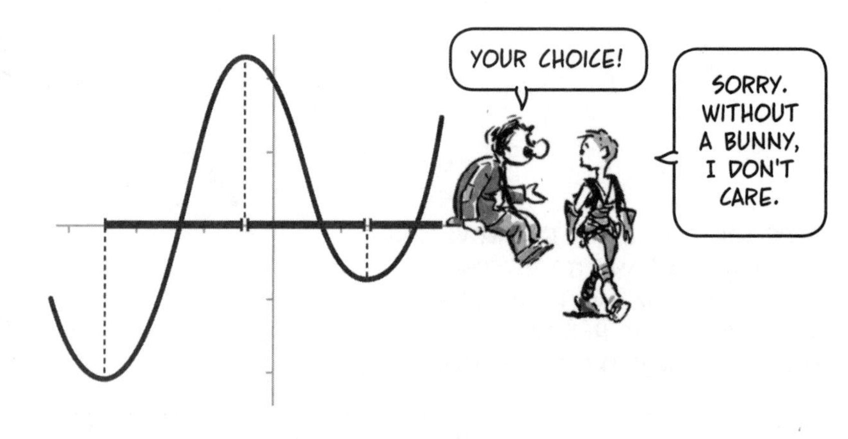
YOUR CHOICE!
SORRY. WITHOUT A BUNNY, I DON'T CARE.

Second Big, Important Example:
Inverse Circular Functions

THE SINE AND COSINE WOBBLE UP AND DOWN, UP AND DOWN... BUT ON SOME SHORT INTERVALS, THEY ARE INCREASING! LET'S CONCENTRATE ON THE SINE, BECAUSE THE COSINE WORKS EXACTLY THE SAME WAY. YOU CAN SEE THAT THE SINE INCREASES ON THE INTERVAL $[-\frac{\pi}{2}, \frac{\pi}{2}]$ WHERE ITS VALUES RISE FROM -1 TO 1.

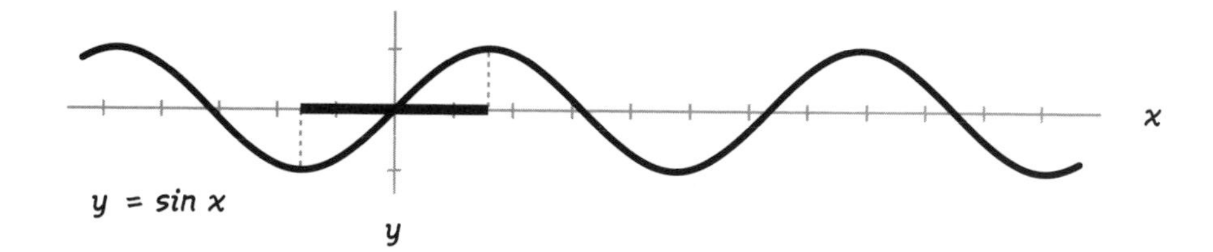
x
y = sin x
y

RESTRICTED TO THAT INTERVAL, THE SINE HAS AN INVERSE FUNCTION, CALLED THE **ARCSINE,** WITH DOMAIN $[-1, 1]$. THE ARCSINE ALWAYS TAKES ON VALUES BETWEEN $-\pi/2$ AND $\pi/2$.

SUCH A LITTLE DOMAIN...

WHY IS IT CALLED THE **ARC**SINE? BECAUSE IT'S THE ARC LENGTH CORRESPONDING TO A GIVEN SINE.

IF $\sin\theta = y$ THEN $\theta = \arcsin y$

θ IS AN ANGLE WHOSE SINE IS y. THIS ANGLE, BEING MEASURED IN RADIANS, IS THE LENGTH OF THE CORRESPONDING ARC ON THE UNIT CIRCLE (SEE P. 35). OTHER ANGLES HAVE THE SAME SINE, BUT θ IS THE **ONLY** ANGLE BETWEEN $-\pi/2$ AND $\pi/2$ WITH $\sin\theta = y$.

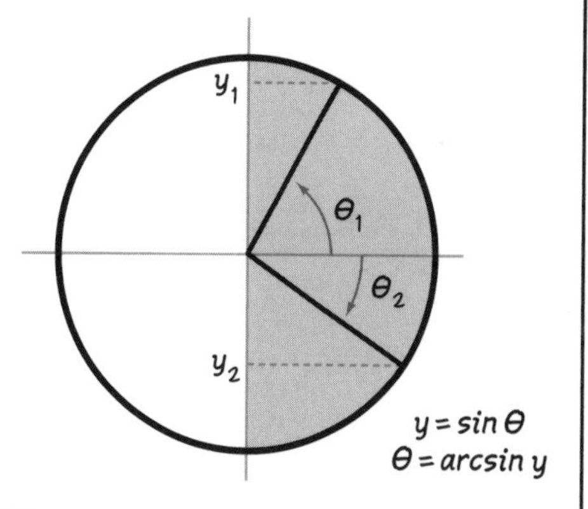
y1
θ1
θ2
y2
y = sin θ
θ = arcsin y

THIS CHAPTER'S FINAL FUNCTION WILL BE THE INVERSE OF THE TANGENT FUNCTION, $f(x) = \tan x$. THE INVERSE IS KNOWN AS THE **ARCTANGENT** FOR THE SAME REASON THE INVERSE SINE IS CALLED ARCSINE, AND IS SYMBOLIZED AS $\arctan x$.

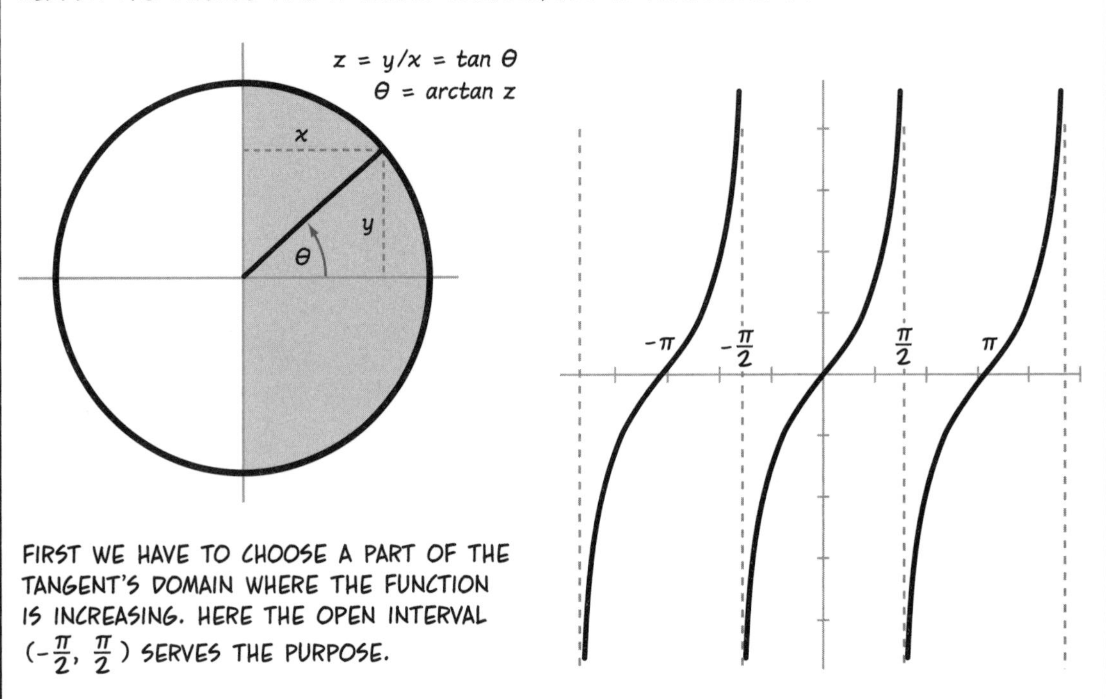

z = y/x = tan θ
θ = arctan z
x
y
θ
-π
-π/2
π/2
π

FIRST WE HAVE TO CHOOSE A PART OF THE TANGENT'S DOMAIN WHERE THE FUNCTION IS INCREASING. HERE THE OPEN INTERVAL $(-\frac{\pi}{2}, \frac{\pi}{2})$ SERVES THE PURPOSE.

THE TANGENT'S VALUES RANGE OVER **ALL REAL NUMBERS,** I.E., THE "INTERVAL" $(-\infty, \infty)$, SO THE ARCTANGENT'S **DOMAIN** IS (∞, ∞). THE FUNCTION IS DEFINED EVERYWHERE, BUT ITS VALUE ALWAYS LIES BETWEEN $-\pi/2$ AND $\pi/2$.

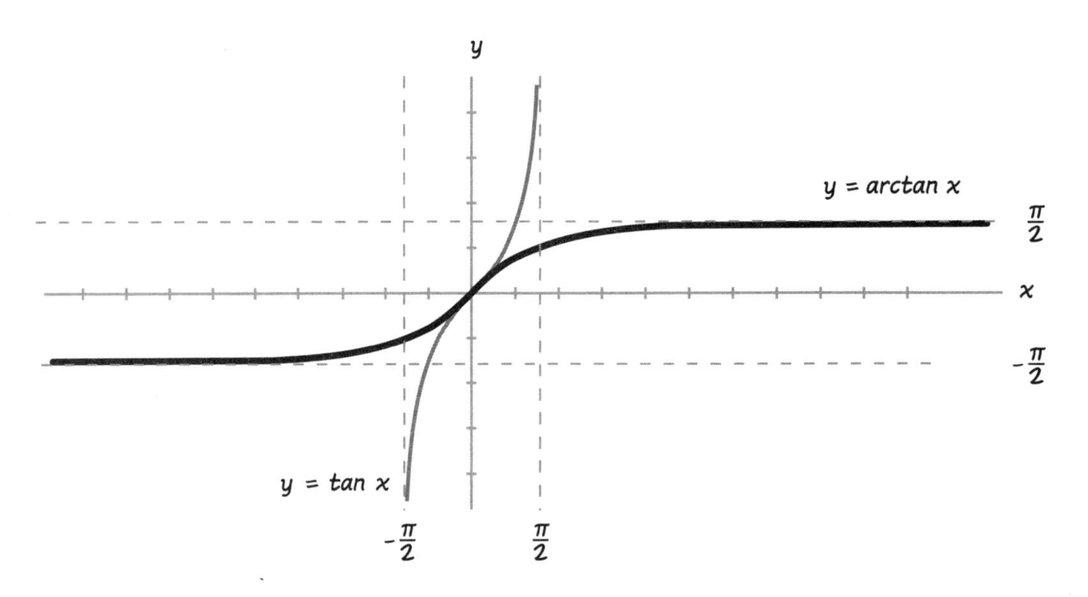

y
y = arctan x
π/2
x
-π/2
y = tan x
-π/2
π/2

THIS COMPLETES OUR TOUR OF THE ELEMENTARY FUNCTIONS! WE'VE SEEN POWER FUNCTIONS (POSITIVE, NEGATIVE, AND FRACTIONAL), THE EXPONENTIAL AND ITS INVERSE THE NATURAL LOGARITHM, AND THE CIRCULAR FUNCTIONS AND THEIR INVERSES. NOT SO MANY, REALLY...

A NICE, MANAGEABLE MENAGERIE!
SST! STOP THAT, YOU TWO...

BUT OF COURSE, WHEN YOU ADD, MULTIPLY, DIVIDE, AND COMPOSE THESE BASIC INGREDIENTS, YOU CAN MAKE MONSTERS LIKE THIS:

$$f(x) = e^{\cos^2[(1 + x^3)^{\frac{1}{2}}(5x - \sin(\ln(\cos x)))^{-\frac{1}{3}}]}$$

AND WHAT ARE WE SUPPOSED TO DO WITH THEM?
CALCULUS, NATURALLY!

Problems

DESCRIBE THE DOMAIN OF EACH OF THE FOLLOWING FUNCTIONS:

1. $Q(t) = \dfrac{3}{1 - 2t}$

2. $f(b) = \dfrac{\sqrt{2b - 1}}{(b - 4)(b + 9)}$

3. $M(x) = \dfrac{1}{1 - |x|}$

4. $V(x) = \sqrt{1 - \left(\frac{x}{2}\right)^2}$

5. $g(\theta) = \dfrac{\tan \theta}{\theta^2 - \frac{\pi}{9}}$

6. $A(x) = (1 - e^{2x})^{-1}$

7. $T(u) = (1 - e^{2u})^{-1/2}$

8. $f(x) = \ln (1 + x^2)$

9. $L(x) = \ln (\ln x)$

HERE IS THE GRAPH OF A FUNCTION $y = f(x)$, A POINT c ON THE x-AXIS, AND A POINT d ON THE y-AXIS.

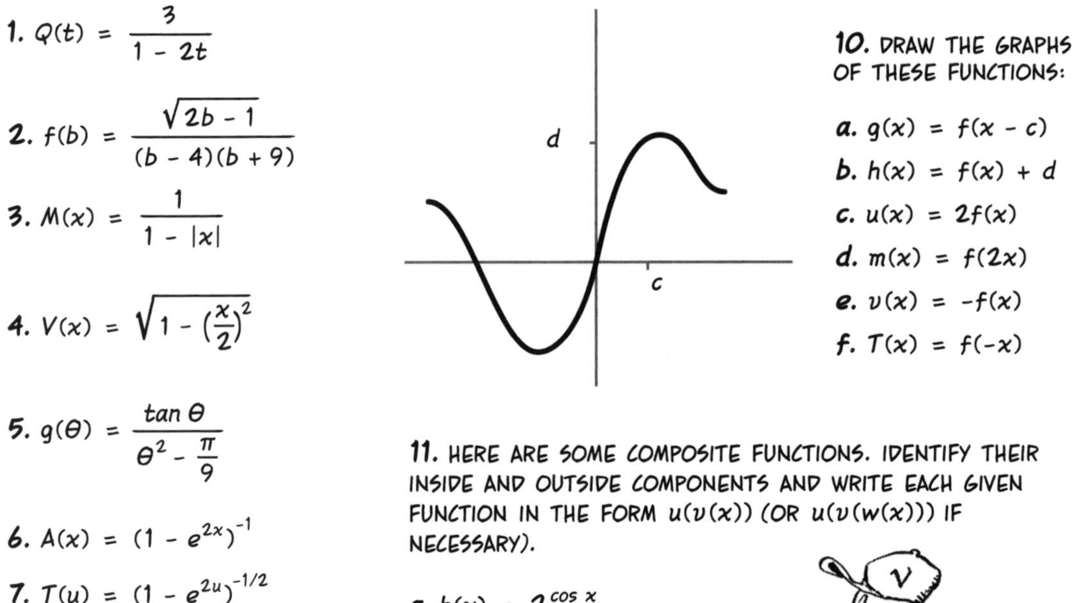

d
c

10. DRAW THE GRAPHS OF THESE FUNCTIONS:

a. $g(x) = f(x - c)$

b. $h(x) = f(x) + d$

c. $u(x) = 2f(x)$

d. $m(x) = f(2x)$

e. $v(x) = -f(x)$

f. $T(x) = f(-x)$

11. HERE ARE SOME COMPOSITE FUNCTIONS. IDENTIFY THEIR INSIDE AND OUTSIDE COMPONENTS AND WRITE EACH GIVEN FUNCTION IN THE FORM $u(v(x))$ (OR $u(v(w(x)))$ IF NECESSARY).

a. $h(x) = 2^{\cos x}$

b. $h(x) = \sqrt{\ln(x^2 - 1)}$

c. $h(x) = 4e^{3x} + e^{2x} + 6e^x - 99$

12. SHOW THAT FOR ANY NUMBER c, A POLYNOMIAL $P(x) = b_0 + b_1x + b_2x^2 + \ldots + b_nx^n$ CAN ALSO BE WRITTEN $P(x) = a_0 + a_1(x - c) + a_2(x - c)^2 + \ldots + a_n(x - c)^n$ WHERE $a_0 = P(c)$. SHOW THAT $a_n \neq 0$ IF $b_n \neq 0$.

13. LET'S DEFINE A FUNCTION f ON THE OPEN INTERVAL $(-1, 1)$ LIKE THIS:

$f(x) = (x + 1)^2$ FOR $-1 < x \leq 0$

$f(x) = x^2 - 1$ FOR $0 < x < 1$

a. IS f AN INCREASING FUNCTION ON ITS WHOLE DOMAIN?

b. IS f ONE-TO-ONE?

c. DRAW THE GRAPH OF f AND ITS INVERSE f^{-1}.

14. SHOW THAT

$$\arctan x = \arccos \frac{1}{\sqrt{1 + x^2}} = \arcsin \frac{x}{\sqrt{1 + x^2}}$$

(HINT: DRAW A TRIANGLE.)

15. IF YOU HAVE A_0 DOLLARS TODAY, AND IT COMPOUNDS SO THAT YOU HAVE $A(t) = A_0e^{rt}$ DOLLARS AFTER t YEARS, HOW LONG DOES IT TAKE TO DOUBLE YOUR MONEY? (r IS ASSUMED FIXED.)

Chapter 1
Limits

A BIG IDEA ABOUT SMALL THINGS

THE LAST CHAPTER WAS ABOUT FUNCTIONS "SITTING STILL," SO TO SPEAK. GIVEN A POINT x, WE FOLLOWED ITS ARROW TO THE **LOCATION** OF $f(x)$.

a
$f(a)$

NOW CALCULUS INTRODUCES A **NEW IDEA:** NOT JUST THE VALUE OF A FUNCTION AT A POINT a, BUT WHAT $f(x)$ LOOKS LIKE **VERY, VERY CLOSE** TO a. IN FACT, WE MAY BE INTERESTED IN THESE VALUES AT NEARBY POINTS x EVEN WHEN f ISN'T DEFINED AT THE POINT a!!

$a + 0.000001$
a
$a - 0.000001$
BUT WHY? JUST SAYIN'...

WHY? THE REASON GOES BACK TO NEWTON'S AND LEIBNIZ'S IDEA ABOUT **VELOCITY.** (SEE PP. 7-8.)

IT PLEASES ME, YOU AGAIN TO SEE, FRÄULEIN!

THEIR IDEA, REMEMBER, WAS THIS: IF $s(t)$ IS POSITION AT TIME t, AND a IS A MOMENT IN TIME, THEN WHEN t IS NEAR a, THE VELOCITY AT TIME a IS VERY CLOSE TO THE "DIFFERENCE QUOTIENT" $D(t)$.

$$D(t) = \frac{s(t) - s(a)}{t - a}$$

D IS A FUNCTION OF t THAT IS NOT DEFINED AT $t = a$, BUT **IS** DEFINED WHEN t IS **NEAR** a. AS t GETS CLOSER TO a, WE EXPECT $D(t)$ TO APPROACH THE INSTANTANEOUS VELOCITY AT a. WE'LL WANT TO WRITE

$$v(a) = \lim_{t \to a} D(t)$$

0/0 GIVES ME THE WORST INDIGESTION...
a
v(a)

AND SAY THAT $v(a)$ IS THE **LIMIT** OF $D(t)$ AS t GOES TO a.

THE KICK LASTED MERE MILLISECONDS, BUT IT CERTAINLY HAD VELOCITY.
CERTAINLY.

FOR EXAMPLE, IT SO HAPPENS THAT ON A RAMP SET AT AN ANGLE OF SLIGHTLY MORE THAN 11.77 DEGREES, A FRICTIONLESS VEHICLE STARTING FROM REST AT $s = 0$ WILL ROLL DOWN ACCORDING TO THE FORMULA

$$s(t) = t^2 \text{ METERS}$$

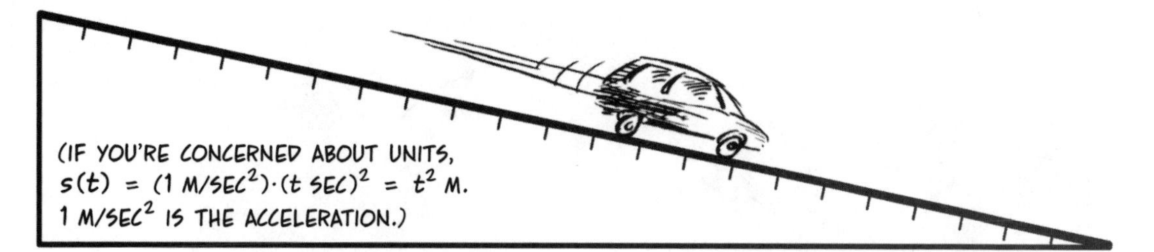
(IF YOU'RE CONCERNED ABOUT UNITS,
s(t) = (1 M/SEC²)·(t SEC)² = t² M.
1 M/SEC² IS THE ACCELERATION.)

THEN NEAR A POINT IN TIME a,

$$D(t) = \frac{t^2 - a^2}{t - a}$$

LET'S SUPPOSE $a = 3$ SEC., AND SEE WHAT HAPPENS TO $D(t)$ WHEN t IS CLOSE TO a:

?

t	$t - 3$	$t^2 - 9$	$D(t)$
2.9	-0.1	-0.59	5.9
2.99	-0.01	-0.0599	5.99
2.999	-0.001	-0.005999	5.999
...	...	...	...
3.001	0.001	0.006001	6.001
3.01	0.01	0.0601	6.01
3.1	0.1	0.61	6.1

$D(t)$ GIVES EVERY APPEARANCE OF APPOACHING A LIMIT OF 6 AS $t \to 3$.

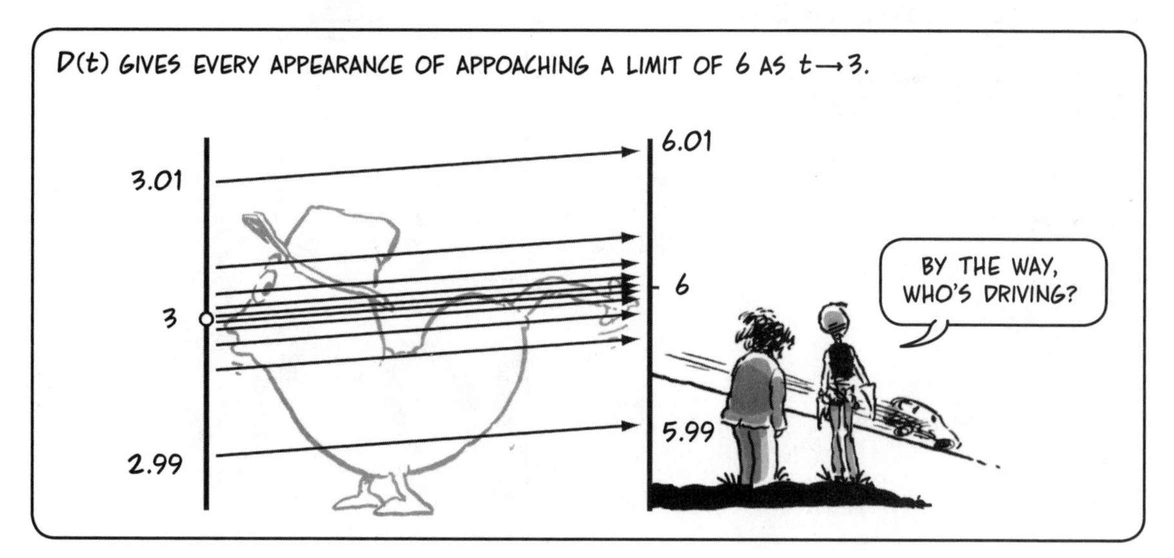
3.01
3
2.99
6.01
6
5.99
BY THE WAY, WHO'S DRIVING?

MAYBE YOU STILL DON'T QUITE BELIEVE IT. YOU CHALLENGE ME TO MAKE $D(t)$ EVEN CLOSER TO 6, WITHIN 0.000001, SAY. THAT IS, YOU REQUIRE

$$|D(t) - 6| < 0.000001$$

I DARE YOU!

I ACCEPT THE CHALLENGE. FIRST, I REWRITE THE EXPRESSION BY LETTING $h = t - 3$ OR $t = 3 + h$. THEN

$$D(t) = \frac{(3 + h)^2 - 3^2}{(3 + h) - 3} = \frac{6h + h^2}{h}$$

$$= 6 + h \quad \text{WHEN } h \neq 0$$

HM! TRUE ENOUGH...

AND I OBSERVE THAT AS LONG AS h IS NON-ZERO AND $|h| < 0.000001$, THEN IT FOLLOWS THAT, SINCE $D(t) = 6 + h$,

$$|D(t) - 6| = |h| < 0.000001$$

SO THERE!

BUT YOU'RE A PERSISTENT SO-AND-SO... YOU CHALLENGE ME AGAIN: NOW YOU WANT $D(t)$ WITHIN 0.0000000001 OF 6.

I'VE GOT A MILLION OF 'EM!

I SATISFY YOUR DEMAND AGAIN: AS LONG AS h IS NON-ZERO AND

$$|h| < 0.0000000001$$

THEN, AS ABOVE,

$$|(D(t)) - 6| = |h| < 0.000000001$$

OR, IF YOU LIKE,

$$5.9999999999 < D(t) < 6.0000000001$$

YOU DECIDE YOU WANT IT EVEN CLOSER, BUT YOU DON'T WANT TO STAND AROUND FEEDING ME SMALL NUMBERS ALL DAY...

I'M PRETTY SURE I HAVE BETTER THINGS TO DO...

SO YOU GIVE ME A **GENERAL CHALLENGE:** "IF I OFFER YOU **ANY** SMALL NUMBER—CALL IT ε, THE GREEK LETTER EPSILON*—CAN YOU MAKE $D(t)$ WITHIN ε OF 6 BY MAKING h SMALL? CAN YOU FORCE $|D(t) - 6| < \varepsilon$?"

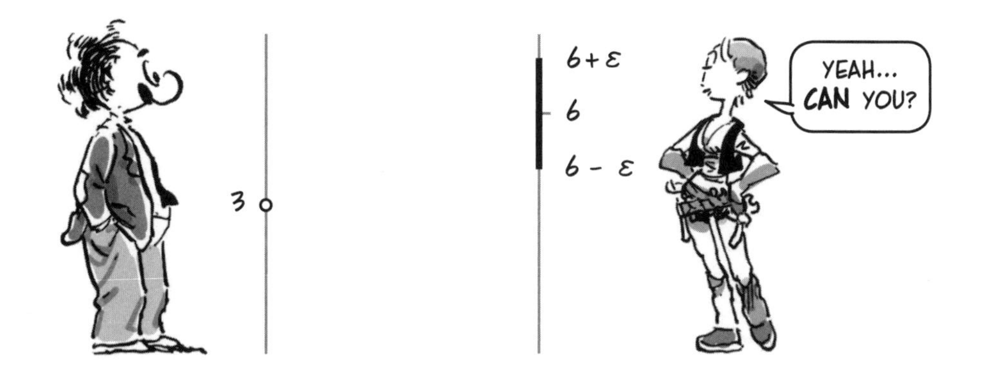

3
6+ε
6
6 - ε
YEAH... CAN YOU?

SIMPLE! I KNOW THAT $D(t) = 6 + h$ WHEN $h \neq 0$, SO I ANSWER THE CHALLENGE BY SAYING, "LET $|h| < \varepsilon$."

TAKE THAT!
3 + ε
3
3 - ε
D
6 + ε
6
6 - ε

IF $|t - 3| = |h| < \varepsilon$,

THEN $|D(t) - 6| =$

$|(6 + h) - 6| =$

$|h| < \varepsilon$.

AND I'VE MET YOUR CHALLENGE.

NOW YOU'RE SATISFIED! I'VE SHOWN THAT $D(t)$ CAN BE MADE WITHIN A HAIR OF 6, NO MATTER HOW SLENDER THE HAIR!!!

UM... CAN WE NOT TALK ABOUT HAIR?

*IT'S TRADITIONAL. SORRY!

BY NOW, YOU MAY BE CONVINCED THAT A FUNCTION REALLY CAN APPROACH A LIMIT AS $x \to a$, EVEN IF THE FUNCTION ISN'T DEFINED AT THE POINT a ITSELF. GRAPHICALLY, IT LOOKS LIKE THIS: $\lim_{x \to a} f(x) = L$ MEANS THAT **THE GRAPH $y = f(x)$ HEADS FOR THE POINT (a, L).**

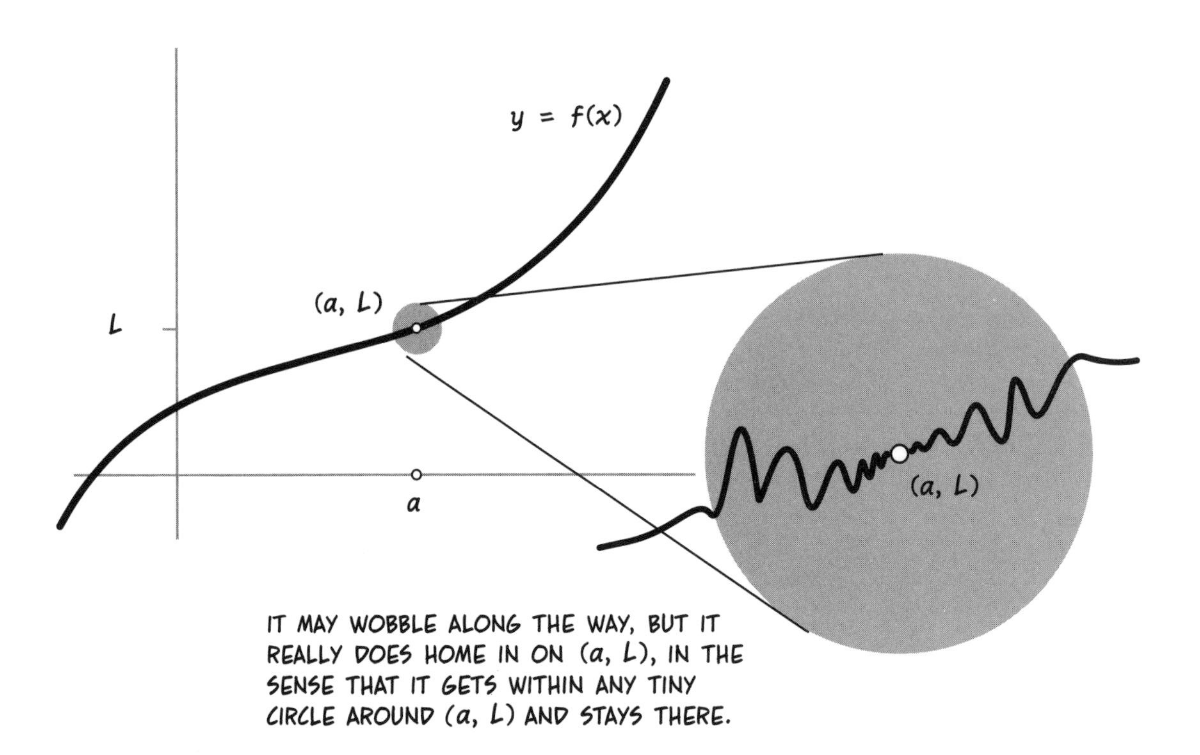

y = f(x)
(a, L)
L
a
(a, L)

IT MAY WOBBLE ALONG THE WAY, BUT IT REALLY DOES HOME IN ON (a, L), IN THE SENSE THAT IT GETS WITHIN ANY TINY CIRCLE AROUND (a, L) AND STAYS THERE.

LIMITS ARE ESPECIALLY EASY WHEN f IS ONE OF OUR **ELEMENTARY FUNCTIONS,** POWER FUNCTIONS, CIRCULAR FUNCTIONS, EXPONENTIALS, AND THEIR INVERSES. WHEN ONE OF THESE FUNCTIONS IS DEFINED AT A POINT a, THE GRAPH GOES WHERE IT OUGHT TO GO, NAMELY

$$\lim_{x \to a} f(x) = f(a)$$

FOR INSTANCE,

$$\lim_{x \to 2} 50x = 100$$

$$\lim_{x \to 9} \frac{1}{x} = \frac{1}{9}$$

$$\lim_{\theta \to \pi/2} \cos \theta = 0$$

TO FIND THE LIMIT AT a, JUST PLUG a INTO THE FUNCTION!
f

ALMOST EVERYTHING ELSE YOU NEED TO KNOW ABOUT LIMITS IS SUMMED UP IN THESE

Basic Limit Facts: SUPPOSE C IS A CONSTANT, AND f AND g ARE TWO FUNCTIONS DEFINED AROUND a*, WITH

$$\lim_{x \to a} f(x) = L \quad \text{AND} \quad \lim_{x \to a} g(x) = M$$

THEN

1a. FOR ANY a, $\lim_{x \to a} C = C$

b. $\lim_{x \to a} Cf(x) = C \lim_{x \to a} f(x)$

c. $\lim_{x \to a} (f(x) + C) = \lim_{x \to a} f(x) + C$

2. $\lim_{x \to a} (f(x) + g(x)) = L + M$

3. $\lim_{x \to a} (f(x)g(x)) = LM$

4. IF $L \neq 0$, THEN $\lim_{x \to a} \frac{1}{f(x)} = \frac{1}{L}$

IN SHORT, YOU CAN TAKE THE LIMIT OF SUMS, PRODUCTS, AND QUOTIENTS TERM BY TERM (WATCHING OUT FOR ZERO DENOMINATORS), AND CONSTANTS "PASS THROUGH" THE LIMIT SYMBOL.

Example: FOR ANY $a \neq 0$,

$$\lim_{x \to a} \left(3x^2 + \frac{e^x \sin x}{x}\right) = 3a^2 + \frac{e^a \sin a}{a}$$

THIS MAKES LIFE **SO** MUCH EASIER!!

*WE'LL USE "DEFINED **AROUND** a" AS SHORTHAND FOR "DEFINED ON AN OPEN INTERVAL CONTAINING a, EXCEPT POSSIBLY AT a ITSELF."

ACTUALLY, THERE ARE A **FEW** MORE THINGS TO KNOW ABOUT LIMITS...

TO BEGIN WITH—THE PRECISE DEFINITION OF A LIMIT! TO UNDERSTAND THIS, LET'S REVIEW WHAT HAPPENED ON PAGES 56 AND 57 WITH THE FUNCTION $D(t)$ NEAR $t = 3$.

O.K., SO I EXAGGERATED...
CAN YOU MAKE IT WITHIN 0.0001?
YES.
WITHIN 0.000000001?
YES.
WITHIN...
YES!!!

IN GENERAL TERMS, IT WENT THIS WAY: YOU CHALLENGED ME TO CONFINE $D(t)$ WITHIN A TINY INTERVAL **I** AROUND L BY MAKING t CLOSE TO a. THE "RADIUS" (HALF-LENGTH) OF THAT INTERVAL WE CALLED ε, EPSILON. YOU DEMANDED THAT I MAKE $L - \varepsilon < D(t) < L + \varepsilon$.

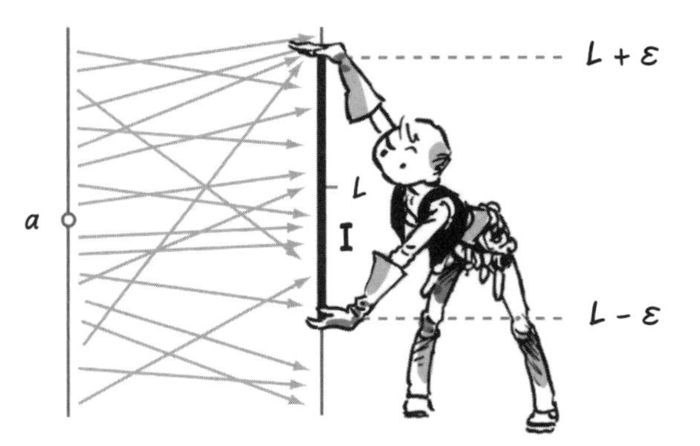

L + ε
a
L
I
L - ε

GIVEN THAT CHALLENGE, I RESPONDED BY FINDING AN INTERVAL **J** AROUND a, WITHIN WHICH THIS WAS TRUE:

IF t IS IN **J**, THEN $D(t)$ IS IN **I**.

J
L
I

AT THAT POINT, YOU CONCEDED THAT THE LIMIT REALLY WAS L.

SIGH... YES... I ADMIT DEFEAT...
THERE, THERE...

WE CAN EXPRESS THIS WITH FORMULAS, TOO. LET'S USE f FOR THE FUNCTION AND x FOR THE VARIABLE INSTEAD OF D AND t, AND I'LL ILLUSTRATE IT WITH A GRAPH, SO YOU CAN SEE THIS PROCESS IN TWO DIFFERENT WAYS. THE MEANING IS IDENTICAL—ONLY THE LANGUAGE IS DIFFERENT.

OH, GREAT... YOU'RE GOING TO MAKE ME ADMIT DEFEAT **TWICE?**
WELL, IT'S FOR, UM... THEM...

SO: GIVEN ANY $\varepsilon > 0$, YOU CHALLENGED ME TO MAKE $|f(x) - L| < \varepsilon$, I.E., TO GET THE GRAPH WITHIN THIS STRIP AROUND L:

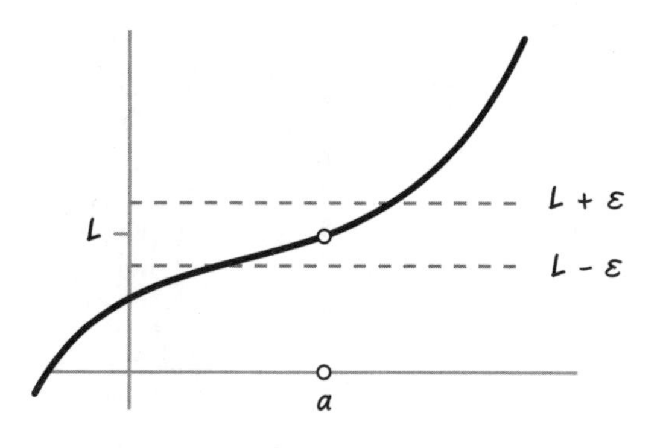

L
L + ε
L − ε
a

I RESPONDED WITH A POSITIVE NUMBER δ (THAT'S THE RADIUS OF THE INTERVAL **J**) WITH THIS PROPERTY:

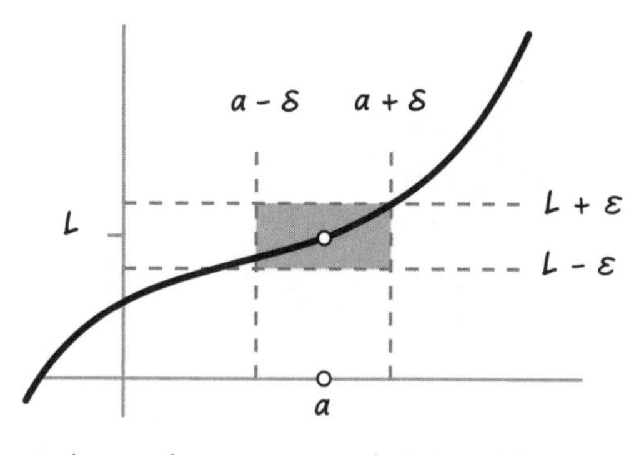

a − δ
a + δ
L
L + ε
L − ε
a

IF $|x - a| < \delta$, THEN $|f(x) - L| < \varepsilon$.

IF I CAN RESPOND TO AN ε CHALLENGE WITH A δ THAT MAKES THAT LAST "IF... THEN" TRUE, THEN YOU AGREE THAT

$$\lim_{x \to a} f(x) = L.$$

YOU O.K.?
ABSOLUTELY! I DECIDED TO TAKE IT AS A VICTORY!

HERE, THEN, ARE TWO WAYS TO EXPRESS THE FORMAL

Definition of the limit: SUPPOSE f IS A FUNCTION DEFINED AROUND POINT a (THOUGH NOT NECESSARILY AT a ITSELF). THEN TO SAY f **HAS THE LIMIT L AS x APPROACHES a** MEANS:

ALGEBRAIC VERSION:

FOR EVERY $\varepsilon > 0$, THERE EXISTS A NUMBER $\delta > 0$, SUCH THAT IF $|x - a| < \delta$ THEN $|f(x) - L| < \varepsilon$.

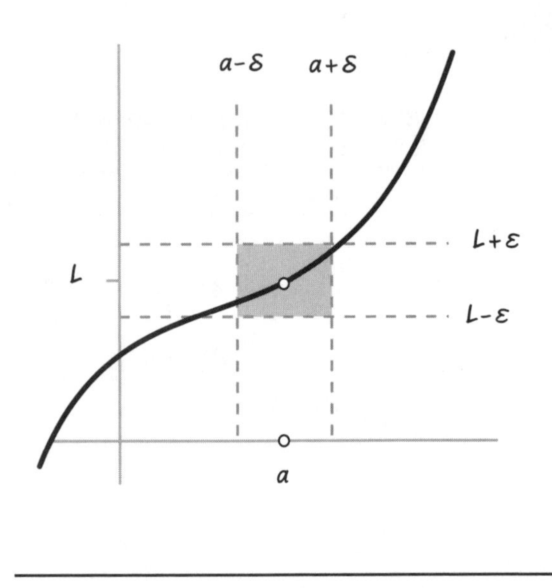

a-δ
a+δ
L+ε
L
L-ε
a

INTERVAL VERSION:

FOR EVERY OPEN INTERVAL **I** AROUND L, THERE IS AN OPEN INTERVAL **J** AROUND a, SUCH THAT IF x IS IN **J**, THEN $f(x)$ IS IN **I**.

ON THE INTERVAL J, f(x) IS "TRAPPED" OR "CAGED" IN I.
a
J
f
I
L

ALTHOUGH I PREFER THE INTERVAL PICTURE, THE ALGEBRAIC VERSION IS THE ONE YOU SEE IN ALL THE TEXTBOOKS, THE ONE RECITED IN A MANTRA-LIKE DRONE BY GENERATIONS OF CALCULUS STUDENTS, UNTIL IT EITHER SINKS IN, OR ELSE, YOU KNOW, IT DOESN'T.

δ
ε
|x|
L
"FOR EVERY EPSILOHHMMMM..."

TO SEE HOW THE DEFINITION WORKS, LET'S PROVE SOME OF THE BASIC LIMIT FACTS ON PAGE 59.

CHECK THIS OUT!
FOR EVERY EPSILON, THERE EXISTS A DELTA... FOR EVERY EPSILON, THERE EXISTS A DELTA... FOR EVERY EPSILON, THERE EXISTS A DELTA... FOR EVERY...

Limit Fact 1b. IF $\lim_{x \to a} f(x) = L$, THEN $\lim_{x \to a} Cf(x) = CL$ WHEN C IS A CONSTANT.

PROOF: GIVEN $\varepsilon > 0$ (THAT'S HOW THESE PROOFS **ALWAYS** START), WE HOPE TO FIND A NUMBER $\delta > 0$ SUCH THAT IF $|x - a| < \delta$, THEN $|Cf(x) - CL| < \varepsilon$. WE NOTICE THAT

$$|Cf(x) - CL| = |C||f(x) - L|$$

SO IF

$$|f(x) - L| < \frac{\varepsilon}{|C|}$$

WE SHOULD GET WHAT WE WANT. BUT CAN WE TRAP $f(x)$ IN THAT $\varepsilon/|C|$ INTERVAL? ANSWER: **OF COURSE WE CAN!** BY DEFINITION OF THE LIMIT, WE CAN TRAP $f(x)$ IN **ANY** SMALL INTERVAL BY USING SOME δ OR OTHER... THIS IS THE KEY TO THE WHOLE CONCEPT!

YES... YES... YESS!

SO TAKE δ SUCH THAT

$$\text{IF } |x - a| < \delta, \text{ THEN } |f(x) - L| < \frac{\varepsilon}{|C|}$$

IN THAT CASE, IF $|x - a| < \delta$, THEN

$$|Cf(x) - CL| = |C||f(x) - L|$$
$$< |C|\frac{\varepsilon}{|C|} = \varepsilon$$

SO $Cf(x)$ IS CAGED WITHIN ε OF CL, AND THE PROOF IS COMPLETE.

Q.-E.-DEEDLY-DUM-DE-DEE!

WARNING!
PROOFS
AHEAD!

SOME FURTHER LIMIT FACTS DEPEND ON THE FOLLOWING PRELIMINARY THEOREM, OR LEMMA, AS MATHEMATICIANS WOULD CALL IT.

Lemma 1: SUPPOSE $\lim_{x \to a} f(x) = \lim_{x \to a} g(x) = L$.

IF **I** IS ANY OPEN INTERVAL AROUND L, THEN THERE IS A **SINGLE** OPEN INTERVAL **J** AROUND a ON WHICH **BOTH** $f(x)$ AND $g(x)$ ARE TRAPPED IN **I**.

WHAT DO YOU DO WITH A LEMMA?
SQUEEZE THEOREMS FROM IT...

PROOF: BY DEFINITION, THERE IS AN OPEN INTERVAL J_f AROUND a WHERE $f(x)$ IS CONFINED TO **I**, AND ANOTHER (POSSIBLY DIFFERENT) OPEN INTERVAL J_g AROUND a WHERE $g(x)$ IS CONFINED TO **I**.

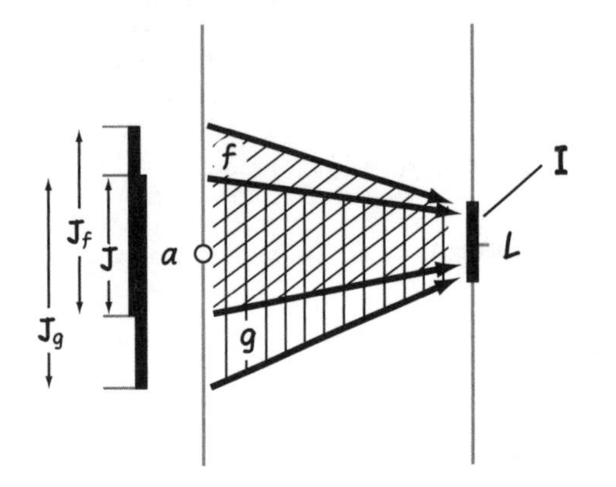
f
I
J_f
J
a
L
J_g
g

THEN THE **INTERSECTION** OF J_f AND J_g, THAT IS, ALL POINTS COMMON TO THE TWO INTERVALS, IS ALSO AN OPEN INTERVAL **J** AROUND a. IF x IS IN **J**, THEN BOTH $f(x)$ AND $g(x)$ ARE IN **I**, AND THE PROOF IS COMPLETE.

Lemma 2: SUPPOSE $\lim_{x \to a} f(x) = \lim_{x \to a} g(x) = 0$. THEN

$$\lim_{x \to a} f(x)g(x) = \lim_{x \to a} f(x) + \lim_{x \to a} g(x) = 0$$

PROOF: GIVEN $\varepsilon > 0$, BY LEMMA 1 THERE IS AN INTERVAL **J** AROUND a SUCH THAT IF x IS IN **J**, THEN

$$|f(x)| < \frac{\varepsilon}{2} \text{ AND } |g(x)| < \frac{\varepsilon}{2}$$

IF x IS IN **J**, THEN,

$$|f(x) + g(x)| \leq |f(x)| + |g(x)| < \frac{\varepsilon}{2} + \frac{\varepsilon}{2} = \varepsilon$$

$$|f(x)g(x)| = |f(x)| \cdot |g(x)| < \frac{\varepsilon^2}{4} < \varepsilon$$

AND THE PROOF IS COMPLETE. (WE ASSUMED $\varepsilon < 1$ HERE, BUT THAT'S O.K.)

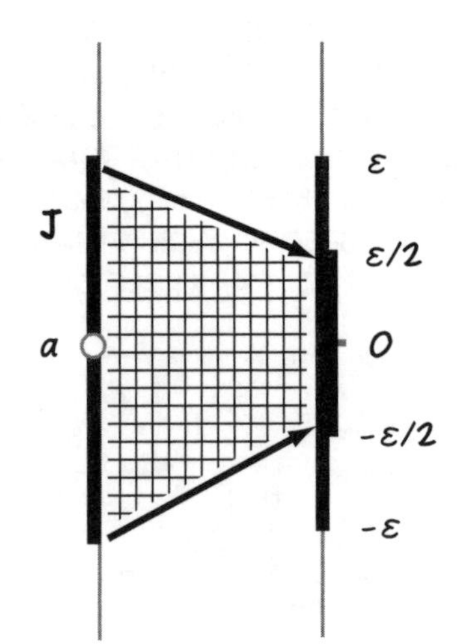
J
a
ε
ε/2
0
-ε/2
-ε

WE LEAVE LIMIT THE PROOF OF FACTS 1*a* AND 1*c* AS AN EASY EXERCISE FOR YOU, READER... ASSUMING THEM TO BE TRUE, WE NOW PROVE FACTS 2 AND 3.

Limit Fact 2. IF $\lim_{x \to a} f(x) = L$ AND $\lim_{x \to a} g(x) = M$, THEN

$$\lim_{x \to a} (f(x) + g(x)) = L + M$$

PROOF: APPLY LEMMA 2 TO THE FUNCTIONS $f - L$ AND $g - M$. THESE BOTH HAVE LIMIT 0 AS $x \to a$, BY FACT 1c. SO

$$0 = \lim_{x \to a} ((f(x) - L) + (g(x) - M)) \quad \text{BY LEMMA 2}$$

$$= \lim_{x \to a} ((f(x) + g(x)) - (L + M)).$$

$$= [\lim_{x \to a} ((f(x) + g(x))] - (L + M) \quad \text{BY FACT 1c, SO}$$

$$\lim_{x \to a} ((f(x) + g(x)) = L + M. \text{ DONE!}$$

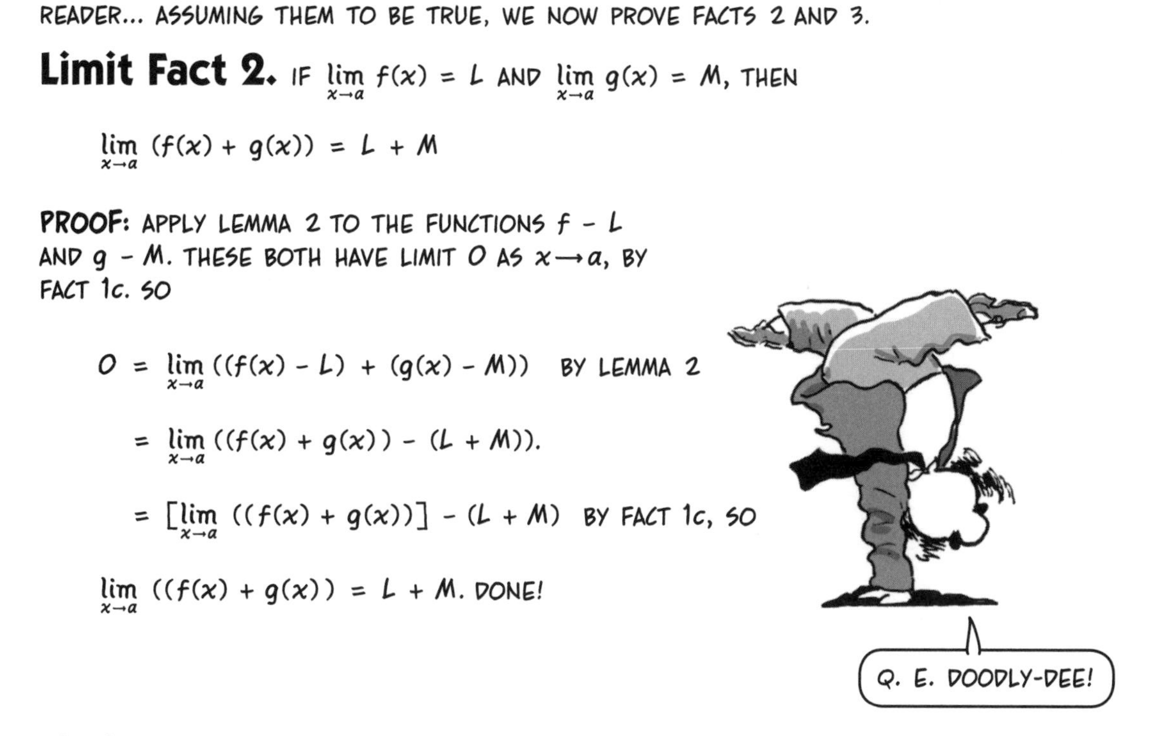
Q. E. DOODLY-DEE!

Limit Fact 3. IF $\lim_{x \to a} f(x) = L$ AND $\lim_{x \to a} g(x) = M$, THEN

$$\lim_{x \to a} (f(x)g(x)) = LM$$

PROOF: AGAIN APPLY LEMMA 2 TO THE FUNCTIONS $f - L$ AND $g - M$, WHICH BOTH HAVE LIMIT 0 AS $x \to a$.

$$0 = \lim_{x \to a} [(f(x) - L)(g(x) - M)] \quad \text{(BY LEMMA 2)}$$

$$= \lim_{x \to a} [f(x)g(x) - Lg(x) - Mf(x) + LM] \quad \text{(JUST ALGEBRA)}$$

$$= \lim_{x \to a} f(x)g(x) - \lim_{x \to a} Lg(x) - \lim_{x \to a} Mf(x) + LM \quad \text{(BY FACTS 2 AND 1a)}$$

$$= \lim_{x \to a} f(x)g(x) - LM - LM + LM \quad \text{(BY FACT 1b)}$$

$$= \lim_{x \to a} f(x)g(x) - LM, \text{ SO}$$

$$\lim_{x \to a} f(x)g(x) = LM. \text{ DONE AGAIN!}$$

THE PROOF OF LIMIT FACT 4 IS LEFT TO THE PROBLEM SETS...

More Limit Facts ABOUT POSITIVE (AND NEGATIVE) FUNCTIONS AND THEIR LIMITS, PLUS SOMETHING ELSE TO CHEW ON...

5a. IF $\lim_{x \to a} f(x) = L > 0$, THEN $f(x) > 0$ ON SOME INTERVAL **J** AROUND a.

PROOF: LET **I** BE ANY OPEN INTERVAL THAT CONTAINS L BUT EXCLUDES 0. BY THE DEFINITION OF A LIMIT, THERE IS AN INTERVAL **J** AROUND a ON WHICH $f(x)$ IS ALWAYS IN **I**. SINCE **I** CONSISTS ENTIRELY OF POSITIVE NUMBERS, THE PROOF IS COMPLETE.

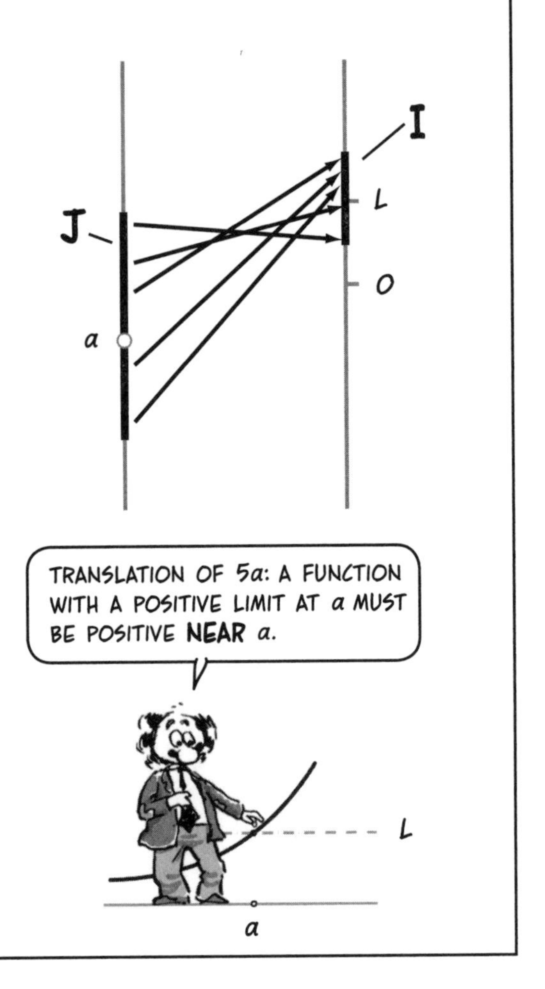
I
L
0
J
a
TRANSLATION OF 5a: A FUNCTION WITH A POSITIVE LIMIT AT a MUST BE POSITIVE NEAR a.

5b. IF $L < 0$, THEN THERE IS AN INTERVAL AROUND a ON WHICH $f(x) < 0$. THIS FOLLOWS BY APPLYING **5a** TO $-f$.

5c. IF $f(x) \geq 0$ FOR ALL x ON SOME INTERVAL AROUND a, THEN $\lim_{x \to a} f(x) \geq 0$ (IF THE LIMIT EXISTS).

PROOF: IF THE LIMIT WERE NEGATIVE, THEN BY **5b**, WE COULD FIND AN INTERVAL AROUND a WHERE $f(x)$ WAS NEGATIVE, CONTRARY TO THE HYPOTHESIS.

5d. SAME AS **5c**, WITH $\geq$ REPLACED THROUGHOUT BY $\leq$.

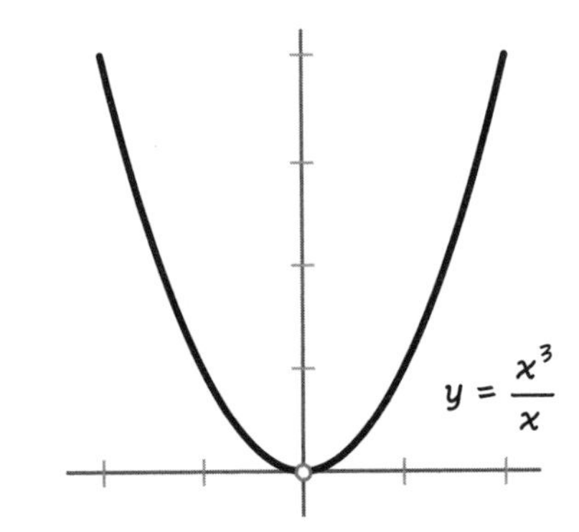
L
a

NOTE: WE CAN **NOT** CONCLUDE THAT A POSITIVE FUNCTION HAS A POSITIVE LIMIT, ONLY A NON-NEGATIVE LIMIT. FOR EXAMPLE,

$$f(x) = x^3/x \quad (x \neq 0)$$

IS ALWAYS POSITIVE, BUT

$$\lim_{x \to 0} f(x) = 0.$$

$y = \frac{x^3}{x}$

AND FINALLY, THIS TASTY RESULT:

Sandwich Theorem: IF $g(x) \le f(x) \le h(x)$ FOR ALL x IN SOME INTERVAL AROUND a, AND $\lim_{x \to a} g(x) = \lim_{x \to a} h(x) = L$, THEN $\lim_{x \to a} f(x) = L$ ALSO.

AS THE BREAD GOES, SO GOES THE PASTRAMI!
DOES THAT WORK WITH VEG TOO?
h(x)
I
x
a
L
f(x)
J
g(x)

PROOF: GIVEN ANY CHALLENGE INTERVAL **I** AROUND L, OUR HELPFUL LEMMA 1 SAYS THERE IS AN INTERVAL **J** AROUND a WHERE **BOTH** $g(x)$ AND $h(x)$ ARE CONFINED TO **I**.

FOR EVERY x IN **J**, THEN, $f(x)$ MUST ALSO BE IN **I**, BECAUSE $f(x)$ LIES BETWEEN $g(x)$ AND $h(x)$. THIS MEANS $\lim_{x \to a} f(x) = L$.

ON A GRAPH, YOU SEE HOW f IS SANDWICHED BETWEEN g AND h, AND SO IS SQUEEZED TOWARD THE POINT (a, L).

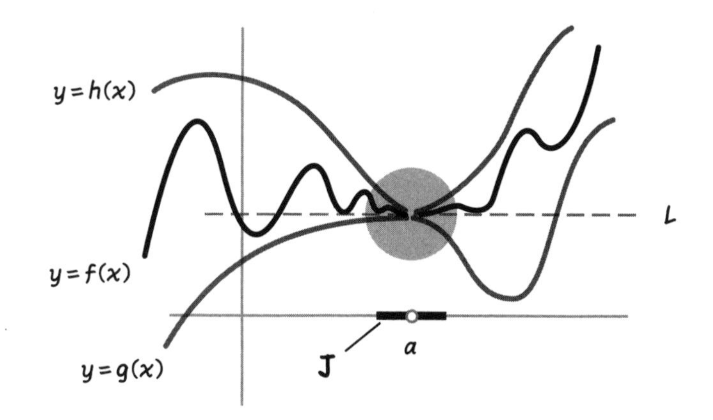
y = h(x)
L
y = f(x)
a
J
y = g(x)

WHOO! ALL THAT THEORY! CAN WE TAKE A BREAK NOW?
YES... LET'S TAKE A SHORT, UM, INTERVAL...

THE SANDWICH THEOREM GIVES US OUR FIRST SURPRISING RESULT INVOLVING ACTUAL, USEFUL FUNCTIONS. LET'S COMPARE AN **ANGLE** WITH ITS **SINE.**

AN ANGLE θ (IN RADIANS!) IS THE LENGTH OF THE ARC IT SWEEPS OUT IN A UNIT CIRCLE, WHILE $\sin\theta$ IS THE VERTICAL LEG OF THE TRIANGLE OAP. AS θ SHRINKS, THE ARC IS LESS CURVED, SO THE DISCREPANCY BETWEEN SINE AND ANGLE SHOULD BE LESS. WHAT HAPPENS WHEN $\theta \to 0$?

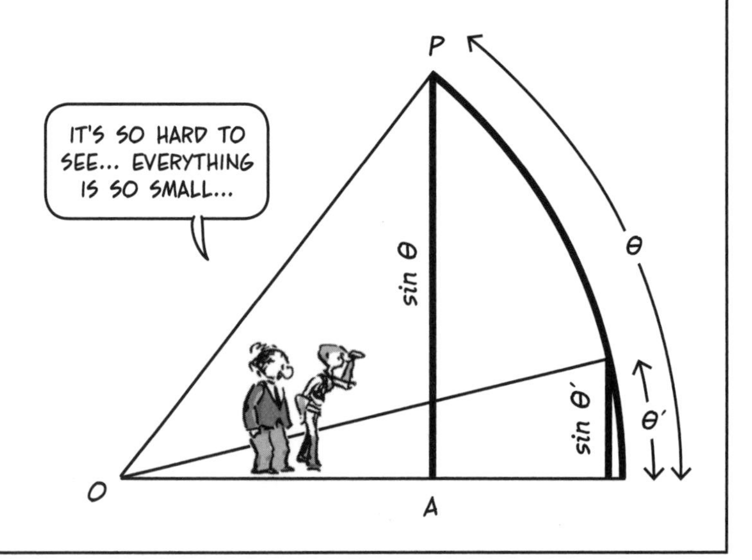
P
IT'S SO HARD TO SEE... EVERYTHING IS SO SMALL...
sin θ
θ
sin θ´
θ´
O
A

IN FACT, THEY BECOME INDISTINGUISHABLE. WE NOW SHOW THIS EXCELLENT RESULT:

$$\lim_{\theta \to 0} \frac{\sin \theta}{\theta} = 1$$

PROOF: SUPPOSE THE ANGLE CUTS THE CIRCLE AT POINT Q. EXTEND THE LINE OQ TO THE POINT Q' DIRECTLY ABOVE P', WHERE THE CIRCLE HITS THE x AXIS. THEN $OP = \cos \theta$, $QP = \sin \theta$, AND $OP' = 1$.

BECAUSE THE TRIANGLES OPQ AND $OP'Q'$ ARE SIMILAR, IT FOLLOWS THAT

$$P'Q' = \frac{P'Q'}{OP'} = \frac{PQ}{OP} = \frac{\sin \theta}{\cos \theta}$$

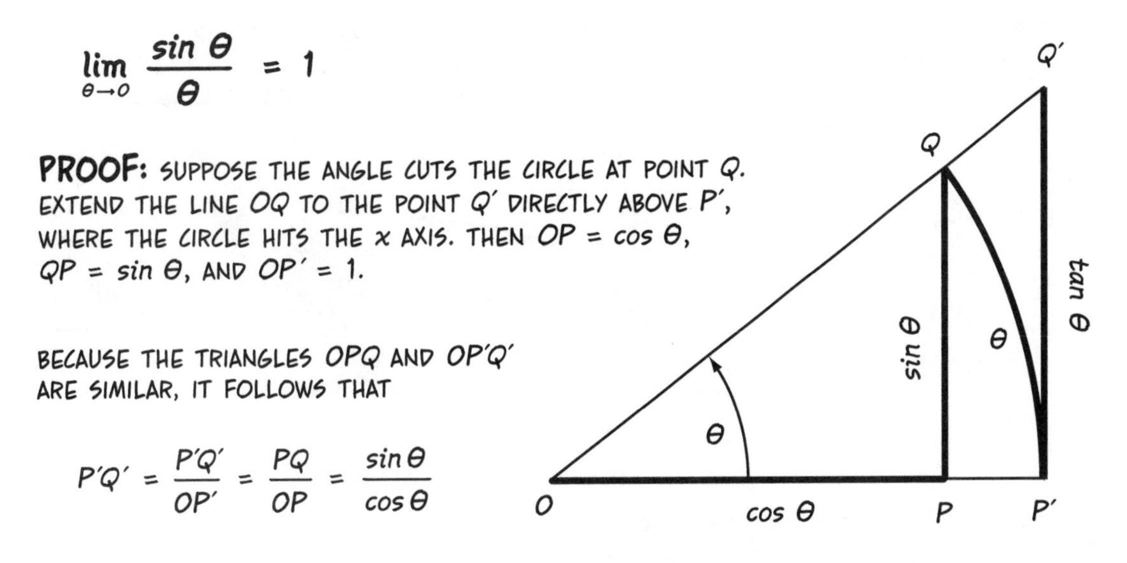

Q′
Q
tan θ
sin θ
θ
θ
O
cos θ
P
P′

NOW THE **AREA** OF THE SECTOR $OP'Q$ IS SIMPLY $\theta/2$ (IN RADIANS, REMEMBER!), SO THE AREAS OF THE SMALL TRIANGLE OPQ, THE SECTOR, AND THE LARGE TRIANGLE $OP'Q'$ FORM THIS SANDWICH OF INEQUALITIES:

$$\tfrac{1}{2} \sin \theta \cos \theta < \tfrac{1}{2}\theta < \tfrac{1}{2} \frac{\sin \theta}{\cos \theta}$$

DIVIDING BY $\frac{1}{2} \sin \theta$ (WHICH IS NOT ZERO!) GIVES

$$\cos \theta < \frac{\theta}{\sin \theta} < \frac{1}{\cos \theta}$$

TURNING EVERYTHING ON ITS HEAD REVERSES THE INEQUALITIES:

$$\cos \theta < \frac{\sin \theta}{\theta} < \frac{1}{\cos \theta}$$

Q-E-DOODLY-DEE!

AS $\theta \to 0$, THE POINT P SLIDES TOWARD P', SO $\cos \theta$ (AND HENCE $1/\cos \theta$) BOTH HAVE LIMIT EQUAL TO 1. THEREFORE, BY THE SANDWICH THEOREM, SO DOES $(\sin \theta)/\theta$, AND WE'RE DONE!

Limits at Infinity, Infinite Limits

SOMETIMES IN CALCULUS WE'RE INTERESTED IN VERY LARGE THINGS AS WELL AS VERY SMALL ONES. WE MAY, FOR EXAMPLE, WANT TO STUDY HOW A FUNCTION BEHAVES IN THE LONG RUN, AS "$x \to \infty$." HERE'S ONE THAT APPROACHES A LIMIT OF 3 AS x GROWS LARGE.

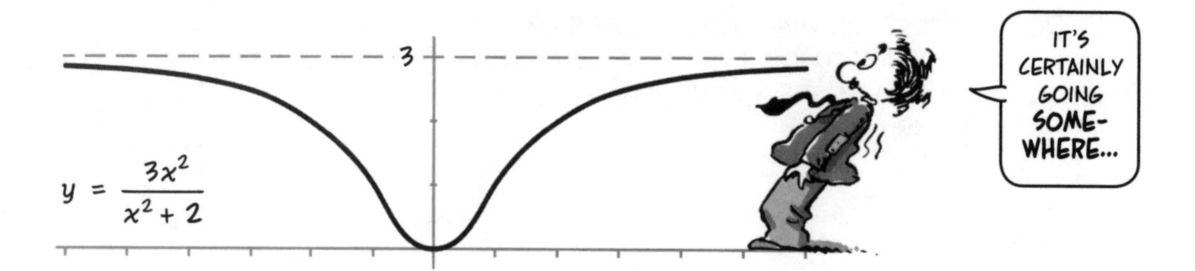

3
$y = \frac{3x^2}{x^2 + 2}$
IT'S CERTAINLY GOING SOME-WHERE...

SOMETIMES A FUNCTION "BLOWS UP TO ∞" AT A POINT a, MEANING THE VALUES OF $f(x)$ GROW WITHOUT BOUND AS $x \to a$. HERE'S ONE THAT BLOWS UP NEAR $x = 2$:

$$f(x) = \frac{1}{(x-2)^2}$$

WE SAY THE LIMIT IS **INFINITE,** AND WE WRITE

$$\lim_{x \to 2} f(x) = \infty$$

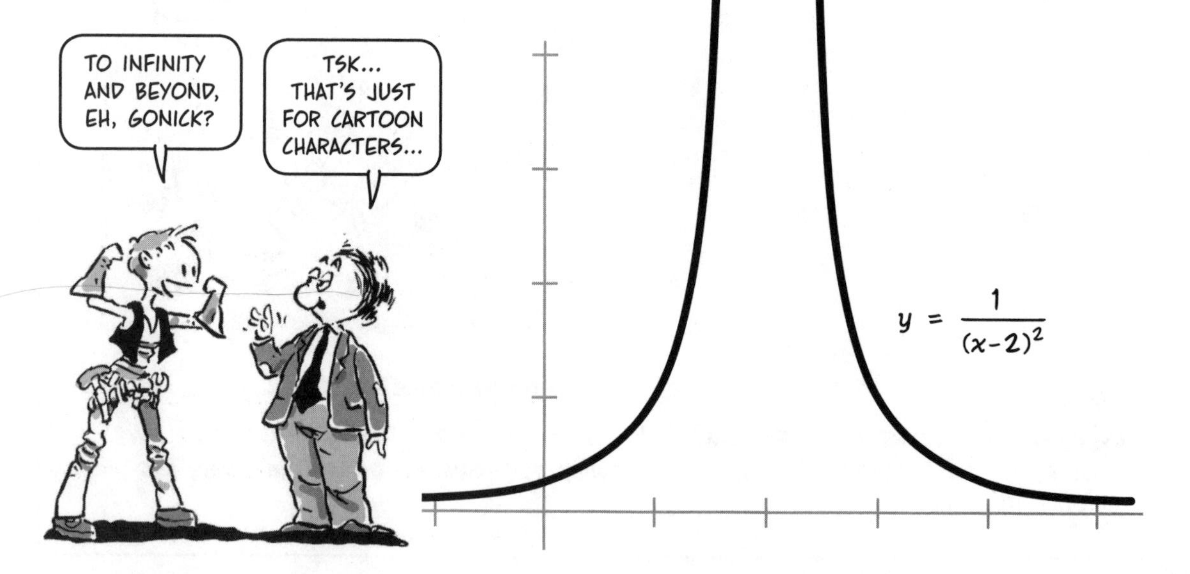

TO INFINITY AND BEYOND, EH, GONICK?
TSK... THAT'S JUST FOR CARTOON CHARACTERS...
$y = \frac{1}{(x-2)^2}$

THE PRECISE MEANING OF $\lim_{x \to a} f(x) = \infty$ IS THIS: GIVEN ANY LARGE NUMBER N, THERE IS AN INTERVAL **J** AROUND a SUCH THAT $f(x) > N$ WHENEVER x IS IN **J**.
REMEMBER, GOING **"TO INFINITY"** REALLY MEANS GOING **AWAY** FROM EVERY OTHER NUMBER!
AK! MY ARM!
I
N
J
a
WE COULD JUST AS WELL SAY THAT FOR EVERY **INTERVAL I AROUND ∞,** THERE IS AN INTERVAL **J** AROUND a SUCH THAT $f(x)$ IS IN **I** WHENEVER x IS IN **J**.
SOUNDS FAMILIAR...

IN A SIMILAR WAY, A FUNCTION'S "LONG-TERM" BEHAVIOR CAN SOMETIMES BE DESCRIBED AS A LIMIT AS $x \to \infty$. FOR EXAMPLE, THE FUNCTION $g(x) = 1/x$ IS DECREASING, AND IN FACT, IT GETS ARBITRARILY CLOSE TO ZERO AS x GROWS WITHOUT BOUND. WE WRITE:

$$\lim_{x \to \infty} \frac{1}{x} = 0$$

MAYBE BY NOW YOU KNOW THE MANTRA TO DEFINE $\lim_{x \to \infty} f(x) = L$:

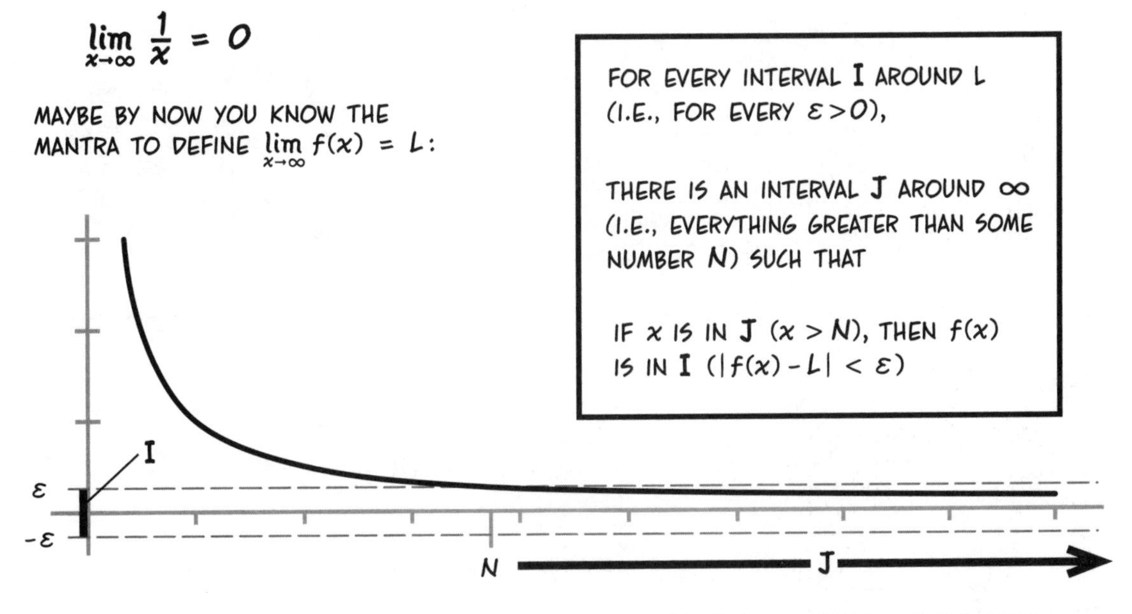
FOR EVERY INTERVAL **I** AROUND L (I.E., FOR EVERY $\varepsilon > 0$),
THERE IS AN INTERVAL **J** AROUND ∞ (I.E., EVERYTHING GREATER THAN SOME NUMBER N) SUCH THAT
IF x IS IN **J** ($x > N$), THEN $f(x)$ IS IN **I** ($|f(x) - L| < \varepsilon$)
I
ε
-ε
N
J

WHEN $x > N$, $f(x)$ IS WITHIN ε OF THE LIMIT.

Polynomials at Infinity

WE CLOSE THIS CHAPTER BY SHOWING HOW POLYNOMIALS GROW AT INFINITY. IN EFFECT, A POLYNOMIAL OF DEGREE n **GROWS AS ITS LEADING TERM** $a_n x^n$ AS $x \to \infty$. ALL THE LOWER-ORDER TERMS BECOME RELATIVELY NEGLIGIBLE.

YOU'RE BENEATH NOTICE...

Polynomial growth theorem: SUPPOSE $P(x)$ AND $Q(x)$ ARE POLYNOMIALS OF DEGREE n AND m, RESPECTIVELY:

$$P(x) = a_n x^n + a_{n-1}x^{n-1} + \dots + a_0$$

$$Q(x) = b_m x^m + b_{m-1}x^{m-1} + \dots + b_0 \quad (a_n, b_m \neq 0)$$

THEN

1. IF $n = m$, THEN $\lim_{x\to\infty} \frac{P(x)}{Q(x)} = \frac{a_n}{b_n}$

2. IF $n < m$, THEN $\lim_{x\to\infty} \frac{P(x)}{Q(x)} = 0$

3. IF $n > m$, AND a_n AND b_m HAVE THE SAME SIGN (I.E., BOTH + OR BOTH −), THEN

   $\lim_{x\to\infty} \frac{P(x)}{Q(x)} = \infty$ AND $-\infty$ WHEN a_n AND b_m HAVE OPPOSITE SIGNS.

IN MATHSPEAK, WE SAY THE POLYNOMIAL OF HIGHER DEGREE **DOMINATES** THE POLYNOMIAL OF LOWER DEGREE.

Examples:

$$\lim_{x\to\infty} \frac{3x^2 + x + 50}{2x^2 + 900x + 1} = \frac{3}{2}$$

(NUMERATOR AND DENOMINATOR HAVE THE SAME DEGREE, 2.)

$$\lim_{x\to\infty} \frac{450x^4 + 8x^3 + 50}{x^8 + x + 1} = 0$$

(DEGREE OF NUMERATOR IS LESS THAN DEGREE OF DENOMINATOR.)

PROOF OF 1: ASSUME $n = m$. BECAUSE A POLYNOMIAL HAS A FINITE NUMBER OF ROOTS, $Q(x) \neq 0$ WHEN x IS LARGE ENOUGH, SO THE FUNCTION P/Q IS DEFINED ON AN INTERVAL AROUND ∞. THEN FOR LARGE x WE CAN WRITE

$$\frac{P(x)}{Q(x)} = \frac{P(x)/x^n}{Q(x)/x^n} = \frac{a_n + \frac{a_{n-1}}{x} + \dots + \frac{a_0}{x^n}}{b_n + \frac{b_{n-1}}{x} + \dots + \frac{b_0}{x^n}}$$

NOW WE CAN TAKE THE LIMIT TERM BY TERM AS $x \to \infty$, AND SINCE EVERYTHING GOES TO ZERO EXCEPT a_n AND b_n, THE RESULT FOLLOWS.

2 IS A CONSEQUENCE OF **1**. IF $n < m$, SAY, THEN FOR LARGE ENOUGH x,

$$\frac{P(x)}{Q(x)} = x^{n-m} \frac{a_n x^m + \dots + a_0 x^{m-n}}{b_m x^m + \dots + b_0}$$

Q-E-DOODLY-D—

DO YOU MIND?

WE JUST SHOWED THAT THE SECOND FACTOR HAS THE FINITE LIMIT a_n/b_m AS $x \to \infty$. SINCE $\lim\limits_{x\to\infty} x^{n-m} = 0$, THE PRODUCT HAS LIMIT 0. PART **3** IS PROVED IN MUCH THE SAME WAY.

THE CASE $Q(x) = 1$ IMPLIES THAT ANY POLYNOMIAL P (I.E., THE NUMERATOR) HAS AN **INFINITE LIMIT AT INFINITY.** POLYNOMIALS CAN'T OSCILLATE (WOBBLE) FOREVER, BUT MUST ZOOM OFF EVENTUALLY.

'BYEE!

$\lim\limits_{x\to\infty} P(x) = \infty$ IF THE LEADING COEFFICIENT IS POSITIVE.

$\lim\limits_{x\to\infty} P(x) = -\infty$ IF THE LEADING COEFFICIENT IS NEGATIVE.

No Limit

FINALLY, I HAVE TO LET YOU IN ON A LITTLE SECRET... SOMETIMES, THERE IS NO LIMIT...
SHHH!

FOR EXAMPLE, NEITHER THE SINE NOR THE COSINE HAS A LIMIT AS $x \to \infty$. BOTH FUNCTIONS OSCILLATE BETWEEN -1 AND 1 FOREVER AS x GETS LARGE. GIVEN ANY SMALL CHALLENGE INTERVAL AROUND ANY NUMBER, THE VALUES $\sin x$ AND $\cos x$ REPEATEDLY ESCAPE THAT INTERVAL... AND SO NEITHER FUNCTION CAN APPROACH A LIMIT AS $x \to \infty$.

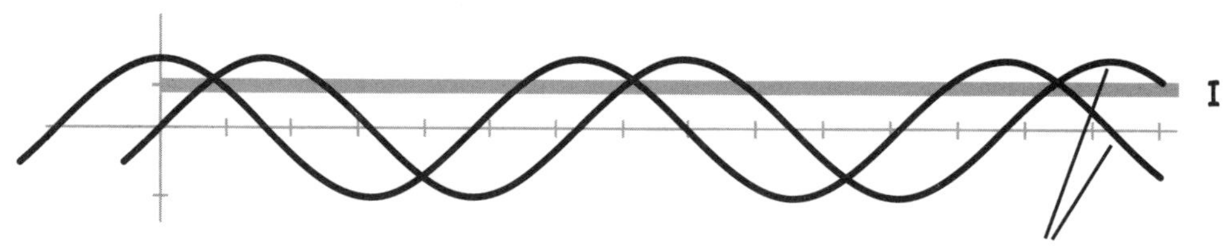
I
POINTS OUTSIDE I

IT IS ALSO POSSIBLE FOR A FUNCTION TO HAVE NO LIMIT AT A FINITE POINT a. THE MONSTER

$$g(x) = \sin\left(\frac{1}{x}\right), \quad x \neq 0$$

WIGGLES UP AND DOWN EVER MORE WILDLY AS $x \to 0$. g HAS NO LIMIT AT $x = 0$.

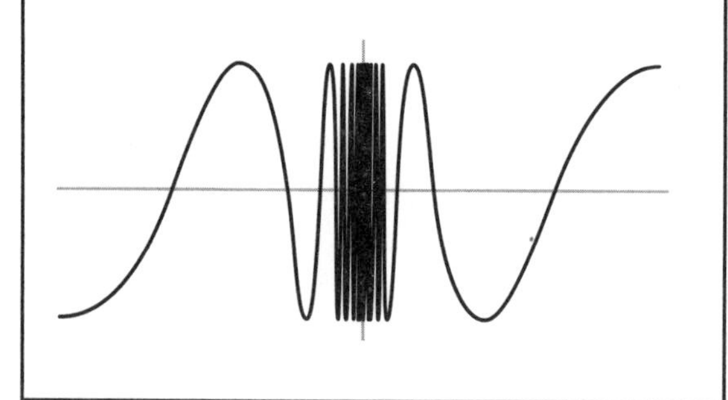

BUT THESE "BAD DOGS" ARE RARE, AT LEAST IN THIS BOOK. CALCULUS IS ALL BASED ON TAKING THINGS TO THE LIMIT, SO WE'LL BE LOOKING AT FUNCTIONS WHERE THE LIMIT EXISTS... YOU CAN EXPECT NOTHING BUT GOOD DOGS FROM NOW ON.

I'LL OUTPUT ON THEIR INPUT!

AND FINDING LIMITS IS EASY, OFTEN ENOUGH. AS WE SAID ON PAGE 58, FINDING $\lim_{x \to a} f(x)$ OFTEN INVOLVES NOTHING MORE THAN PLUGGING a INTO f:

a
f

$$\lim_{x \to 3} e^x = e^3$$

$$\lim_{x \to 9} \frac{1}{x} = \frac{1}{9}$$

$$\lim_{\theta \to 4} \sin \theta = \sin 4$$

AND SO ON...

THE MORE CHALLENGING EXAMPLES IN THIS CHAPTER WERE THESE TWO:

$$\lim_{x \to 0} \frac{\sin x}{x}$$

$$\lim_{x \to \infty} \frac{P(x)}{Q(x)}$$

UM... I WOULDN'T...
sin x / x
0

BOTH OF THESE FUNCTIONS, NOT COINCIDENTALLY, ARE QUOTIENTS... THE DENOMINATOR GOES TO ZERO OR INFINITY... NO WONDER THEY'RE CHALLENGING! YOU CAN'T SIMPLY PLUG IN!!

sin x / x
0
0/0 WILL DO THAT TO YOU...

IN THE NEXT CHAPTER, WE LOOK AT NOTHING BUT LIMITS OF QUOTIENTS...

Problems

FIND THE LIMITS:

1. $\lim_{x \to 2} 3x$

2. $\lim_{x \to 2} (3x + C)$, C A CONSTANT

3. $\lim_{x \to \infty} \frac{x^3 + x + 1}{4x^3 + 17}$

4. $\lim_{x \to -\infty} \frac{x^3 + x^2 + 1}{9x^2 + 8}$

5. $\lim_{t \to e^3} 2 \ln t$

6. $\lim_{x \to \infty} \frac{\cos x}{x - 1}$

7. $\lim_{x \to 1} \frac{x^2 + x - 2}{x - 1}$

HINT: SUBSTITUTE $y = 1/(x - 1)$ AND FIND THE LIMIT AS $y \to \infty$. ALTERNATIVELY, LET $h = x - 1$ AND FIND THE LIMIT AS $h \to 0$.

8. $\lim_{x \to 0} \frac{\sin 2x}{x}$

HINT: USE A TRIG IDENTITY FOR $\sin 2x$.

9. $\lim_{x \to 0} \frac{\sin x}{x^2}$

10. $\lim_{x \to 0} x \sin \left(\frac{1}{x}\right)$

HINT: USE THE SANDWICH THEOREM.

11. ON P. 19, WE DEFINED THE FUNCTION $f(x) = [x]$ TO BE THE WHOLE NUMBER PART OF x, THAT IS, THE LARGEST INTEGER $\leq x$. HERE IS THE GRAPH OF THE FUNCTION $g(x) = x - [x]$. DOES $\lim_{x \to 2} (x - [x])$ EXIST? HOW ABOUT $\lim_{x \to n} (x - [x])$ FOR ANY INTEGER n?

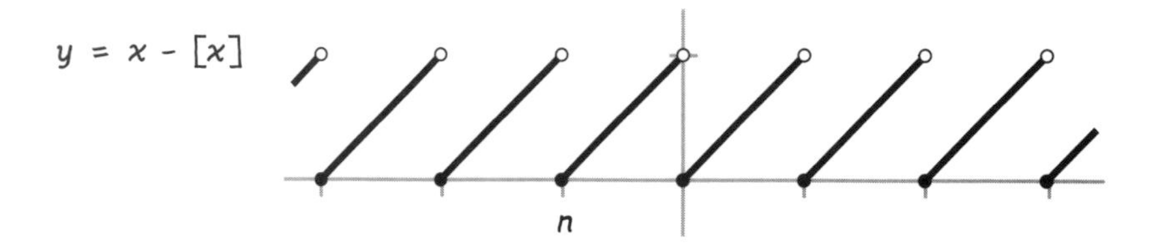
y = x - [x]
n

IF WE APPROACH n FROM THE **LEFT,** $g(x) \to 1$. IF WE APPROACH n FROM THE **RIGHT,** $g(x) \to 0$. THIS SUGGESTS THE IDEA OF HAVING **RIGHT-HAND** AND **LEFT-HAND LIMITS.** DO YOU THINK THIS IS A GOOD IDEA? MATHEMATICIANS DO... AND THEY WRITE THEM LIKE THIS:

$\lim_{x \to a^-} g(x)$ THE LIMIT FROM THE LEFT.

$\lim_{x \to a^+} g(x)$ THE LIMIT FROM THE RIGHT.

OPTIONAL PROBLEM: WORK OUT THE DETAILED DEFINITIONS!

12. SUPPOSE f IS ANY FUNCTION, WITH $\lim_{x \to a} f(x) = L$ AND $L \neq 0$. USING THE DEFINITION OF THE LIMIT, PROVE THAT THERE IS AN OPEN INTERVAL **J** AROUND a SUCH THAT IF x IS IN **J**, THEN $|f(x)| > |L/2|$.

13. SHOW THAT THIS IMPLIES THAT IF x IS IN **J**, THEN

$$\left| \frac{1}{f(x)} - \frac{1}{L} \right| < \frac{2|f(x) - L|}{L^2}$$

SHOW HOW THIS IMPLIES THAT

$$\lim_{x \to a} \frac{1}{f(x)} = \frac{1}{L}$$

Chapter 2
The Derivative

PICKING UP SPEED

NOW WE COME TO THE HEART OF CALCULUS: A FUNCTION'S **RATE OF CHANGE.** AS AN EXAMPLE, TAKE THE FUNCTION $s(t) = t^2$, WHICH DESCRIBES A CAR ROLLING DOWN A RAMP.

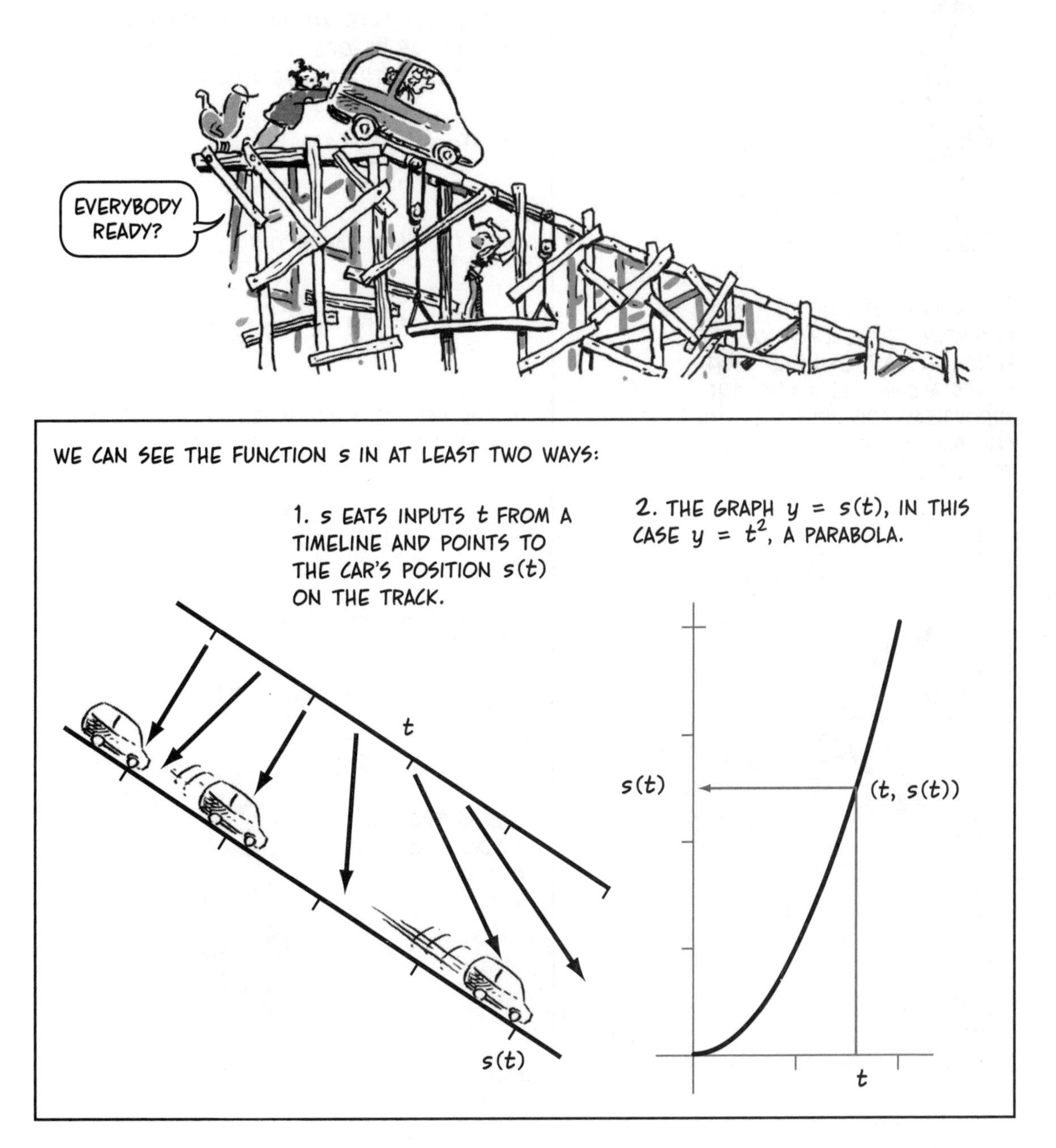

EVERYBODY READY?
WE CAN SEE THE FUNCTION s IN AT LEAST TWO WAYS:
1. s EATS INPUTS t FROM A TIMELINE AND POINTS TO THE CAR'S POSITION s(t) ON THE TRACK.
2. THE GRAPH y = s(t), IN THIS CASE y = t², A PARABOLA.
t
s(t)
s(t)
(t, s(t))
t

HERE ARE THREE WAYS TO THINK OF THE CAR'S VELOCITY IN TERMS OF THE FUNCTION s.

1. IN THE TIMELINE PICTURE, IT IS SIMPLY THE VELOCITY OF THE FUNCTION'S **ARROWHEAD** AS IT MOVES ALONG THE s AXIS! THE ARROWHEAD COINCIDES WITH THE CAR, SO THEY HAVE THE SAME VELOCITY.

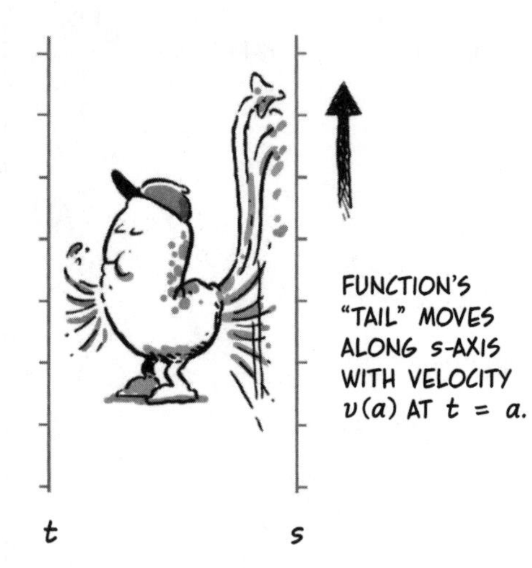
FUNCTION'S "TAIL" MOVES ALONG s-AXIS WITH VELOCITY $v(a)$ AT $t = a$.
t
s

2. AT TIME $t = a$, THE VELOCITY $v(a)$ IS

$$v(a) = \lim_{t \to a} \frac{s(t) - s(a)}{t - a}$$

AS WE SAW ON PAGE 54. THE **AVERAGE** VELOCITY ON THE INTERVAL (a, t) APPROACHES THE **INSTANTANEOUS** VELOCITY AS THE TIME INTERVAL GETS SHORTER AND SHORTER. AS BEFORE, WE SET $h = t - a$ AND REWRITE THE DIFFERENCE QUOTIENT:

$$\frac{s(a + h) - s(a)}{h}$$

THEN THE LIMIT TAKES THE FORM

$$\boldsymbol{v(a) = \lim_{h \to 0} \frac{s(a + h) - s(a)}{h}}$$

IN THE CASE AT HAND, WHEN $s(t) = t^2$, WE CAN ACTUALLY EVALUATE THIS EXPRESSION:

$$v(a) = \lim_{h \to 0} \frac{(a + h)^2 - a^2}{h}$$

$$= \lim_{h \to 0} \frac{a^2 + 2ah + h^2 - a^2}{h}$$

$$= \lim_{h \to 0} (2a + h)$$

$$= \mathbf{2a}$$

THIS IS THE CAR'S VELOCITY AT TIME $t = a$.

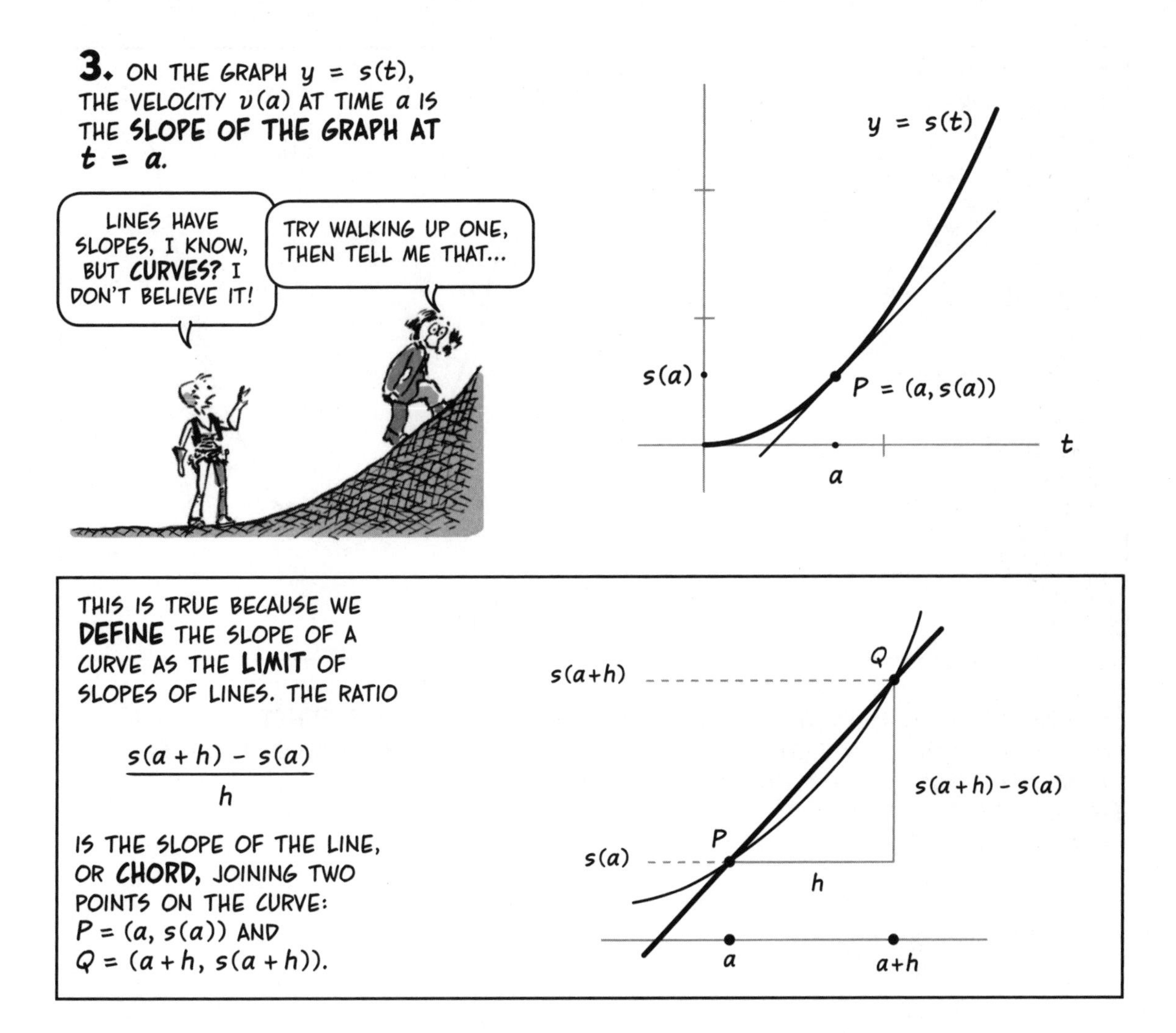
LINES HAVE SLOPES, I KNOW, BUT **CURVES?** I DON'T BELIEVE IT!
TRY WALKING UP ONE, THEN TELL ME THAT...
$y = s(t)$
$s(a)$
$P = (a, s(a))$
t
a
$s(a+h)$
Q
$s(a+h) - s(a)$
P
$s(a)$
h
a
$a+h$

3. ON THE GRAPH $y = s(t)$, THE VELOCITY $v(a)$ AT TIME a IS THE **SLOPE OF THE GRAPH AT** $t = a$.

THIS IS TRUE BECAUSE WE **DEFINE** THE SLOPE OF A CURVE AS THE **LIMIT** OF SLOPES OF LINES. THE RATIO

$$\frac{s(a+h) - s(a)}{h}$$

IS THE SLOPE OF THE LINE, OR **CHORD,** JOINING TWO POINTS ON THE CURVE: $P = (a, s(a))$ AND $Q = (a+h, s(a+h))$.

AS $h \to 0$, Q SLIDES TOWARD P, AND THE SLOPES OF THE CHORDS PQ, PQ', PQ'', ETC., APPROACH A LIMITING VALUE, WHICH WE INTERPRET AS THE **SLOPE OF THE CURVE** AT THE POINT P. IF $s(t) = t^2$, WE JUST FOUND THAT THIS SLOPE IS $v(a) = 2a$.

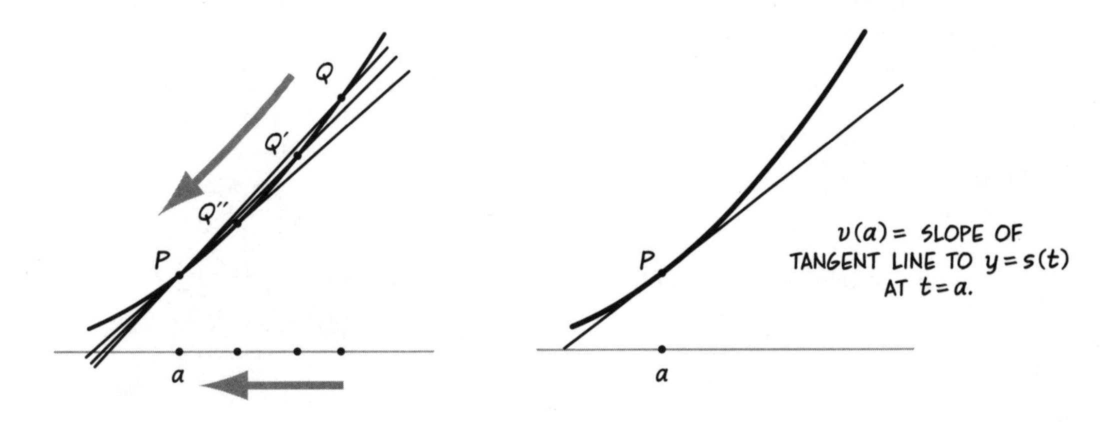
Q
Q'
Q''
P
a
P
a
$v(a)$ = SLOPE OF TANGENT LINE TO $y = s(t)$ AT $t = a$.

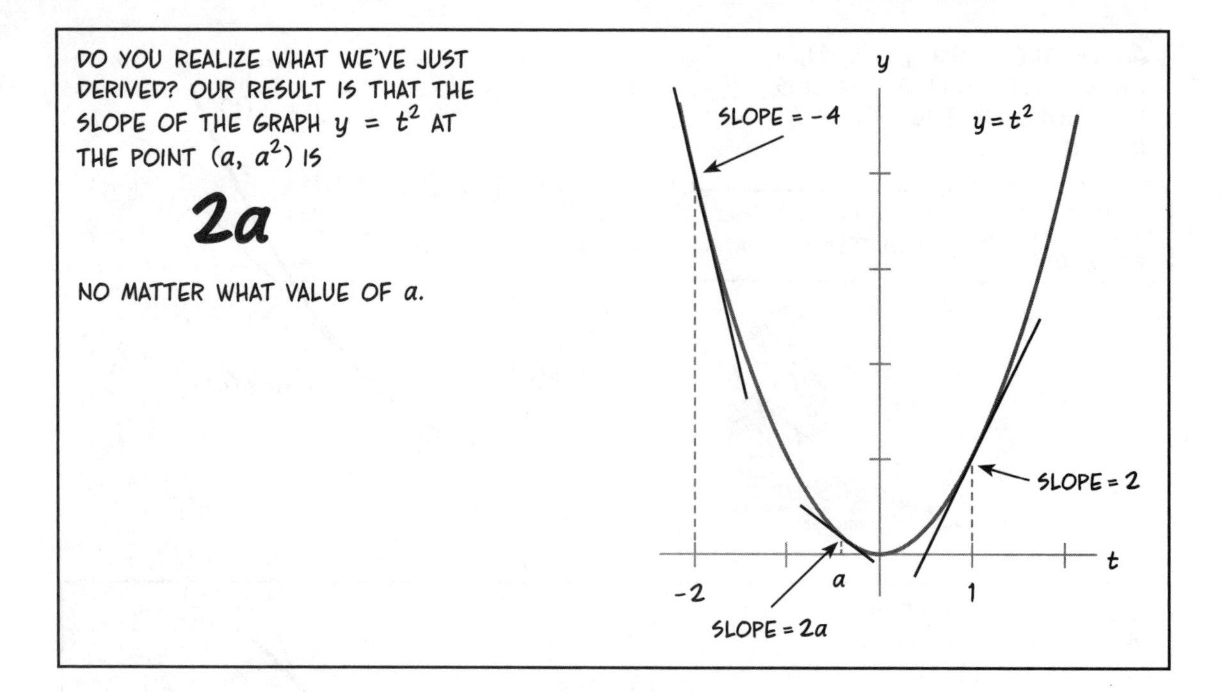
DO YOU REALIZE WHAT WE'VE JUST DERIVED? OUR RESULT IS THAT THE SLOPE OF THE GRAPH $y = t^2$ AT THE POINT (a, a^2) IS
$2a$
NO MATTER WHAT VALUE OF a.
y
SLOPE = −4
$y = t^2$
SLOPE = 2
t
−2
a
1
SLOPE = 2a

SIMILAR REASONING FINDS THE SLOPE OF THE GRAPH OF ANY **POWER FUNCTION** $y = t^n$ (n BEING A POSITIVE INTEGER) AT A POINT $P = (a, a^n)$. A CHORD BETWEEN P AND A NEARBY POINT $Q = (a + h, (a + h)^n)$ HAS SLOPE

$$\frac{(a+h)^n - a^n}{h}$$

DOES THIS HAVE A LIMIT AS $h \to 0$?
BY ALGEBRA, WE CAN EXPAND:

$$(a+h)^n = a^n + na^{n-1}h + C_2h^2 + C_3h^3 + \dots + h^n$$

WHERE THE COEFFICIENTS C_i ARE CONSTANTS INVOLVING POWERS OF a. SUBTRACTING a^n AND DIVIDING BY h, WE GET

$$\frac{(a+h)^n - a^n}{h} = na^{n-1} + C_2h + C_3h^2 + \dots + h^{n-1}$$

ALL TERMS AFTER THE FIRST HAVE LIMIT 0 AS $h \to 0$, SO

$$\lim_{h \to 0} \frac{(a+h)^n - a^n}{h} = na^{n-1}$$

NOTE: THE VERY LAST STEP USED LIMIT FACT 2: THE LIMIT OF A SUM IS THE SUM OF THE LIMITS!

AS WE'VE JUST SEEN, THIS SLOPE CAN BE INTERPRETED AS A VELOCITY. FOR EXAMPLE, IF A ROCKET CAN BLAST AHEAD SO FAST THAT $s(t) = t^5$, THEN AT ANY TIME a, THE ROCKET HAS VELOCITY $v(a) = 5a^4$.

a	$s(a) = a^5$	$v(a) = 5a^4$
-2	-32	$5(-2)^4 = (5)\cdot(16) = 80$
-1	-1	$5(-1)^4 = 5$
0	0	$5(0)^4 = 0$
$\frac{1}{2}$	$\frac{1}{32}$	$5\cdot(\frac{1}{2})^4 = \frac{5}{16}$
3	243	$5\cdot(3)^4 = (5)\cdot(81) = 405$
...	...	...

OR, IF $g(t) = t^4$, THEN $v(a) = 4a^3$ FOR ANY a:

a	$g(a)$	$v(a) = 4a^3$
-10	10,000	$4(-10)^3 = -4{,}000$
-2	16	$4(-2)^3 = -32$
-1	1	$4(-1)^3 = -4$
0	0	$(4)(0) = 0$
1	1	$4(1)^3 = 4$
2	16	$4(2)^3 = 32$
10	10,000	$4(10)^3 = 4{,}000$
...	...	...

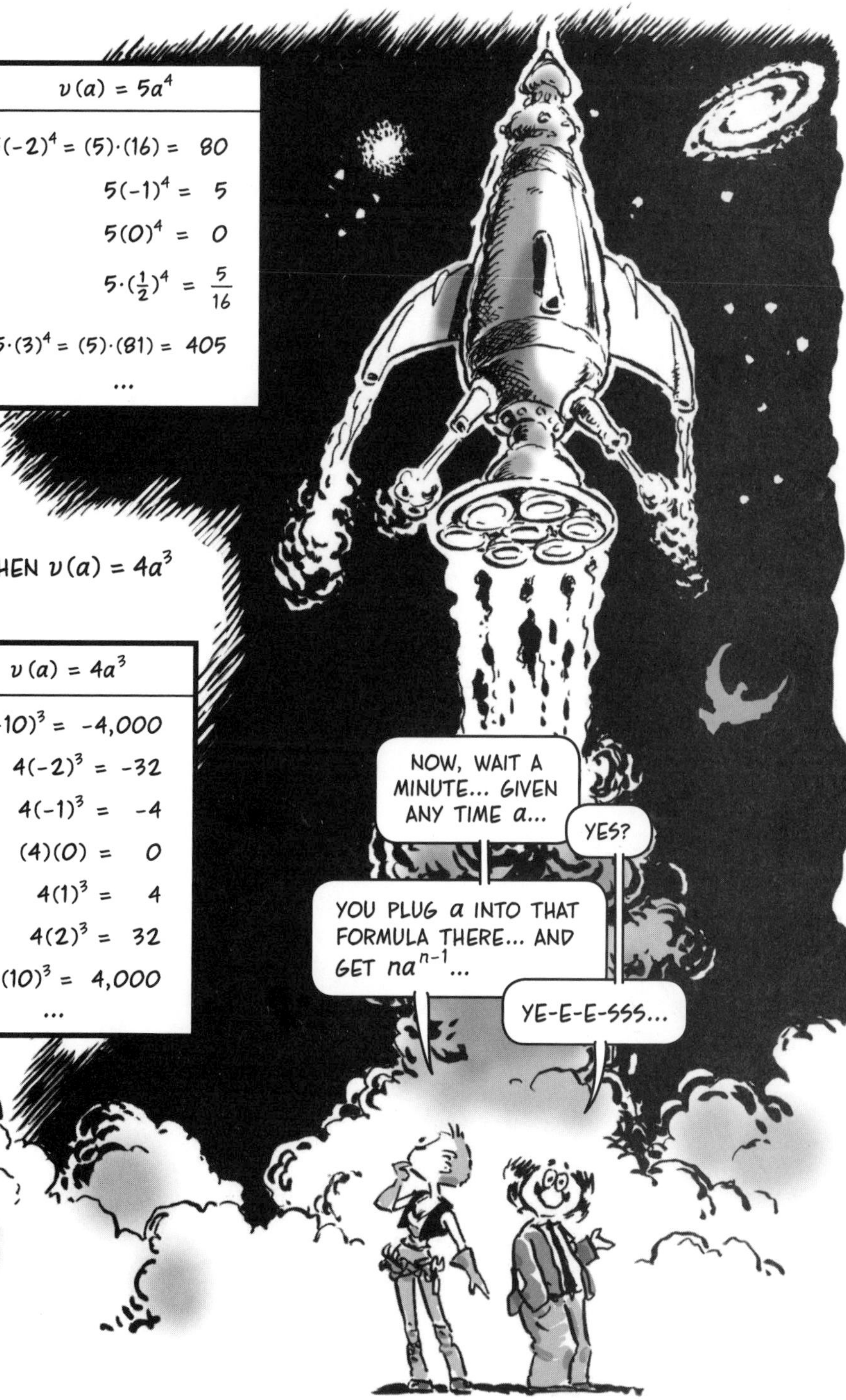

NOW, WAIT A MINUTE... GIVEN ANY TIME a...
YES?
YOU PLUG a INTO THAT FORMULA THERE... AND GET na^{n-1}...
YE-E-E-SSS...

SO WHY ISN'T v A FUNCTION OF t?

READER, IT IS! WE KEPT SAYING "FOR ANY TIME a," BUT WE COULD JUST AS WELL HAVE SAID "FOR ANY TIME t." VELOCITY, AFTER ALL, IS OBVIOUSLY A FUNCTION OF TIME: AT ANY TIME, THE CAR (OR ROCKET) HAS A VELOCITY! IN FACT, WE HAVE NOW PROVED THAT IF THE CAR'S POSITION AT TIME t IS t^n, THEN **ITS VELOCITY AT THAT TIME,** $v(t)$, IS nt^{n-1}.

t s t^n

I'M A FUNCTION TOO!

t v nt^{n-1}

WE HAVE DERIVED A **NEW FUNCTION** FROM s: THIS DERIVED FUNCTION, OR **DERIVATIVE,** GIVES THE SLOPE OF THE GRAPH $y = s(t)$ AT EACH POINT t, A SLOPE EQUAL TO THE VELOCITY AT TIME t.

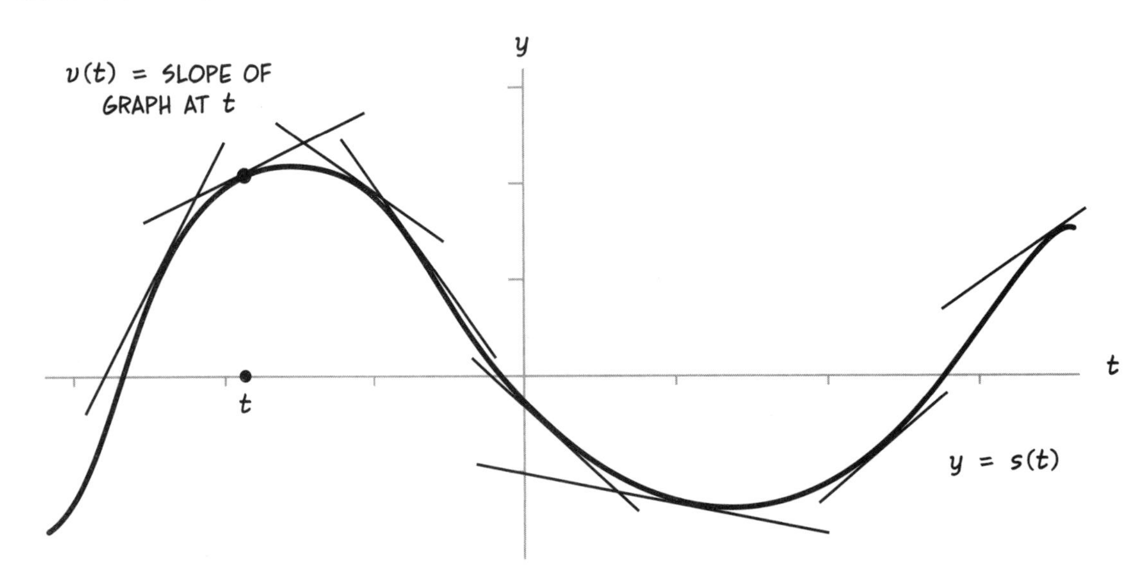

v(t) = SLOPE OF GRAPH AT t
y
t
t
y = s(t)

THIS DERIVED FUNCTION IS SO ASTOUNDINGLY AND WIDELY USEFUL, IN CONTEXTS FAR BEYOND CARS ROLLING DOWN RAMPS, THAT IT DESERVES ITS OWN NAME, DEFINITION, AND NOTATION:

THIS IS **"ONLY"** THE CENTRAL CONCEPT OF CALCULUS!!!

The Derivative Defined:

IF f IS ANY FUNCTION, AND x IS ANY POINT IN ITS DOMAIN, THE **DERIVATIVE** OF f, WRITTEN f' AND READ "EFF-PRIME," IS THE FUNCTION DEFINED BY

$$f'(x) = \lim_{h \to 0} \frac{f(x+h) - f(x)}{h}$$

FOR EACH x WHERE THIS LIMIT EXISTS.

FINDING THE DERIVATIVE f' IS CALLED **DIFFERENTIATING** THE FUNCTION f. $f'(x)$ IS THE SLOPE OF THE GRAPH $y = f(x)$ AT THE POINT $(x, f(x))$. FROM NOW ON, WE DISPENSE WITH THE LETTER v FOR VELOCITY, AND WRITE $s'(t)$ INSTEAD. IN THIS NEW TERMINOLOGY, THE RESULTS OF THE PREVIOUS PAGES ARE KNOWN AS THE **POWER RULE:**

$$\text{IF } f(x) = x^n, \text{ THEN } f'(x) = nx^{n-1}$$

KIND OF A SIMPLE FORMULA...
THAT'S WHAT MAKES IT SO COOL!!!

YOU CAN EASILY CHECK THAT IT AGREES WITH WHAT WE FOUND WHEN $n = 2$. WHAT DOES IT SAY WHEN $n = 1$? WHEN $n = 0$?

KNOWING THE DERIVATIVE OF $f(x) = x^n$, WE ALSO IMMEDIATELY KNOW THE DERIVATIVE OF **ANY POLYNOMIAL**, THANKS TO

Derivative Fact 1: Sums and Constants are Easy!

1a. IF C IS A CONSTANT AND f IS A FUNCTION WITH DERIVATIVE f', THEN $(Cf)' = Cf'$. TAKING THE DERIVATIVE "PASSES THROUGH" A CONSTANT.

1b. IF f AND g ARE TWO FUNCTIONS, THEN

$$(f + g)' = f' + g'.$$

THE DERIVATIVE OF A SUM IS THE SUM OF THE DERIVATIVES.

THESE FOLLOW FROM LIMIT FACTS 1b AND 2 ON PAGE 59.
IF YOU SAY SO...

WOULD YOU LIKE TO SEE THE PROOF?
CAN I STOP YOU?

$$(f + g)'(x) =$$

$$\lim_{h \to 0} \frac{f(x+h) + g(x+h) - (f(x) + g(x))}{h} =$$

$$\lim_{h \to 0} \frac{f(x+h) - f(x)}{h} + \lim_{h \to 0} \frac{g(x+h) - g(x)}{h} =$$

$$f'(x) + g'(x)$$

KEW-EE-DOODLY—

SIGH...

THIS MEANS WE CAN DIFFERENTIATE (TAKE THE DERIVATIVE OF) A POLYNOMIAL ONE TERM AT A TIME.

$g(x) = x^9 + x^8 + 2x^2 \qquad g'(x) = 9x^8 + 8x^7 + 4x$

$f(x) = 3x^4 + 6x^2 + 5 \qquad f'(x) = 12x^3 + 12x$

ETC.

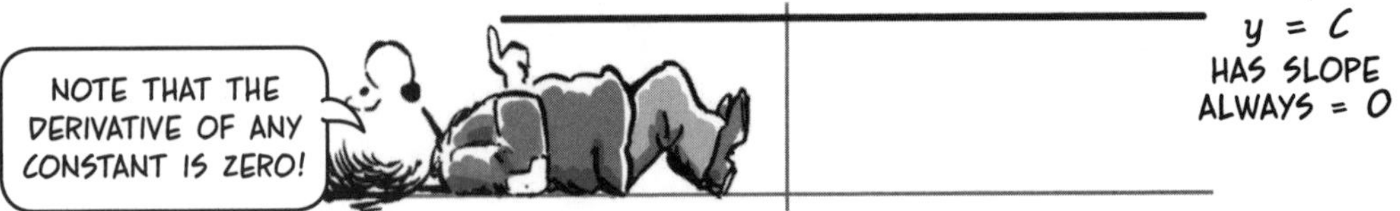
NOTE THAT THE DERIVATIVE OF ANY CONSTANT IS ZERO!
y = C
HAS SLOPE
ALWAYS = 0

Real-Life Example:

ISAAC NEWTON IS BOUNCING ON A VERY SPRINGY TRAMPOLINE WITH A MEMBRANE 1 METER OFF THE GROUND. IF IT FLINGS ISAAC UPWARD AT AN INITIAL VELOCITY OF 100 METERS PER SECOND, THEN HIS HEIGHT s ABOVE THE GROUND (VERTICAL POSITION, WITH UPWARD BEING THE POSITIVE DIRECTION), MEASURED IN METERS, IS GIVEN BY

$$s(t) = 1 + 100t - 4.9t^2$$

HOW FAST IS HE MOVING AFTER 10 SECONDS? IN WHAT DIRECTION?

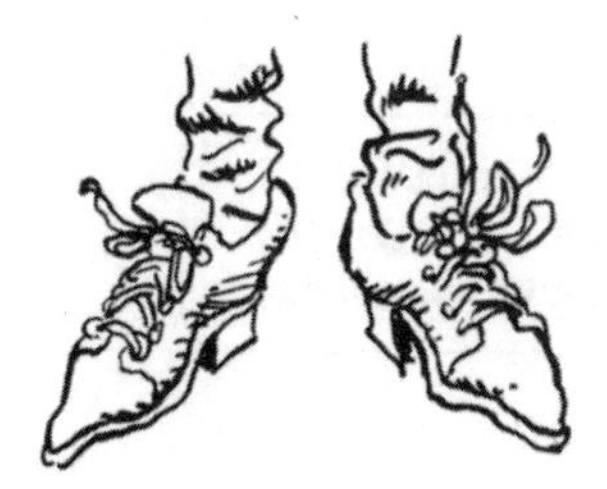

Solution: THE DERIVATIVE OF s GIVES THE VELOCITY AT ANY TIME. DIFFERENTIATE s TERM BY TERM:

$$s'(t) = 100 - (4.9)(2t)$$

$$= 100 - 9.8t \text{ M/SEC}$$

THAT IS THE GENERAL FORMULA FOR NEWTON'S VELOCITY AT TIME t. PLUG IN $t = 10$ SECONDS FOR THE ANSWER:

$$s'(10) = 100 - (9.8)(10)$$

$$= \mathbf{2} \text{ METERS PER SECOND.}$$

THE POSITIVE VELOCITY MEANS NEWTON IS STILL GOING UP AT THAT TIME!

WHOA! AFTER 10 SECONDS?
THAT IS CALCULUS-STRENGTH ELASTIC...

LET'S PAUSE HERE A MOMENT TO CONTEMPLATE THE DERIVATIVE... ALL THOSE PAGES ABOUT LIMITS WERE JUST A LEAD-IN TO THIS ONE KEY IDEA, THE SIMPLE ACT OF CROWNING AN f WITH A LITTLE TICK MARK, OR PRIME.

IT WAS THE FIRST BRILLIANT INSIGHT OF NEWTON AND LEIBNIZ TO SEE THAT THIS DERIVATIVE FUNCTION COULD HAVE A SIMPLE AND EXACT FORMULA, WHICH, WITH A STROKE, UNLOCKS THE SECRETS OF MOTION AND CHANGE. TAKE THAT, ZENO!

nx^{n-1}
THE ILLUSION OF MOTION IS OVERWHELMING!

AND ALTHOUGH NEWTON HAPPENED TO BE THINKING ABOUT VELOCITY WHEN HE DREAMED UP HIS "FLUXIONS," THE DERIVATIVE'S IMPORTANCE EXTENDS FAR BEYOND VELOCITY.

REGARDLESS OF WHAT f AND x STAND FOR, THE FRACTION

$$\frac{f(x+h) - f(x)}{h}$$

IS THE CHANGE IN THE VALUE OF f RELATIVE TO A SMALL CHANGE IN THE VARIABLE x. IN THE LIMIT, THEN, f' IS THE **INSTANTANEOUS RATE OF CHANGE OF** f WITH RESPECT TO x.

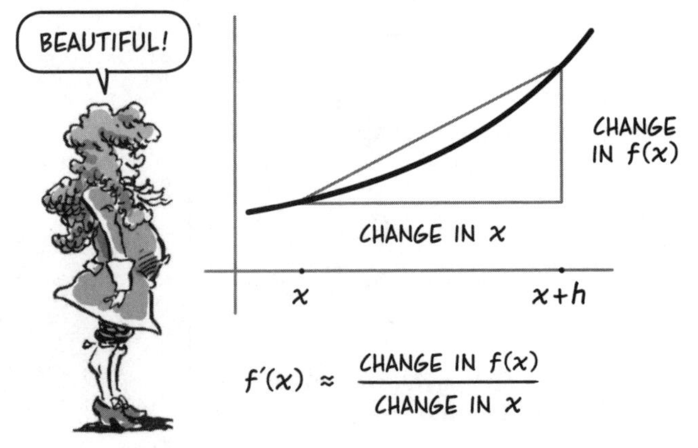
BEAUTIFUL!
CHANGE IN $f(x)$
CHANGE IN x
x
$x+h$
$f'(x) \approx \frac{\text{CHANGE IN } f(x)}{\text{CHANGE IN } x}$

For Example:

SUPPOSE SOME FLUID IS FLOWING INTO OR OUT OF A STORAGE TANK. IF $V(t)$ IS THE VOLUME IN LITERS PRESENT AT TIME t MINUTES, THEN

$$V'(t) = \lim_{h \to 0} \frac{V(t+h) - V(t)}{h}$$

IS THE (INSTANTANEOUS) **RATE OF FLOW,** MEASURED IN LITERS PER MINUTE.

NOTE: THIS IS NOT VELOCITY, BECAUSE IT DOESN'T REFER TO POSITION!

IF $C(t)$ IS THE COST OF LIVING AT TIME t, THEN

$$C'(t) = \lim_{h \to 0} \frac{C(t+h) - C(t)}{h}$$

IS THE RATE AT WHICH THE COST IS CHANGING AT TIME t.

THIS IS THE RATE OF INFLATION!

MANY REAL-WORLD FUNCTIONS DEPEND ON VARIABLES OTHER THAN TIME. FOR INSTANCE, AIR THINS OUT AT HIGHER ALTITUDE. IF $P(x)$ IS THE PRESSURE AT ALTITUDE x, THEN

$$P'(x) = \lim_{h \to 0} \frac{P(x+h) - P(x)}{h}$$

IS THE RATE OF CHANGE AT ALTITUDE x OF PRESSURE PER UNIT OF ALTITUDE (PASCALS PER METER, SAY), THE SO-CALLED **PRESSURE GRADIENT.**

A STRAIGHT ROAD GOES INTO THE MOUNTAINS. IF $A(x)$ IS THE ALTITUDE AT POSITION x, THEN

$$A'(x) = \lim_{h \to 0} \frac{A(x+h) - A(x)}{h}$$

IS THE ACTUAL SLOPE OR **GRADE** OF THE ROAD AT POINT x. (THERE ARE NO UNITS, SINCE WE HAVE DIVIDED METERS BY METERS. GRADE IS USUALLY GIVEN IN PERCENTAGE TERMS.)

NOW WE'RE READY TO START DIFFERENTIATING THE ELEMENTARY FUNCTIONS, BUT FIRST...

A Note on Notation (Leibniz-Style)

WRITING f' FOR THE DERIVATIVE OF f MAKES TWO THINGS CLEAR:

a) THE DERIVATIVE IS A FUNCTION

b) f' IS DERIVED FROM THE FUNCTION f

BUT YOU'LL OFTEN SEE THE DERIVATIVE WRITTEN IN AN ENTIRELY DIFFERENT WAY, LIKE THIS:

$$\frac{dy}{dx} \quad \text{OR} \quad \frac{df}{dx}$$

WHY SO MANY DIFFERENT SYMBOLS?
HEY, THE DERIVATIVE IS A ROCK STAR! IT CAN HAVE AS MANY AS IT LIKES!
THE FUNCTION FORMERLY KNOWN AS x

THIS WIDELY USED NOTATION EMPHASIZES OTHER ASPECTS OF THE DERIVATIVE:

c) ITS ORIGIN AS A QUOTIENT

d) THE VARIABLE x WITH RESPECT TO WHICH THE DERIVATIVE IS TAKEN

LEIBNIZ INVENTED THE dy/dx SCRIBBLE BASED ON THIS DIAGRAM. Δx, PRONOUNCED "DELTA-EKS," MEANS THE CHANGE IN x, OR WHAT WE'VE BEEN CALLING h. Δf OR Δy IS THE RESULTING CHANGE IN THE VALUE OF THE FUNCTION, I.E., $\Delta y = f(x + \Delta x) - f(x)$. THE SYMBOL Δ (GREEK CAPITAL DELTA) SIMPLY MEANS "THE CHANGE IN..."

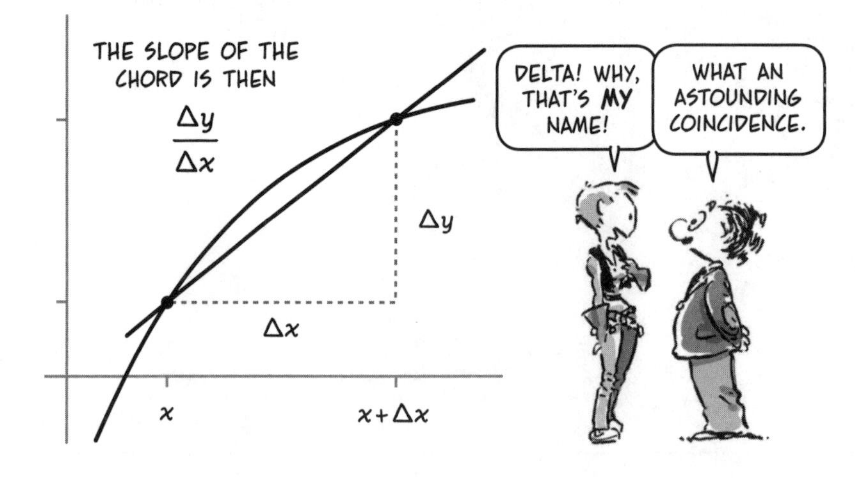
THE SLOPE OF THE CHORD IS THEN
Δy/Δx
Δy
Δx
x
x+Δx
DELTA! WHY, THAT'S MY NAME!
WHAT AN ASTOUNDING COINCIDENCE.

IN THIS NOTATION, WE WOULD WRITE:

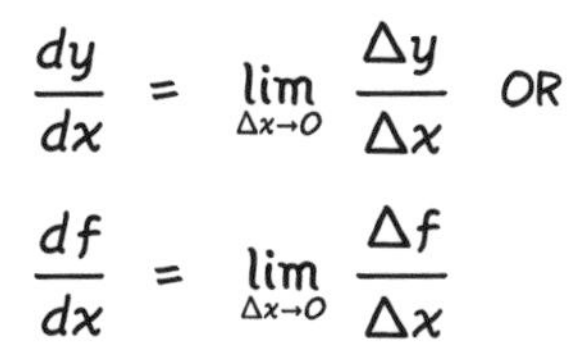

$$\frac{dy}{dx} = \lim_{\Delta x \to 0} \frac{\Delta y}{\Delta x} \quad \text{OR}$$

$$\frac{df}{dx} = \lim_{\Delta x \to 0} \frac{\Delta f}{\Delta x}$$

WHAT A WEIRD, UNNECESSARY IDEA! WHERE'D YOU GET IT?
YOU HAVE AN INFINITESIMAL IMAGINATION...

LEIBNIZ BELIEVED THAT dx AND dy WERE SOME KIND OF "INFINITELY SMALL" VERSIONS OF Δx AND Δy AND THAT THE DERIVATIVE WAS THE QUOTIENT OF THESE "INFINITESIMALS."

ALTHOUGH THIS IDEA WAS EVENTUALLY ABANDONED BY MOST MATHEMATICIANS, IT'S ACTUALLY PRETTY HELPFUL TO THINK OF THE DERIVATIVE, FOR ALL PRACTICAL PURPOSES, AS A LITTLE BIT OF y DIVIDED BY A LITTLE BIT OF x...

DOGGONE CURVE LOOKS PRETTY MUCH LIKE A STRAIGHT LINE FROM UP CLOSE ANYWAY...
dy
dx

THE LEIBNIZ WAY IS OFTEN MORE CONVENIENT—BEGINNING NOW, AS WE WRITE

$$\frac{d}{dx}(x^n), \quad \frac{d}{dx}(\sin x), \quad \text{AND} \quad \frac{d}{dx}(e^x)$$

TO REFER TO THE DERIVATIVES OF THE INDIVIDUAL FUNCTIONS. IT'S GREAT NOTATION!

ARE YOU A MATHEMATICIAN OR A STENOGRAPHER?
JEALOUS, ARE WE????!!!!

SO... ARE WE READY TO FIND $\frac{d}{dx}(\sin x)$?

CRANK IT UP!

Derivative of the Sine:

$$\frac{d}{d\theta}(\sin\theta) = \cos\theta$$

THE DERIVATIVE OF THE SINE IS THE COSINE.

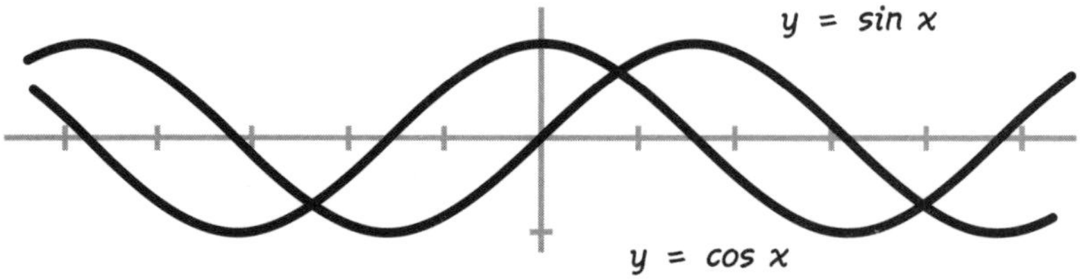

$y = \sin x$
$y = \cos x$

PROOF: BY DEFINITION OF THE DERIVATIVE, THE SINE'S DERIVATIVE IS

$$(1) \quad \lim_{h\to 0} \frac{\sin(\theta + h) - \sin\theta}{h}$$

IF THE LIMIT EXISTS.

EXPANDING $\sin(\theta+h)$ BY A TRIG IDENTITY, THE NUMERATOR BECOMES:

$$(\sin\theta\cos h + \sin h\cos\theta) - \sin\theta$$

SO THE DIFFERENCE QUOTIENT IN (1) IS

$$(2) \quad \cos\theta\,\frac{\sin h}{h} + \sin\theta\,\frac{\cos h - 1}{h}$$

IN THE LAST CHAPTER, WE SHOWED THAT

$$\lim_{h\to 0} \frac{\sin h}{h} = 1,$$

SO THE LIMIT OF (2) AS $h \to 0$ WILL BE

$$(3) \quad \cos\theta + (\sin\theta)\lim_{h\to 0}\frac{\cos h - 1}{h}$$

NOW WE SHOW THAT THE LAST FACTOR IS ZERO.

AH, "TRICKE-NOMETRY!"

$$\lim_{h\to 0} \frac{\cos h - 1}{h} = 0$$

BECAUSE

$$\frac{\cos h - 1}{h} = \left(\frac{\cos h - 1}{h}\right)\left(\frac{\cos h + 1}{\cos h + 1}\right)$$

$$= \frac{\cos^2 h - 1}{h((\cos h) + 1)} = \frac{-\sin^2 h}{h(\cos h + 1)}$$

$$= \left(\frac{-\sin h}{h}\right)\left(\frac{\sin h}{\cos h + 1}\right)$$

$\cos h$ HAS LIMIT 1 AS $h \to 0$, SO THE PRODUCT HAS LIMIT

$$(-1)\left(\frac{0}{2}\right) = 0 \text{ AS } h \to 0.$$

PUTTING THAT INTO (3) GIVES THE RESULT.

$$\lim_{h\to 0} \frac{\sin(\theta + h) - \sin\theta}{h} = \boldsymbol{\cos\theta}$$

WHAT THIS SAYS: TO FIND THE **SLOPE** OF THE SINE CURVE AT A POINT x, LOOK AT THE **VALUE** OF THE COSINE THERE.

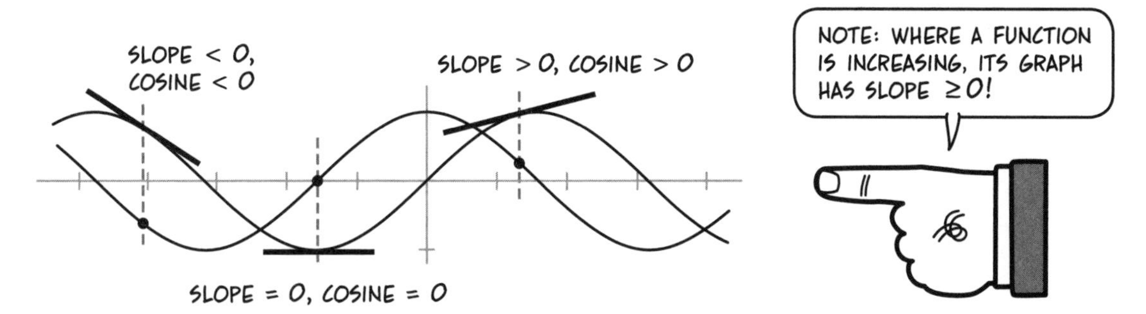
SLOPE < 0,
COSINE < 0
SLOPE > 0, COSINE > 0
SLOPE = 0, COSINE = 0
NOTE: WHERE A FUNCTION IS INCREASING, ITS GRAPH HAS SLOPE ≥0!

WHERE THE SINE IS INCREASING AND ITS CURVE IS RISING (BETWEEN $-\pi/2$ AND $\pi/2$, SAY), IT HAS POSITIVE SLOPE AND THE COSINE IS POSITIVE. WHERE THE SINE IS DECREASING AND ITS CURVE IS FALLING, THE SLOPE IS NEGATIVE AND SO ARE THE VALUES OF $\cos x$.

Derivative of the Cosine:

$$\frac{d}{d\theta}(\cos\theta) = -\sin\theta$$

THE DERIVATIVE OF THE COSINE IS THE NEGATIVE OF THE SINE.

RATHER THAN SUFFER MORE TRIG TORTURE, LET'S SIMPLY NOTICE THAT THE COSINE CURVE IS IDENTICAL TO THE SINE'S, BUT SHIFTED TO THE LEFT BY $\pi/2$. THEREFORE, THE COSINE'S DERIVATIVE MUST BE THE **COSINE ITSELF,** SHIFTED OVER ANOTHER $\pi/2$ TO THE LEFT!

THAT, IN TURN, IS THE SINE SHIFTED LEFTWARD BY A FULL π UNITS, OR $\sin(x + \pi)$. THIS IS THE SAME AS $-\sin x$, AS THE GRAPH MAKES CLEAR (OR YOU CAN WORK OUT WITH TRIG IDENTITIES OR ON THE UNIT CIRCLE).

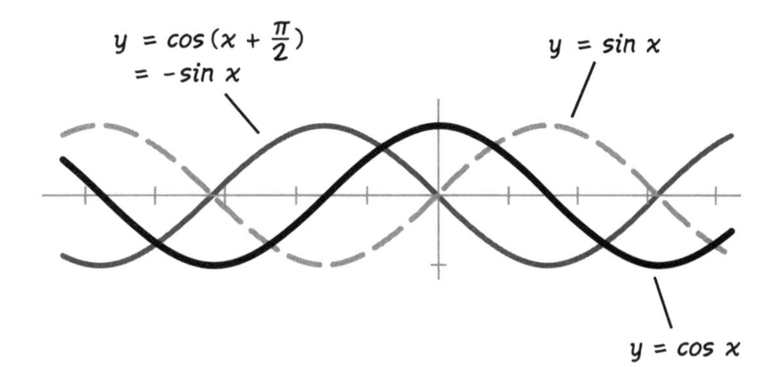
$y = \cos(x + \frac{\pi}{2})$
$= -\sin x$
$y = \sin x$
$y = \cos x$

Derivative of the Exponential:

THE SINE AND COSINE ARE EACH OTHER'S DERIVATIVES (UP TO A MINUS SIGN, ANYWAY). THE EXPONENTIAL'S DERIVATIVE IS—**ITSELF!**

$$\frac{d}{dx}e^x = e^x$$

THIS FOLLOWS FROM THE EQUATION $e^{x+h} = e^x e^h$ AND THE DEFINITION OF THE DERIVATIVE:

$$\frac{d}{dx}e^x = \lim_{h\to 0}\frac{e^{x+h} - e^x}{h} = \lim_{h\to 0}\frac{e^x e^h - e^x}{h}$$

$$= \lim_{h\to 0} e^x \frac{e^h - 1}{h} = e^x \lim_{h\to 0}\left(\frac{e^h - 1}{h}\right)$$

RECALL FROM THE COMPOUND INTEREST DISCUSSION ON PAGE 30 THAT $e \approx (1 + h)^{1/h}$ WHEN h IS SMALL. (THINK OF h AS $1/n$ IN THE ORIGINAL EXAMPLE.) RAISING BOTH SIDES TO THE hTH POWER GIVES $e^h \approx 1 + h$, SO

$$\frac{e^h - 1}{h} \approx \frac{(1 + h) - 1}{h} = 1$$

THAT IS, THE LIMIT OF THIS RATIO IS 1 AS $h \to 0$, AND SO THE DERIVATIVE IS $e^x \cdot (1) = e^x$.

THE **RATE OF INCREASE** OF THE EXPONENTIAL FUNCTION $Exp(x) = e^x$ IS EQUAL TO THE **VALUE** OF THE FUNCTION AT THAT POINT!!

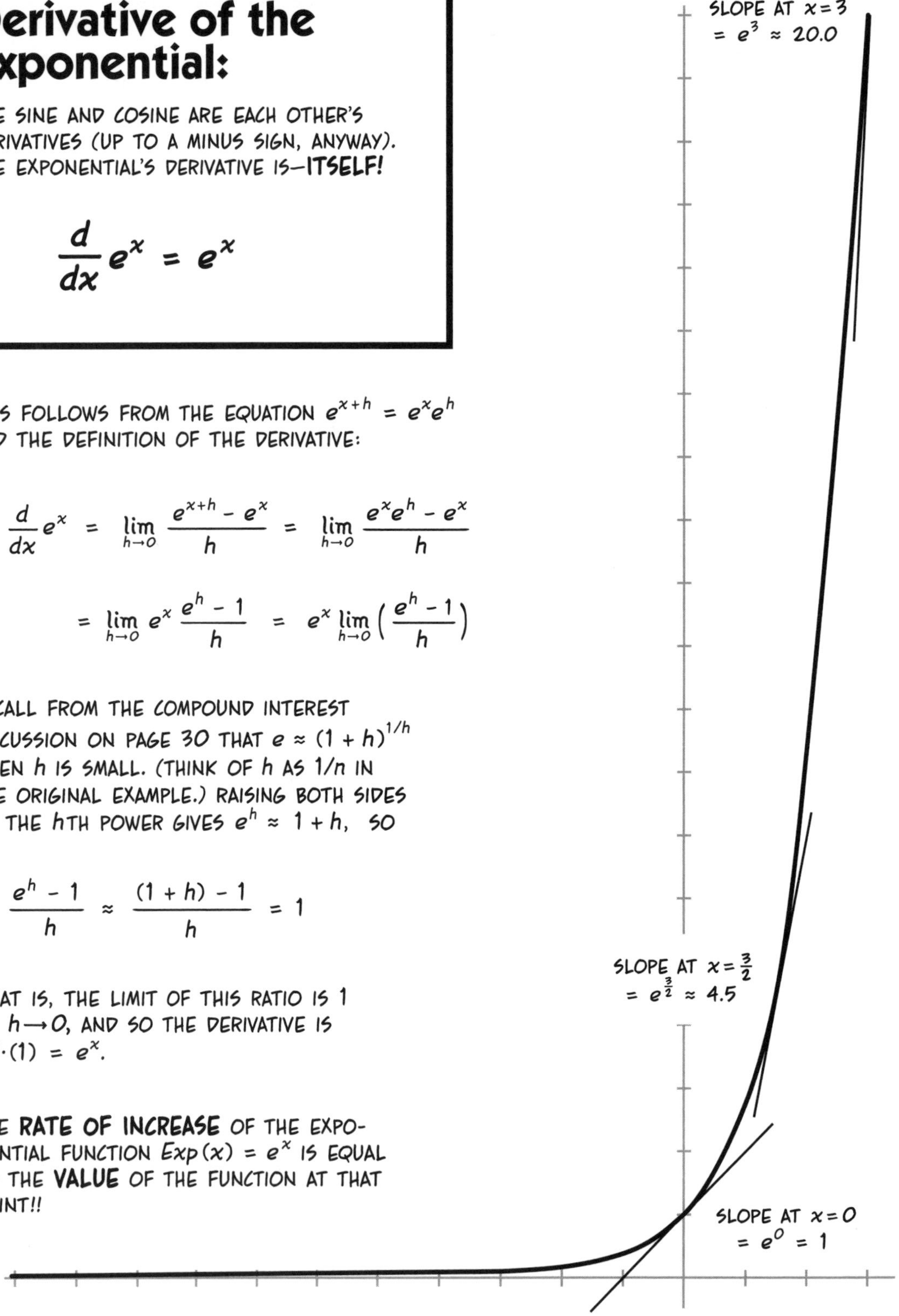

SLOPE AT $x = 3$
$= e^3 \approx 20.0$
SLOPE AT $x = \frac{3}{2}$
$= e^{\frac{3}{2}} \approx 4.5$
SLOPE AT $x = 0$
$= e^0 = 1$

THIS MAY SEEM COMPLETELY BIZARRO, A BIT OF MATHEMATICAL MAGIC, OR ELSE THE OPPOSITE—WHO KNOWS? MAYBE THERE ARE PLENTY OF FUNCTIONS THAT HAVE THEMSELVES AS DERIVATIVE...

WELL... NO, THERE AREN'T. THE EXPONENTIAL e^x AND ITS CONSTANT MULTIPLES Ae^x ARE THE **ONLY** FUNCTIONS WITH THIS PROPERTY. (YOU'LL PROVE THIS YOURSELF AS AN EXERCISE ON P. 160.)

HAVEN'T I SEEN YOU SOMEWHERE BEFORE?

SECOND, IT'S NOT REALLY THAT WEIRD, WHEN YOU THINK ABOUT COMPOUND INTEREST. THE **INTEREST ADDED PER YEAR** IS A FIXED PERCENTAGE OF THE **AMOUNT** OF MONEY IN THE ACCOUNT.

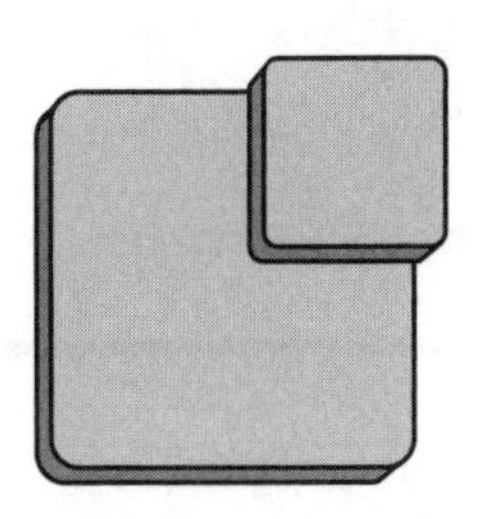

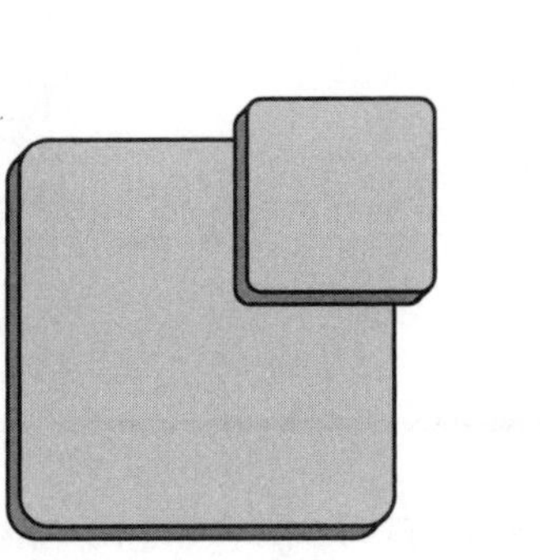

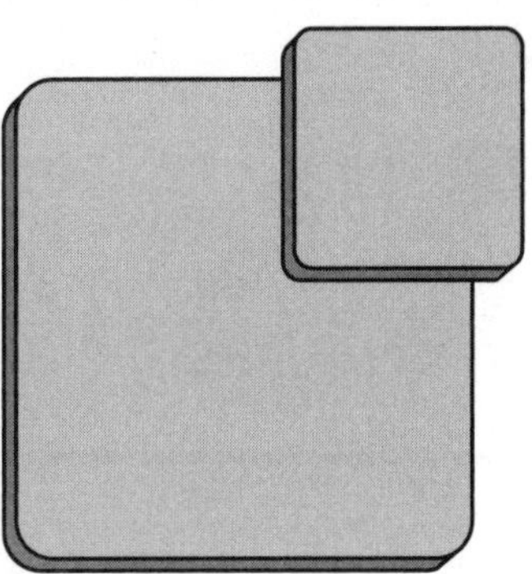

IN OTHER WORDS, THE **RATE OF CHANGE** IN VALUE, IN DOLLARS PER YEAR, IS PROPORTIONAL TO THE **VALUE** ITSELF. IF THE INTEREST IS COMPOUNDED CONTINUOUSLY, WE SHOULD EXPECT THAT THE **INSTANTANEOUS RATE OF CHANGE** OF THE VALUE V IS PROPORTIONAL TO V: $V'(t) = CV(t)$ FOR SOME CONSTANT C.

YOU SEE? WHEN THE INTEREST RATE IS 100% PER YEAR, THE CONSTANT IS 1, PROVIDED t IS MEASURED IN YEARS.
I STILL THINK THE BANK OWES ME $0.00000000127...

Derivatives of Products and Quotients

TAKING DERIVATIVES OF SUMS AND CONSTANT MULTIPLES IS STILL EASY: JUST GO TERM BY TERM. (SEE PAGE 84.) FOR EXAMPLE,

$$\frac{d}{dx}(5x^2 + \sin x) = 10x + \cos x$$

$$\frac{d}{dt}(e^x + \cos x - 2\sin x) = e^x - \sin x - 2\cos x$$

BUT—

HOW AWFULLY INCONVENIENT!
CAN'T BE HELPED!
IS WHAT IT IS!
SORRY!

Derivative Fact 2: Products are Trickier

THE DERIVATIVE OF A PRODUCT fg IS MOST EMPHATICALLY **NOT** THE PRODUCT OF THE DERIVATIVES. THE **PRODUCT RULE** IS:

$$(fg)' = f'g + fg' \quad \text{OR}$$

$$\frac{d}{dx}(fg) = f\frac{dg}{dx} + g\frac{df}{dx}$$

TO SEE WHY THIS IS TRUE, LET'S IMAGINE $f(x)$ AND $g(x)$ AS THE SIDES OF A RECTANGLE WITH AREA $f(x)g(x)$. THEN A SMALL CHANGE h IN x PRODUCES CHANGES Δf AND Δg IN f AND g, THAT IS, $f(x+h) = f(x) + \Delta f$ AND $g(x+h) = g(x) + \Delta g$:

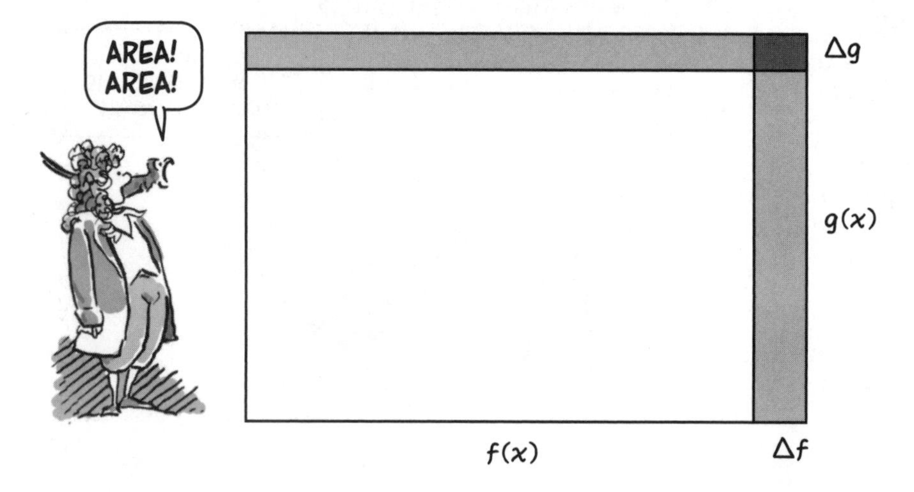
AREA! AREA!
Δg
g(x)
f(x)
Δf

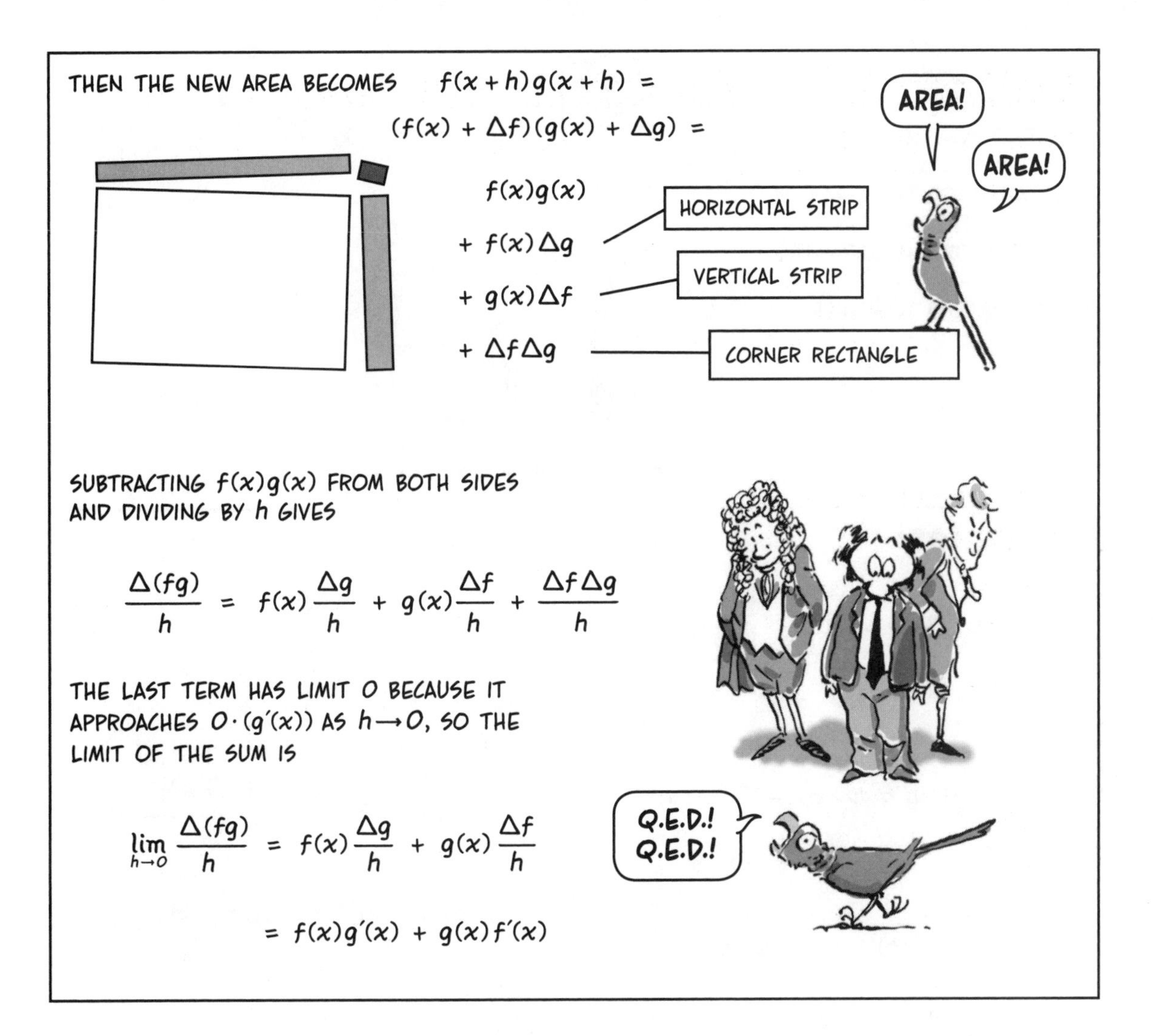
THEN THE NEW AREA BECOMES $f(x+h)g(x+h) =$
$(f(x) + \Delta f)(g(x) + \Delta g) =$
$f(x)g(x)$
$+ f(x)\Delta g$
$+ g(x)\Delta f$
$+ \Delta f \Delta g$
HORIZONTAL STRIP
VERTICAL STRIP
CORNER RECTANGLE
AREA!
AREA!
SUBTRACTING $f(x)g(x)$ FROM BOTH SIDES AND DIVIDING BY h GIVES
$$\frac{\Delta(fg)}{h} = f(x)\frac{\Delta g}{h} + g(x)\frac{\Delta f}{h} + \frac{\Delta f \Delta g}{h}$$
THE LAST TERM HAS LIMIT 0 BECAUSE IT APPROACHES $0 \cdot (g'(x))$ AS $h \to 0$, SO THE LIMIT OF THE SUM IS
$$\lim_{h \to 0} \frac{\Delta(fg)}{h} = f(x)\frac{\Delta g}{h} + g(x)\frac{\Delta f}{h}$$
$$= f(x)g'(x) + g(x)f'(x)$$
Q.E.D.! Q.E.D.!

LEIBNIZ WOULD SAY THAT

$$d(fg) = f\,dg + g\,df$$

IN THE LIMIT, THE "DIFFERENTIAL" OF fg—THE TINY BIT ADDED TO fg—CONSISTS OF THE TWO SIDE STRIPS OF SIZE $f\,dg$ AND $g\,df$, WHILE THE CORNER PIECE OF SIZE $df\,dg$ IS NEGLIGIBLE.

f
dg
g
df

IN OTHER WORDS, TO DIFFERENTIATE THE PRODUCT OF TWO FUNCTIONS, MULTIPLY THE FIRST FUNCTION BY THE DERIVATIVE OF THE SECOND, MULTIPLY THE SECOND FUNCTION BY THE DERIVATIVE OF THE FIRST, AND ADD THE TWO NUMBERS TOGETHER.

WHICH DO YOU PREFER, WORDS OR FORMULAS?
BIRDSEED! BIRDSEED!

Examples:

1. $\frac{d}{dx}(x^2e^x) = (\frac{d}{dx}(x^2))e^x + x^2\frac{d}{dx}(e^x)$

$$= \mathbf{2xe^x + x^2e^x}$$

2. $\frac{d}{d\theta}(\sin\theta\cos\theta) = (\frac{d}{d\theta}(\sin\theta))\cos\theta + \sin\theta\frac{d}{d\theta}(\cos\theta)$

$$= \mathbf{\cos^2\theta - \sin^2\theta}$$

CRANK!
CRANK!

3. $\frac{d}{dt}(\sin^2 t) = \frac{d}{dt}((\sin t)\cdot(\sin t))$

$$= \sin t\cos t + \cos t\sin t$$

$$= \mathbf{2\sin t\cos t}$$

TO DIFFERENTIATE THE PRODUCT OF MORE THAN TWO FUNCTIONS, FOLLOW THE SAME SORT OF RULE:

$$(fgh)' = f'gh + fg'h + fgh'$$

FOR INSTANCE,

$$\frac{d}{dx}(x\sin x\cos x) = 1\cdot\sin x\cos x + x\cos x\cos x + x\sin x(-\sin x)$$

$$= \mathbf{\sin x\cos x + x(\cos^2 x - \sin^2 x)}$$

Derivative Fact 3: Quotients are Weird.

3a. IF f IS DIFFERENTIABLE AT x AND $f(x) \neq 0$, THEN $1/f$ IS ALSO DIFFERENTIABLE AT x, AND

$$\left(\frac{1}{f}\right)'(x) = \frac{-f'(x)}{(f(x))^2}$$

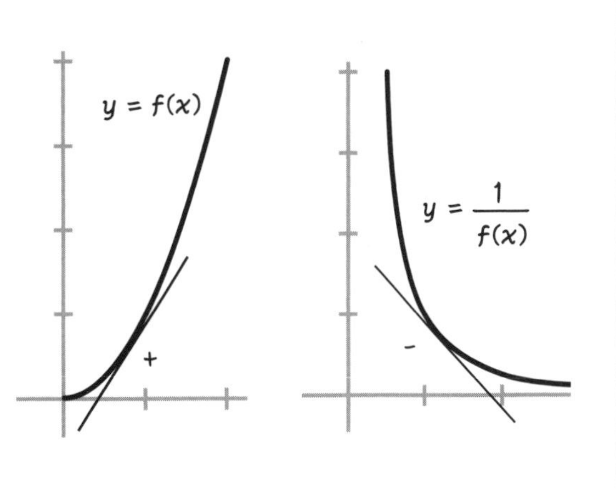

$y = f(x)$
+
$y = \frac{1}{f(x)}$
-

WHERE DID THAT MINUS SIGN COME FROM? WELL... f IS INCREASING WHEREVER $1/f$ IS DECREASING, AND VICE VERSA, SO THEIR DERIVATIVES MUST HAVE OPPOSITE SIGNS AT ANY POINT.

IT'S SIMPLY ALGEBRA:

$$\frac{1}{f(x+h)} - \frac{1}{f(x)} = \frac{f(x) - f(x+h)}{f(x+h)\,f(x)}$$

OR

$$\Delta\left(\frac{1}{f}\right) = \frac{-\Delta f}{f(x)\,f(x+h)}$$

ALGEBRA! ALGEBRA!

DIVIDING BOTH SIDES BY h AND TAKING THE LIMIT AS $h \to 0$ GIVES THE RESULT.*

3b. Quotient Rule: IF f AND g ARE BOTH DIFFERENTIABLE AT A POINT x, AND $g(x) \neq 0$, THEN f/g IS DIFFERENTIABLE AT x, AND

$$\left(\frac{f}{g}\right)'(x) = \frac{f'(x)g(x) - f(x)g'(x)}{g(x)^2}$$

THIS FOLLOWS BY TAKING THE DERIVATIVE OF THE PRODUCT $f \cdot (1/g)$ AND APPLYING **3a**.

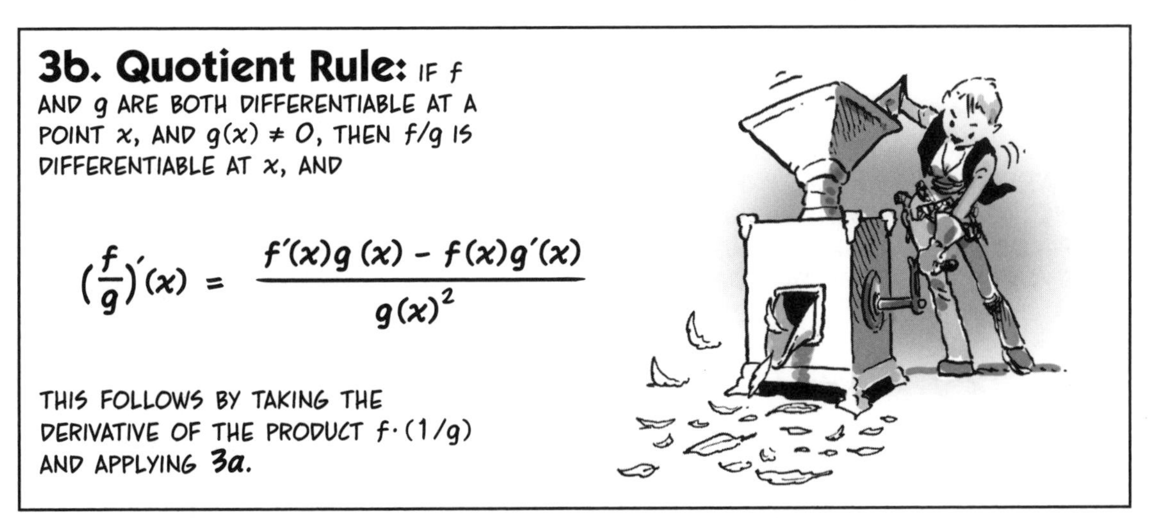

*NOTE THAT WE HAVEN'T DIVIDED BY ZERO ANYWHERE HERE: $f(x+h) \neq 0$ WHEN h IS SMALL ENOUGH, BECAUSE $f(x) \neq 0$, AND $f(x+h)$ GETS ARBITRARILY CLOSE TO $f(x)$.

Example: Negative Powers

WHEN $f(x) = 1/x^n = x^{-n}$, THEN THE FORMULA SAYS

$$\frac{d}{dx}(x^{-n}) = \frac{d}{dx}\left(\frac{1}{x^n}\right) =$$

$$\frac{-\frac{d}{dx}(x^n)}{x^{2n}} = \frac{-nx^{n-1}}{x^{2n}} = \frac{-n}{x^{n+1}}$$

$$= \mathbf{-nx^{-n-1}}$$

$f(x)$	$f'(x)$
$\frac{1}{x}$	$-\frac{1}{x^2}$
$\frac{1}{x^2}$	$-\frac{2}{x^3}$
$\frac{1}{x^3}$	$-\frac{3}{x^4}$
$\frac{1}{x^4}$	$-\frac{4}{x^5}$
$\frac{1}{x^5}$	$-\frac{5}{x^6}$

OR

$f(x)$	$f'(x)$
x^{-1}	$-x^{-2}$
x^{-2}	$-2x^{-3}$
x^{-3}	$-3x^{-4}$
x^{-4}	$-4x^{-5}$
x^{-5}	$-5x^{-6}$
x^{-6}	$-6x^{-7}$

ETC.

NEGATIVE POWERS FOLLOW THE SAME POWER RULE AS POSITIVE POWERS: TO DIFFERENTIATE, MAKE THE EXPONENT A COEFFICIENT AND REDUCE THE POWER BY 1:

$$\frac{d}{dx}(x^p) = px^{p-1}$$

WHETHER p IS A POSITIVE OR NEGATIVE INTEGER. WE'LL SEE IN THE NEXT CHAPTER THAT THIS RULE WORKS FOR FRACTIONAL POWERS AS WELL.

Example: Tangent Function

$$\frac{d}{d\theta}\tan\theta = \sec^2\theta$$

PROOF: WE APPLY THE QUOTIENT FORMULA TO

$$\tan\theta = \frac{\sin\theta}{\cos\theta}$$

HERE $f = \sin\theta$, $g = \cos\theta$, SO

$$\frac{f'g - fg'}{g^2} =$$

$$\frac{\cos\theta\cos\theta - \sin\theta(-\sin\theta)}{\cos^2\theta} =$$

$$\frac{\cos^2\theta + \sin^2\theta}{\cos^2\theta} = \frac{1}{\cos^2\theta}$$

$$= \sec^2\theta$$

MY HEAD IS BURSTING WITH DERIVATIVES!!
YEAH, ISN'T IT GREAT??

SOMEONE ONCE SAID THAT THE PURPOSE OF SCIENCE IS TO SAVE US FROM UNNECESSARY THINKING, AND THAT'S WHAT CALCULUS DOES. HAVING ONCE PENETRATED THE MYSTERIES OF LIMITS AND CHANGE, CALCULUS POPS OUT A BUNCH OF SIMPLE FORMULAS DESCRIBING THE RATES OF CHANGE OF COMMON FUNCTIONS. HALF THE SUBJECT IS **USING THESE FORMULAS!**

WHAT'S THE OTHER HALF?
A HEALTHY BREAKFAST!
ADULT FORMULA
FLAX

$\frac{d}{dx}(x^n) = nx^{n-1} \quad n = 0, \pm 1, \pm 2, \ldots$

$\frac{d}{dx}(e^x) = e^x$

$\frac{d}{dx}\sin x = \cos x$

$\frac{d}{dx}\cos x = -\sin x$

$\frac{d}{dx}\tan x = \sec^2 x \quad (\cos x \neq 0)$

$\frac{d}{dx}(C) = 0$ IF C IS CONSTANT

$(Cf)' = Cf'$, C A CONSTANT

$(f + g)' = f' + g'$

$(fg)' = f'g + fg'$

$\left(\frac{f}{g}\right)' = \frac{f'g - fg'}{g^2}$ WHEREVER $g(x) \neq 0$

A GOOD LIST, BUT STILL MISSING A FEW... WE CAN'T YET DIFFERENTIATE A **COMPOSITE** FUNCTION, NOT EVEN ONE AS SIMPLE AS $h(x) = e^{2x}$... NOR **INVERSE** FUNCTIONS SUCH AS THE LOGARITHM, ARCSINE, AND ARCTANGENT... THOSE COME IN THE NEXT CHAPTER...

AND THEN ON TO THE PROMISED LAND OF PRACTICAL APPLICATIONS!
MILK
HONEY
AND ABOUT TIME, TOO...

BUT FIRST, WHY NOT DO SOME

Problems?

FIND THE DERIVATIVES OF THE GIVEN FUNCTIONS:

1. $f(x) = x^3 + 5x + 1$
2. $f(x) = x^3 + 5x + 1{,}000{,}000$
3. $P(x) = (x^2 + 1)^{-1}$
4. $g(x) = 7$
5. $h(x) = \cos x - \dfrac{5}{\sqrt[3]{x}}$
6. $R(x) = \dfrac{x + 1}{x - 1}$
7. $u(x) = \dfrac{\cos x}{e^x}$
8. $v(t) = \sec t$
9. $F(x) = \dfrac{1}{xe^x}$
10. $B(\theta) = \tan^2\theta$
11. $Q(x) = \dfrac{529x}{x^3 - x^2 - x - 1}$
12. $F(p) = \dfrac{\cos p + pe^p}{p^{10} + p^{-2}}$

13. A PROJECTILE HURLED STRAIGHT UPWARD FROM GROUND LEVEL AT AN INITIAL VELOCITY OF v_0 M/SEC HAS AN ALTITUDE AT TIME t OF

$$A(t) = -4.9t^2 + v_0 t$$

a. IF A BALL IS THROWN VERTICALLY AT AN INITIAL VELOCITY OF 30 M/SEC, WHAT IS ITS VELOCITY AFTER 3 SECONDS? AFTER 5 SECONDS?

b. THE FASTEST AN UNAIDED HUMAN CAN THROW A BALL UPWARD IS AROUND 45 M/SEC. ESTIMATE HOW HIGH THE BALL WILL GO, AND HOW LONG IT TAKES TO RETURN TO EARTH. (HINT: VELOCITY IS POSITIVE BEFORE THE TOP AND NEGATIVE AFTERWARD.)

14. A TRAIL LEADING INTO A MOUNTAIN RANGE HAS ALTITUDE

$$A(x) = x + 0.3\sin x \text{ METERS,}$$

WHERE x IS THE HORIZONTAL DISPLACEMENT FROM THE TRAILHEAD.

a. WHAT IS THE SLOPE OF THE TRAIL AT $x = \pi$ METERS? AT $x = 25\pi$ METERS?

b. DOES THE TRAIL EVER GO DOWNHILL? DRAW A PICTURE OF THE TRAIL.

USE THE DEFINITION OF THE DERIVATIVE TO SHOW THE FOLLOWING:

15. IF f IS INCREASING ON AN INTERVAL (a, b), AND x IS ANY POINT IN THE INTERVAL, THEN $f'(x) \geq 0$.

16. A FUNCTION f IS CALLED **EVEN** IF $f(-x) = f(x)$ FOR ANY x. THE COSINE IS AN EXAMPLE. f IS **ODD** IF $f(-x) = -f(x)$. THE SINE IS AN EXAMPLE.

SHOW THAT AN EVEN FUNCTION HAS AN ODD DERIVATIVE, AND VICE VERSA.

Chapter 3
Chain, Chain, Chain

COMPOSITE FUNCTIONS, ELEPHANTS, MICE, AND FLEAS

NOW WE'RE ON A ROLL... OR MAYBE IT'S A CRAWL... A FORMULA CRAWL... SO LET'S KEEP CRAWLING, SHALL WE? THIS CHAPTER BEGINS BY FINDING THE DERIVATIVES OF ALL THE REMAINING ELEMENTARY FUNCTIONS, AND NICE, SIMPLE FORMULAS THEY ARE...

FORMULA...
FORMULA...

THE KEY TO DERIVING THESE FORMULAS (AND MUCH ELSE BESIDES) IS SOMETHING CALLED THE **CHAIN RULE.** WE'LL START BY SAYING WHAT IT IS, THEN WE'LL USE IT, AND FINALLY WE'LL EXPLAIN WHY IT'S TRUE.

THE CHAIN RULE IS A PROCEDURE FOR DIFFERENTIATING **COMPOSITE** FUNCTIONS, FUNCTIONS MADE BY PLUGGING ONE FUNCTION INTO ANOTHER. [SEE PP. 38-39.] FOR EXAMPLE,

$$h(x) = e^{2x}$$

HERE THE INSIDE FUNCTION IS $u(x) = 2x$, WHILE THE OUTSIDE FUNCTION IS $v(u) = e^u$.

x
$2x$
e^{2x}

The Chain Rule:

TO DIFFERENTIATE A COMPOSITION $h(x) = v(u(x))$, FOLLOW THESE STEPS:

1. DIFFERENTIATE THE INSIDE FUNCTION. THAT IS, FIND $u'(x)$.
2. TREATING THE ENTIRE INSIDE FUNCTION u AS A VARIABLE, DIFFERENTIATE THE OUTSIDE FUNCTION WITH RESPECT TO u: I.E., FIND $v'(u)$.
3. MULTIPLY THE RESULTS OF **1** AND **2**.
4. FINALLY, REPLACE u BY $u(x)$ IN $v'(u)$.

IN SYMBOLS,

$$h'(x) = u'(x) \cdot v'(u(x))$$

THIS IS THE KEY TO EVERYTHING!

I DON'T NEED A KEY, I NEED, UM, FORMULA...
GOOD. THIS IS THE KEY TO THE FRIDGE, TOO...

THIS PROBABLY LOOKS WORSE THAN IT REALLY IS. IN ESSENCE, THE CHAIN RULE SIMPLY SAYS TO MULTIPLY THE DERIVATIVE OF THE INSIDE FUNCTION BY THE DERIVATIVE OF THE OUTSIDE FUNCTION.

Example: AS ABOVE, SUPPOSE $h(x) = e^{2x}$. WE GO STEP BY STEP:

1. $u'(x) = 2$
2. $v'(u) = e^u$
3. THE PRODUCT IS $2e^u$
4. WE REPLACE u BY $u(x) = 2x$ TO GET THE FINAL RESULT:

$$h'(x) = \mathbf{2e^{2x}}$$

REMEMBER: ALWAYS TREAT THE ENTIRE INSIDE FUNCTION AS A VARIABLE IN STEP 2!!

Example: $G(x) = \sin(x^2)$. THE INSIDE FUNCTION IS $u(x) = x^2$. THE OUTSIDE FUNCTION IS $v(u) = \sin u$.

1. $u'(x) = 2x$
2. $v'(u) = \cos u$
3. THE PRODUCT IS $2x \cos u$
4. WRITING $u(x) = x^2$ FOR u GIVES THE DERIVATIVE:

$$G'(x) = \mathbf{2x \cos(x^2)}$$

WHAT'S WRONG?
I'M TIRED OF BEING TREATED LIKE A VARIABLE...

One More Example!

$f(x) = (2x^3 + 3)^8$.
INSIDE FUNCTION: $u(x) = 2x^3 + 8$.
OUTSIDE FUNCTION: $v(u) = u^8$

$$f'(x) = u'(x)g'(u)$$
$$= (6x^2)(8u^7)$$
$$= (6x^2)(8(2x^3 + 3)^7)$$
$$= \mathbf{48x^2(2x^3 + 3)^7}$$

HERE THE CHAIN RULE LETS US DIFFERENTIATE A MONSTER 24TH-DEGREE POLYNOMIAL WITHOUT HAVING TO EXPAND IT FIRST.

Derivatives of Inverse Functions

THE CHAIN RULE CAN ALSO HELP US FIND THE DERIVATIVE OF AN INVERSE f^{-1} WHEN WE KNOW THE DERIVATIVE OF f.

f(x)
x

Example: SUPPOSE $u(x) = \sqrt{x}$ OR $x^{\frac{1}{2}}$, THE INVERSE OF $v(u) = u^2$. THEN THE COMPOSITION $f(x) = v(u(x)) = x$, SO OBVIOUSLY,

$$f'(x) = 1$$

BUT THE CHAIN RULE GIVES ANOTHER FORMULA FOR $f'(x)$:

$$f'(x) = u'(x)v'(u(x))$$

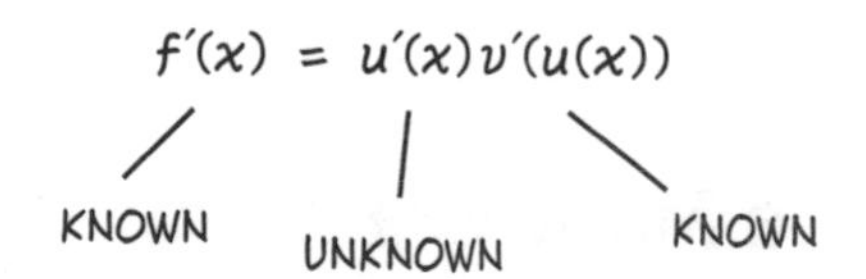
KNOWN
UNKNOWN
KNOWN

ERHM... WHAT HAPPENS HERE IF ONE FACTOR IS 0?
AHEM... COUGH!... JUST ASSUME IT ISN'T, OKAY?

EQUATING THESE, WE GET:

$$1 = \frac{d}{dx}(x^{\frac{1}{2}})\frac{d}{du}(u^2) = 2u\frac{d}{dx}(x^{\frac{1}{2}})$$

$$= 2x^{\frac{1}{2}}\frac{d}{dx}(x^{\frac{1}{2}})$$

NOW DIVIDE BY $2x^{\frac{1}{2}}$ TO SOLVE FOR THE DERIVATIVE:

$$\frac{d}{dx}(x^{\frac{1}{2}}) = \frac{1}{2x^{\frac{1}{2}}}$$

OR

$$\boxed{\frac{1}{2}x^{-\frac{1}{2}}}$$

WHEN $x \neq 0$!!

YOU CAN RUN THROUGH THE SAME SET OF STEPS FOR $u(x) = x^{1/n}$ AND $v(u) = u^n$: THEN $f(x) = v(u(x)) = x$, AND SO

$$1 = u'(x)v'(u(x)) \quad \text{PROVIDED } v'(u(x)) \neq 0$$

$$= u'(x) \cdot n(x^{1/n})^{n-1} \quad \text{SO}$$

$$u'(x) = \frac{1}{n}(x^{1/n})^{1-n} = \frac{1}{n}x^{\frac{1-n}{n}}$$

$$= \frac{1}{n}x^{\frac{1}{n}-1}$$

$$\boxed{\frac{d}{dx}(x^{\frac{1}{n}}) = \frac{1}{n}x^{(\frac{1}{n}-1)}}$$

IF $x \neq 0$

WHAT WE JUST DID FOR $x^{\frac{1}{n}}$ AND u^n, WE CAN DO FOR **ANY** PAIR OF INVERSE FUNCTIONS f AND f^{-1}: TO FIND $(f^{-1})'$, THE DERIVATIVE OF THE INVERSE, IN TERMS OF f':

$$x = f(f^{-1}(x))$$

$$1 = \frac{d}{dx}(f(f^{-1}(x))$$

$$= (f^{-1})'(x) \cdot f'(f^{-1}(x)) \quad \text{SO}$$

$$\boxed{(f^{-1})'(x) = \frac{1}{f'(f^{-1}(x))}}$$

IF $f'(f^{-1}(x)) \neq 0$

HERE'S HOW IT LOOKS ON A GRAPH. BECAUSE THE INVERSE SWITCHES x AND y, THE SLOPE $\Delta y/\Delta x$ OF THE GRAPH OF f BECOMES $\Delta x/\Delta y$ ON THE GRAPH OF f^{-1}. YOU HAVE TO CHASE AROUND THE GRAPH A BIT TO FIND THE RIGHT POINT AT WHICH TO EVALUATE $(f^{-1})'$ ... BUT DON'T WORRY! SOON WE'LL SEE A DIFFERENT DIAGRAM THAT MAKES THINGS MUCH CLEARER.

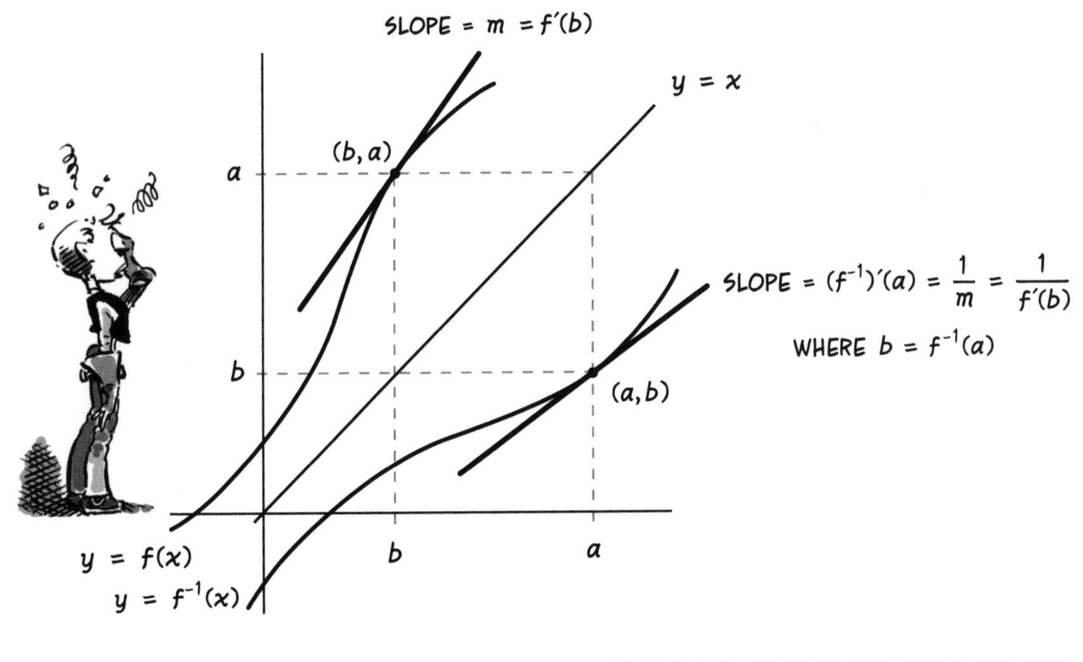

SLOPE = m = f′(b)
y = x
(b, a)
a
b
(a, b)
SLOPE = (f⁻¹)′(a) = 1/m = 1/f′(b)
WHERE b = f⁻¹(a)
b
a
y = f(x)
y = f⁻¹(x)

FOR NOW, LET'S JUST USE THE FORMULA BLINDLY, PLUGGING IN INVERSE FUNCTIONS TO FIND THEIR DERIVATIVES. THE SIMPLICITY OF THE RESULTS MAY SURPRISE YOU...

WE APPLY THE INVERSE DERIVATIVE FORMULA TO THREE FUNCTIONS: THE **LOGARITHM,** THE **ARCSINE,** AND THE **ARCTANGENT.**

1. TAKE $f(u) = e^u$ AND $f^{-1}(x) = \ln x$. THEN $f'(u) = e^u$, AND

$$\frac{d}{dx}\ln x = \frac{1}{e^{\ln x}} = \boxed{\frac{1}{x}}$$

2. $f(u) = \sin u$, $f^{-1}(x) = \arcsin x$. $f'(u) = \cos u$

$$\frac{d}{dx}(\arcsin x) = \frac{1}{\cos(\arcsin x)}$$

HOW DO WE EVALUATE THE COSINE OF $\arcsin x$? BY REMEMBERING THAT $\sin^2 u + \cos^2 u = 1$.

$$\cos u = \sqrt{1 - \sin^2 u} \quad \text{SO}$$

$$\cos(\arcsin x) = \sqrt{1 - \sin^2(\arcsin x)}$$

$$= \sqrt{1 - x^2} \quad \text{SO}$$

$$\frac{d}{dx}(\arcsin x) = \boxed{\frac{1}{\sqrt{1 - x^2}}}$$

NOTE THAT IT WAS O.K. TO TAKE THE POSITIVE SQUARE ROOT HERE: VALUES OF THE ARCSINE LIE BETWEEN $-\pi/2$ AND $\pi/2$, AND ON THIS INTERVAL THE COSINE IS POSITIVE.

3. $f(u) = \tan u$, $f^{-1}(x) = \arctan x$. $f'(u) = \sec^2 x$

$$\frac{d}{dx}(\arctan x) = \frac{1}{\sec^2(\arctan x)}$$

THE TRIG IDENTITY $\sec^2 x = 1 + \tan^2 x$ GIVES
$\sec^2(\arctan x) = 1 + \tan^2(\arctan x) = 1 + x^2$!!!

$$\frac{d}{dx}\arctan x = \boxed{\frac{1}{1 + x^2}}$$

MIND-BOGGLING, AREN'T THEY?

IT'S VERY STRANGE... TRIG FUNCTIONS AND EXPONENTIALS TAKE THEIR DERIVATIVES FROM AMONG THEMSELVES... BUT THEIR **INVERSES** HAVE DERIVATIVES MADE OF ORDINARY **POLYNOMIALS** AND **SQUARE ROOTS.** HOW DID **THAT** HAPPEN?

$\frac{d}{dx} \ln x = \frac{1}{x}$
$\frac{d}{dx} \arcsin x = \frac{1}{\sqrt{1-x^2}}$
$\frac{d}{dx} \arctan x = \frac{1}{1+x^2}$
IT'S A MYSTERY!
The Case of the Inverse Function
A GIFT!

THE LOGARITHM'S DERIVATIVE IS PERHAPS MOST SURPRISING: x^{-1} LOOKS LIKE THE DERIVATIVE OF A POWER FUNCTION. BUT THE POWER RULE $\frac{d}{dx}(x^n) = nx^{n-1}$ CAN PRODUCE DERIVATIVES ONLY WITH EXPONENTS **OTHER THAN −1,** SINCE $\frac{d}{dx}(x^0) = 0$.

THE NATURAL LOG PERFECTLY FILLS THAT ONE HOLE IN THE POWER LIST:

$f(x)$	$f'(x)$
x^2	$2x$
x	1
$x^0 = 1$	0
$\ln x$	**x^{-1}**
x^{-1}	$-x^{-2}$
x^{-2}	$-2x^{-3}$
ETC.	

Examples of Derivatives Found by the Chain Rule:

1. $h(x) = x^{\frac{m}{n}}$, m AND n INTEGERS.

$x^{\frac{m}{n}} = (x^{\frac{1}{n}})^m$, SO

INSIDE FUNCTION: $u(x) = x^{\frac{1}{n}}$, $u'(x) = \frac{1}{n}x^{\frac{1}{n}-1}$

OUTSIDE FUNCTION: $v(u) = u^m$, $v'(u) = mu^{m-1}$

$$h'(x) = u'(x)v'(u(x)) = (\tfrac{1}{n}x^{\frac{1}{n}-1})(mu^{m-1})$$

$$= (\tfrac{1}{n}x^{\frac{1}{n}-1})(m(x^{\frac{1}{n}})^{m-1})$$

$$= \frac{m}{n}x^{(\frac{1-n}{n} + \frac{m-1}{n})}$$

$$= \mathbf{\frac{m}{n}x^{\frac{m}{n}-1}}$$

YESS!! THE POWER RULE AGAIN!

2. $f(x) = \arctan(3x)$

INSIDE: $u(x) = 3x$, $u'(x) = 3$

OUTSIDE: $v(u) = \arctan u$, $v'(u) = \dfrac{1}{1+u^2}$

$$f'(x) = u'(x)v'(u(x)) = \frac{3}{1+u^2}$$

$$= \frac{3}{1+(3x)^2} = \mathbf{\frac{3}{1+9x^2}}$$

3. $g(x) = f(ax)$, a A CONSTANT

INSIDE: $u(x) = ax$, OUTSIDE f, SO

$$\mathbf{g'(x) = af'(ax)}$$

4. $F(x) = \sqrt{1-x^2}$

INSIDE: $u(x) = 1 - x^2$, $u'(x) = -2x$

OUTSIDE: $v(u) = u^{\frac{1}{2}}$, $v'(u) = \frac{1}{2}u^{-\frac{1}{2}}$

$$F'(x) = -2x \cdot (\tfrac{1}{2}u^{-\frac{1}{2}}) = -2x(\tfrac{1}{2})(1-x^2)^{-\frac{1}{2}}$$

$$= \mathbf{\frac{-x}{\sqrt{1-x^2}}}$$

5. $G(x) = \ln(x^2 + x)$

INSIDE: $u(x) = x^2 + x$, $u'(x) = 2x + 1$

OUTSIDE: $v(u) = \ln u$, $v'(u) = 1/u$

$$G'(x) = (2x + 1)(1/u)$$

$$= \mathbf{\frac{2x+1}{x^2+x}}$$

6. $P(t) = (2 + t + 2t^3)^{5/6}$

INSIDE: $u(x) = 2 + t + 2t^3$, $u'(x) = 1 + 6t^2$

OUTSIDE: $v(u) = u^{5/6}$, $v'(u) = \frac{5}{6}u^{-1/6}$

$$P'(t) = (1 + 6t^2)(\tfrac{5}{6}u^{-1/6})$$

$$= \mathbf{\frac{5}{6}(1 + 6t^2)(2 + t + 2t^3)^{-1/6}}$$

7. $U(x) = (f(x))^n$ FOR ANY DIFFERENTIABLE FUNCTION f, ANY RATIONAL n

INSIDE: $f(x)$, DERIVATIVE $= f'(x)$

OUTSIDE: $v(u) = u^n$, $v'(u) = nu^{n-1}$

$$U'(x) = f'(x)(nu^{n-1})$$

$$= \mathbf{nf'(x)(f(x))^{n-1}}$$

WE HAVE NOW FOUND DERIVATIVES OF ALL THE ELEMENTARY FUNCTIONS... FROM THESE WE CAN BUILD THE DERIVATIVE OF **ANY** FUNCTION MADE BY PILING UP THE ELEMENTARIES BY COMBINATIONS OF ADDITION, MULTIPLICATION, DIVISION, AND COMPOSITION. WE'RE EMPOWERED!

sin x² x
2e^cos x³
e^e^sin x
EMPOWERED, BABY!

AND YES, WE DO KNOW HOW TO DIFFERENTIATE CHAINS LONGER THAN TWO FUNCTIONS: JUST MULTIPLY ALL THE DERIVATIVES!

$$\frac{d}{dt}v(u(y(x(t)))) = \frac{dv}{du}\frac{du}{dy}\frac{dy}{dx}\frac{dx}{dt}$$

OR, IF YOU PREFER THE OTHER NOTATION:
IF $f(t) = v(u(y(x(t))))$, THEN

$$f'(t) = x'(t)y'(x(t))u'(y(x(t)))v'(u(y(x(t))))$$

PIECE OF CAKE!
READY FOR ANYTHING!!
HUP!
YEAH!

Three-function example:

$$\frac{d}{dx}\sin(e^{x^2}) = 2xe^{x^2}\cos(e^{x^2})$$

(INNER: $u(x) = x^2$, MIDDLE: $v(u) = e^u$,
OUTER: $g(v) = \sin v$)

ARE YOU READY FOR THIS?
REMEMBER, WE STILL HAVEN'T SHOWN WHY THE CHAIN RULE IS TRUE! TO SEE THIS, LET'S LOOK AT THE DERIVATIVE IN THE PARALLEL-AXES PICTURE OF A FUNCTION f.
f
UH... WHERE'D THE DERIVATIVE GO, ANYWAY?
$x+h$
$f(x+h)$
h
Δf
$f(x)$
x

WHAT IS $\Delta f/h$ IN THIS PICTURE? ON THE LEFT ARE THE POINTS x AND $x + h$, AND ON THE RIGHT ARE THE "TARGET" OR "IMAGE" POINTS $f(x)$ AND $f(x + h)$.

HERE, THEN, THE DIFFERENCE QUOTIENT $\Delta f/h$ IS A **SCALING FACTOR** THAT MULTIPLIES THE AMOUNT h TO GET Δf.

$$\Delta f = \left(\frac{\Delta f}{h}\right) h$$

THIS FACTOR MAY ENLARGE, SHRINK, AND/OR INVERT THE SPACE BETWEEN THE POINTS x AND $x + h$.

$\frac{\Delta f}{h} > 1$ $\quad$ $0 < \frac{\Delta f}{h} < 1$ $\quad$ $\frac{\Delta f}{h} < 0$

WHAT HAPPENS AS $h \to 0$? IT ISN'T EASY TO SEE... EVERYTHING IS SO SMALL... SO LET'S TALK ABOUT **SMALLNESS...**

SMALLNESS IS **RELATIVE...** SOMETHING IS SMALL ONLY IN COMPARISON WITH SOMETHING ELSE. NEXT TO AN **ELEPHANT,** A **MOUSE** IS SMALL, BUT THAT SAME MOUSE INSPIRES AWE IN A **FLEA...** THE MOUSE, MEANWHILE, SEES THE FLEA AS SMALL, WHILE TO THE ELEPHANT A FLEA IS COMPLETELY BENEATH NOTICE.

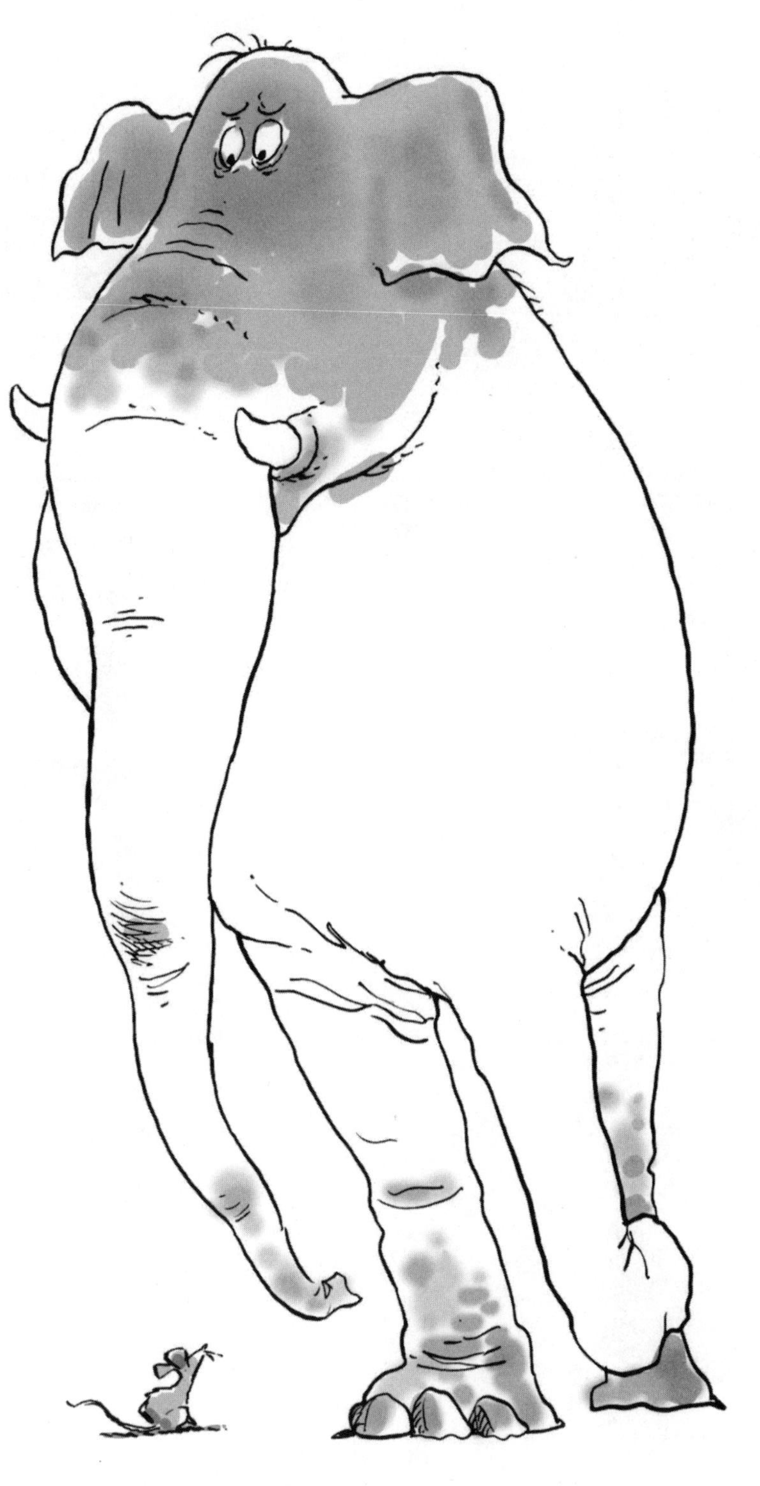

IT'S THE SAME WITH NUMBERS... WE WANT TO THINK OF ORDINARY NUMBERS, LIKE a AND $f(a)$, AS ELEPHANTS, PART OF THE MACRO-WORLD. (I KNOW, THEY CAN BE ZERO SOMETIMES, BUT NOT USUALLY!)

THE INCREMENT h IS ASSUMED TO BE SMALL COMPARED TO AN ELEPHANTINE NUMBER LIKE, SAY, 1. IN GENERAL, WE'LL CALL SOMETHING A **MOUSE** IF IT SHRINKS WITH h, THAT IS, IF

$$\lim_{h \to 0} (\text{MOUSE}) = 0$$

A MATHEMATICAL **FLEA** IS SOMETHING SMALL EVEN COMPARED TO h. FOR INSTANCE, h^2 IS A FLEA: IF $h = \frac{1}{1000}$, THEN $h^2 = \frac{1}{1000}$ OF $\frac{1}{1000}$, AS SMALL COMPARED TO h AS h IS TO 1. WE'LL CALL SOMETHING A FLEA IF

$$\lim_{h \to 0} \frac{\text{FLEA}}{h} = 0$$

WOW...

SO h^2, h^3, AND $h^{3/2}$ ARE ALL FLEAS. EVENTUALLY, AS $h \to 0$, THEY ALL LOOK SMALL COMPARED TO h.

$$\lim_{h \to 0} \frac{h^{3/2}}{h} = \lim_{h \to 0} h^{1/2} = 0$$

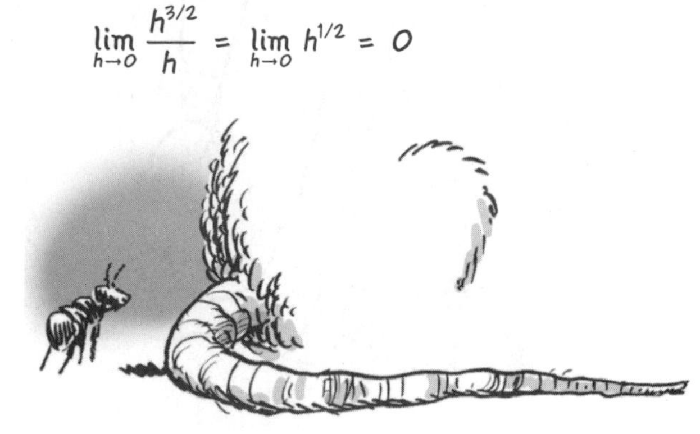

FROM THE DEFINITIONS, IT FOLLOWS IMMEDIATELY THAT

$\frac{\text{FLEA}}{h}$ IS A MOUSE

$h \cdot (\text{MOUSE})$ IS A FLEA

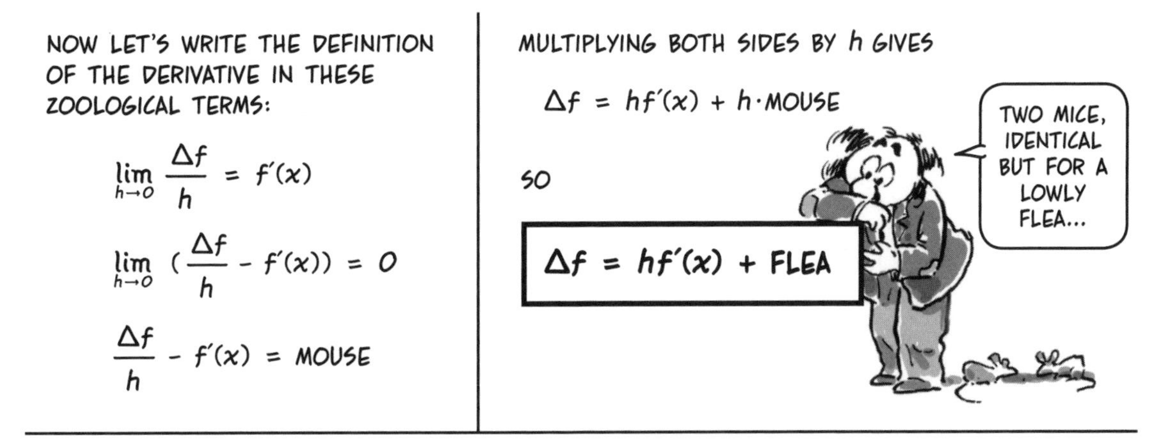
TWO MICE, IDENTICAL BUT FOR A LOWLY FLEA...

NOW LET'S WRITE THE DEFINITION OF THE DERIVATIVE IN THESE ZOOLOGICAL TERMS:

$$\lim_{h \to 0} \frac{\Delta f}{h} = f'(x)$$

$$\lim_{h \to 0} \left(\frac{\Delta f}{h} - f'(x)\right) = 0$$

$$\frac{\Delta f}{h} - f'(x) = \text{MOUSE}$$

MULTIPLYING BOTH SIDES BY h GIVES

$$\Delta f = hf'(x) + h \cdot \text{MOUSE}$$

SO

$$\boxed{\Delta f = hf'(x) + \text{FLEA}}$$

I CALL THIS LAST EQUATION THE **FUNDAMENTAL EQUATION OF CALCULUS.** ('COURSE, NOBODY ELSE DOES, SO DON'T EXPECT TO SEE IT ON THE TEST...) I LIKE IT BECAUSE EVERYTHING IN IT IS SMALL: IT GIVES US A "MOUSE-SCALE" VIEW OF FUNCTIONS ON VERY SHORT INTERVALS. IN FACT, I LIKE IT SO WELL, I'M GOING TO WRITE IT AGAIN, REALLY LARGE:

$\Delta f = hf'(x) + \text{FLEA}$
A LARGE EQUATION ABOUT SMALL THINGS!

ON A GRAPH, IT MEANS THIS: AS h GETS SMALL, THE DISCREPANCY BETWEEN THE CURVE $y = f(x)$ AND ITS TANGENT LINE BECOMES NEGLIGIBLE, A MERE FLEA—SMALL COMPARED TO h. IF WE ZOOM IN CLOSE ENOUGH, IN OTHER WORDS, **THE CURVE BECOMES VIRTUALLY INDISTINGUISHABLE FROM A STRAIGHT LINE.**

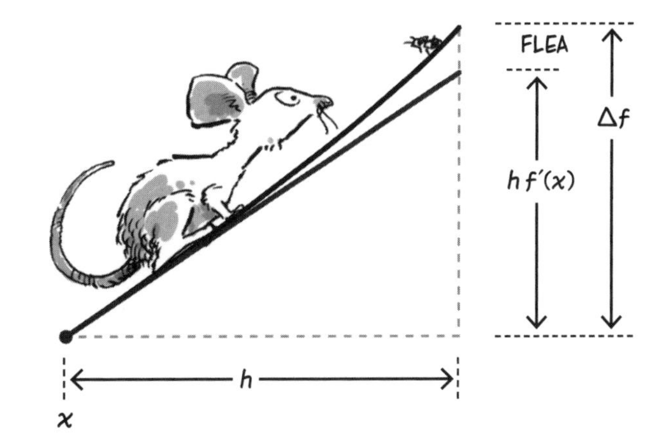
FLEA
Δf
hf′(x)
h
x

IN THE PARALLEL-AXES VIEW, IT MEANS THIS: IN THE LIMIT, AS $h \to 0$, WE CAN REPLACE THE SCALING FACTOR $\Delta f/h$ BY $f'(x)$. THAT IS, THE FUNCTION f **SCALES A SMALL CHANGE IN x BY A FACTOR OF $f'(x)$,** ASIDE FROM A DISCREPANCY THAT BECOMES NEGLIGIBLE.

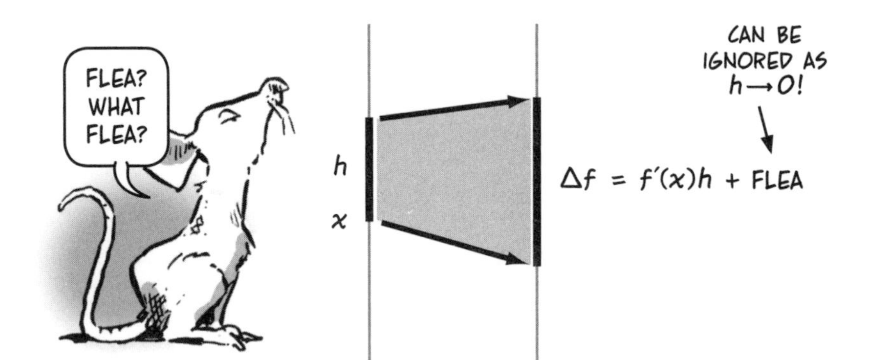

FLEA? WHAT FLEA?
CAN BE IGNORED AS h→0!
h
x
Δf = f′(x)h + FLEA

THIS IMMEDIATELY SHOWS WHY AN INVERSE FUNCTION'S DERIVATIVE IS WHAT IT IS: THE INVERSE f^{-1} **REVERSES THE ARROWS OF f.** WHATEVER SCALING IS DONE BY f IS **UNSCALED** BY f^{-1}.

f SCALES A SMALL CHANGE IN t BY A FACTOR OF $f'(t)$ (ASSUME $f'(t) \neq 0$)

h
t
f
Δf
f(t)

$$\Delta f \approx f'(t)h$$

REVERSING THE ARROWS THEN "UNSCALES" BY A FACTOR OF $1/f'(t)$.

Δ(f⁻¹)
t
f⁻¹
k
x

$$\Delta(f^{-1}) \approx \frac{1}{f'(t)}k$$

BUT THE DERIVATIVE IS THE SCALING FACTOR! SO THE DERIVATIVE $(f^{-1})'(x)$ HAS TO BE $1/f'(t)$, AND, SINCE $t = f^{-1}(x)$, WE GET THE FORMULA OF PAGE 105:

$$(f^{-1})'(x) = \frac{1}{f'(f^{-1}(x))}$$

FOR THE CHAIN RULE, THE PICTURE IS SIMILAR. NOW WE HAVE TWO FUNCTIONS u AND v. THE INSIDE FUNCTION u IS ON THE LEFT, SINCE IT COMES FIRST, AND WE WANT TO SEE THE DERIVATIVE OF THE FUNCTION f DEFINED BY $f(x) = v(u(x))$.

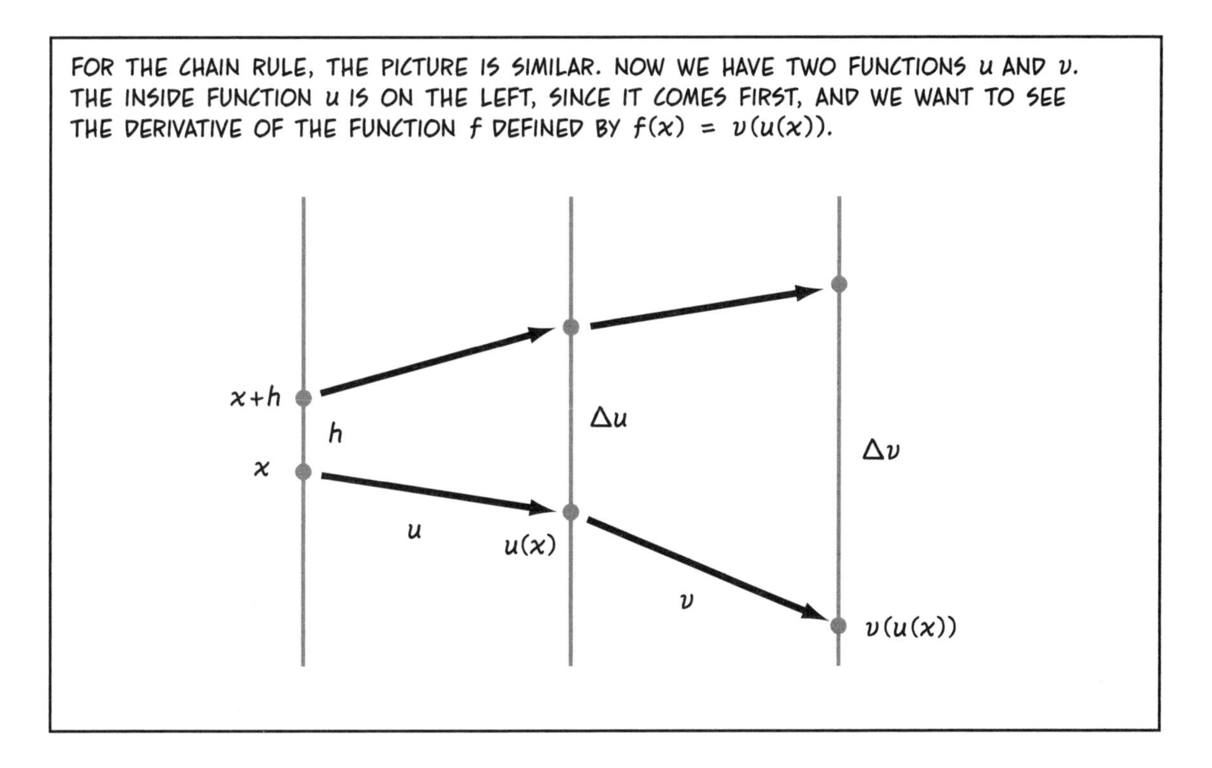

x+h
h
x
Δu
u
u(x)
Δv
v
v(u(x))

HERE THE QUANTITY h IS SCALED TWICE: FIRST BY A FACTOR $u'(x)$ AND THEN BY A FACTOR OF v' EVALUATED AT $u(x)$. THE NET EFFECT OF BOTH FUNCTIONS, THEN, IS TO SCALE h BY THE **PRODUCT** $u'(x)\,v'(u(x))$, SO THIS MUST BE THE DERIVATIVE OF f AT THE POINT x. (IMAGINE FIRST DOUBLING, THEN TRIPLING; THE EFFECT WOULD BE TO MULTIPLY BY SIX.)

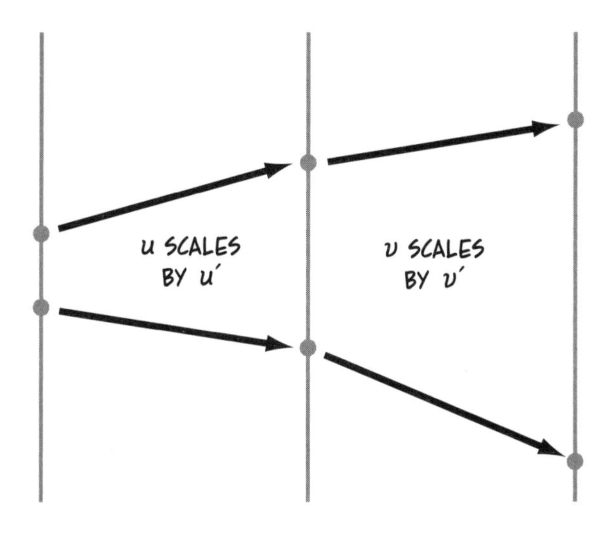

u SCALES BY u′
v SCALES BY v′

$$\Delta u \approx u'(x)h$$

$$\Delta v \approx v'(u(x))\Delta u$$

$$\approx v'(u(x))\,u'(x)h$$

WHICH SAYS THAT THE SCALING FACTOR, AND THEREFORE THE DERIVATIVE, OF THE COMPOSITION IS $u'(x)v'(u(x))$. AND THAT IS THE CHAIN RULE!

$$\mathbf{f'(x) = u'(x)\,v'(u(x))}$$

Q.E.D., SORT OF!

Problems

1. SUPPOSE $f(x) = x^2$ AND $g(u) = \cos u$. WHAT IS $f(g(u))$? WHAT IS $g(f(x))$? GRAPH BOTH COMPOSITE FUNCTIONS. WHAT ARE THEIR DERIVATIVES?

2. SUPPOSE $u(x) = -x^2$ AND $v(u) = e^u$. SAME QUESTIONS AS PROBLEM 1.

3. DIFFERENTIATE:

a. $f(t) = \sqrt{1 + t + t^2}$

b. $g(x) = (\cos x - \sin x)^{25}$

c. $h(\theta) = \tan^2\theta$

d. $P(r) = (r^2 + 7)^{10}$

e. $Q(r) = (r^2 + 7)^{-10}$

f. $f(y) = \cos(\sqrt{y})$

g. $E(x) = e^{x-a}$

h. $F(x) = e^{\left(\frac{x-a}{2}\right)}$

i. $v(z) = (\sin(z)^2 + 2)^{-1/3}$

---

4. IF f IS DIFFERENTIABLE, SHOW THAT

$$\frac{d}{dx}\ln(f(x)) = f'(x)/f(x)$$

THIS FORMULA, TOGETHER WITH THE FACT THAT $\ln(ab) = \ln a + \ln b$, CAN SIMPLIFY DIFFERENTIATION WHEN THE FUNCTION INVOLVES PRODUCTS AND QUOTIENTS. FOR EXAMPLE, SUPPOSE

$$y = x^2\cos x \quad \text{SO}$$

$$\ln y = 2\ln x + \ln(\cos x)$$

DIFFERENTIATING WITH RESPECT TO x GIVES

$$\frac{y'}{y} = \frac{2}{x} - \frac{\sin x}{\cos x}$$

BUT $y = x^2\cos x$ (IT'S WHERE WE STARTED!). MULTIPLY THROUGH BY THIS TO FIND y':

$$y' = \left(\frac{2}{x} - \frac{\sin x}{\cos x}\right)x^2\cos x$$

$$= 2x\cos x - x^2\sin x$$

5. USE THIS **LOGARITHMIC DIFFERENTIATION** TECHNIQUE ON THESE FUNCTIONS:

a. $f(x) = x^5e^x(1 + x)^{-1/3}$

b. $g(x) = x^{\sqrt{x}}$

6. SHOW THAT IF $F_1(h)$ AND $F_2(h)$ ARE BOTH FLEAS, THEN SO IS $F_1 + F_2$.

7a. IF $f(x) = 2 + \sin x$, WHAT IS THE INVERSE FUNCTION f^{-1}? DRAW ITS GRAPH ON A SUITABLE DOMAIN, AND FIND $(f^{-1})'(x)$.

HINT: WRITE $y = 2 + \sin x$ AND SOLVE FOR x.

b. SAME THING FOR $f(x) = \sqrt{x^2 + 1}$.

8. A POTATO AT ROOM TEMPERATURE (25° C) IS PUT INTO A 275° OVEN. THE POTATO'S TEMPERATURE T, IN DEGREES CELSIUS, AFTER t MINUTES IS

$$T(t) = 25 + 250(1 - e^{-0.46t})$$

a. DRAW A GRAPH OF THIS FUNCTION. HOW FAST IS THE POTATO HEATING UP, IN DEGREES PER MINUTE, AFTER 10 MINUTES? 20 MINUTES? 60 MINUTES? 100 MINUTES?

b. HOW MANY MINUTES DOES IT TAKE THE POTATO TO REACH 274°?

9. WHICH OF THESE FUNCTIONS IS A FLEA? A MOUSE? NEITHER?

a. $h^{3/2}$

b. $h^{1/2}$

c. $\dfrac{1 - h^2}{h}$

d. $\sin h$

e. $h\cos h$

f. $\cos h - 1$

g. $\Delta f \Delta g$ WHEN f AND g ARE DIFFERENTIABLE.

Chapter 4
Using Derivatives, Part 1: Related Rates

IN WHICH WE ACTUALLY TALK ABOUT THE REAL WORLD

THE CHAIN RULE IS MORE THAN A FORMULA FOR TAKING DERIVATIVES: IT ALSO HELPS US **SOLVE PROBLEMS.**

AND ABOUT TIME, DON'T YOU THINK?
OF COURSE I THINK.

Example 1:

AN AIRPLANE TRAVELING AT A CONSTANT ALTITUDE OF 3 KM IS BEING TRACKED BY A GROUND-BASED RADAR STATION. AT A CERTAIN TIME t_0, THE RADAR CREW MEASURES THE PLANE TO BE 5 KM DISTANT, AND THIS DISTANCE IS FALLING AT A RATE OF 320 KM/HR. HOW FAST IS THE PLANE FLYING AT TIME t_0?

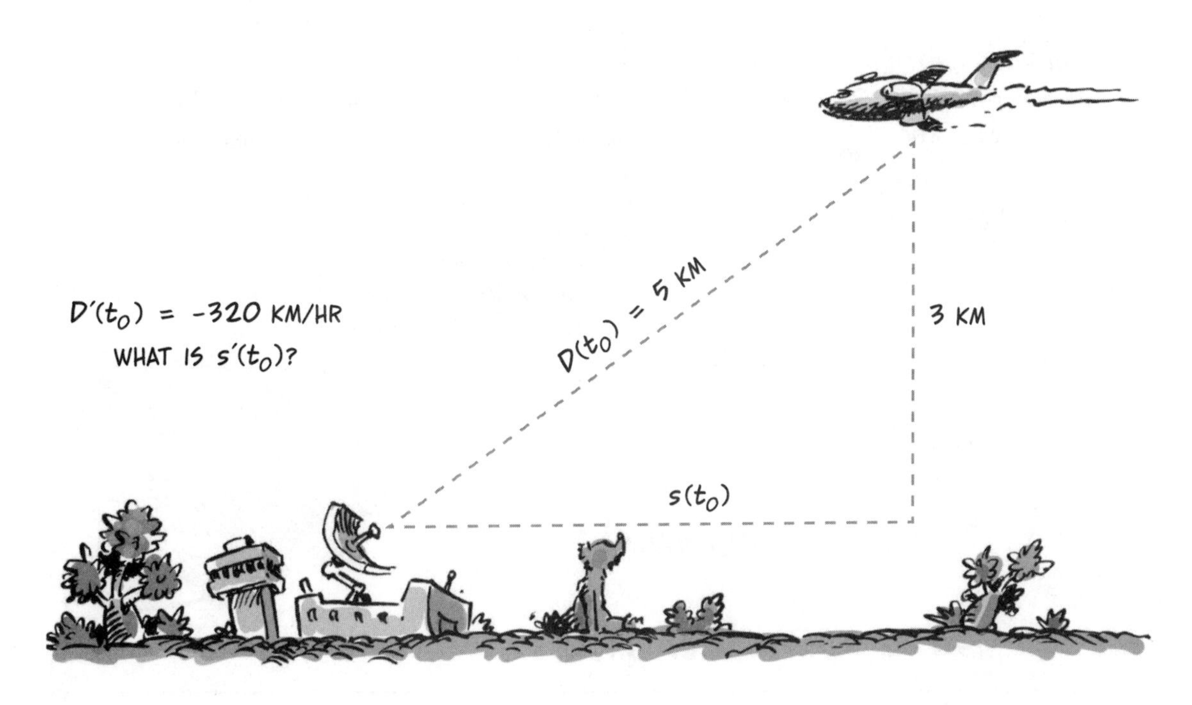

$D'(t_0) = -320$ KM/HR
WHAT IS $s'(t_0)$?
$D(t_0) = 5$ KM
3 KM
$s(t_0)$

AT ANY TIME t, THE RADAR SITS AT ONE CORNER OF A RIGHT TRIANGLE OPQ WITH HYPOTENEUSE $D(t)$. IF $s(t)$ IS THE PLANE'S **HORIZONTAL** DISPLACEMENT AT TIME t, WE ARE ASKING: WHAT IS $s'(t)$, THE DERIVATIVE OF s?

YOU MIGHT WONDER HOW CAN WE FIND $s'(t)$ WHEN WE HAVE **NO IDEA** WHAT THE FUNCTION s LOOKS LIKE. THE PILOT COULD BE ACCELERATING AND DECELERATING LIKE A DRUNKEN AVIATOR!

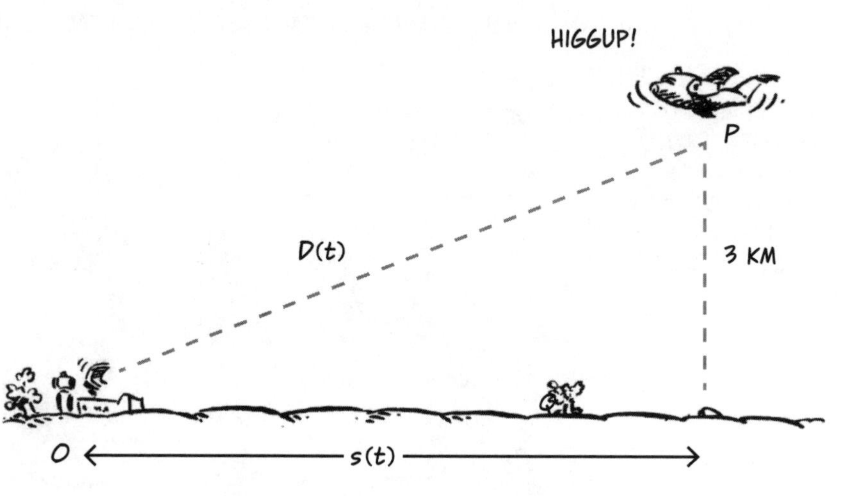
HIGGUP!
P
D(t)
3 KM
O
s(t)

WHAT WE DO KNOW IS THIS:

$$D^2 - s^2 = 3^2 \quad \text{AND ALSO}$$

$$D(t_0) = 5 \qquad s(t_0) = 4 \qquad D'(t_0) = -320$$

EVEN WITHOUT KNOWING THE FUNCTIONS $s(t)$ AND $D(t)$, THE FIRST EQUATION IMPLIES A RELATIONSHIP BETWEEN THEIR DERIVATIVES. BY THE CHAIN RULE, WE CAN DIFFERENTIATE THE SQUARE OF A FUNCTION: $\frac{d}{dx}(f)^2 = 2f'f$. (SEE EXAMPLE 7, P. 108.) SO WE DIFFERENTIATE:

$$2DD' - 2ss' = 0$$

SO

$$s' = \frac{DD'}{s} \qquad \text{WHENEVER } s(t) \neq 0$$

AT TIME t_0, THEN,

$$s'(t_0) = \frac{5}{4}(-320) = \mathbf{-400\ km/hr}$$

FROM A GROUND-BASED OBSERVATION, WE GET THE PLANE'S AIRSPEED!

THE DERIVATIVES s' AND D' ARE **RELATED RATES.**

Implicit Differentiation

IN THE PREVIOUS EXAMPLE, THE EQUATION $D^2 - s^2 = 9$ **IMPLIED** A RELATIONSHIP BETWEEN THE DERIVATIVES OF D AND s. THE PROCESS OF FINDING THIS RELATIONSHIP IS CALLED **IMPLICIT DIFFERENTIATION**. WE DIFFERENTIATE WITHOUT EVER WRITING AN EXPLICIT FORMULA FOR EITHER FUNCTION.

SOMETIMES IT'S BETTER NOT TO BE TOO EXPLICIT..

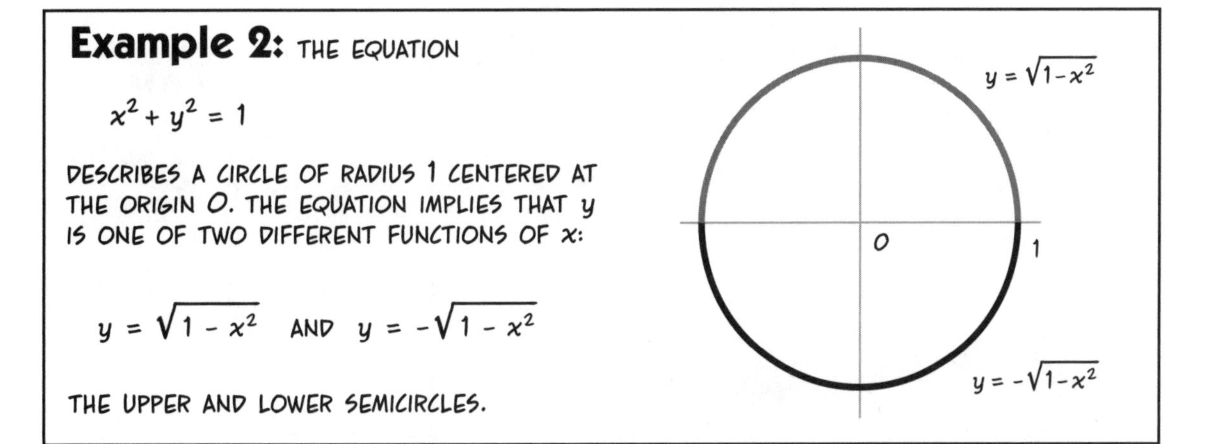
$y = \sqrt{1-x^2}$
0
1
$y = -\sqrt{1-x^2}$

Example 2: THE EQUATION

$$x^2 + y^2 = 1$$

DESCRIBES A CIRCLE OF RADIUS 1 CENTERED AT THE ORIGIN O. THE EQUATION IMPLIES THAT y IS ONE OF TWO DIFFERENT FUNCTIONS OF x:

$$y = \sqrt{1 - x^2} \quad \text{AND} \quad y = -\sqrt{1 - x^2}$$

THE UPPER AND LOWER SEMICIRCLES.

WE COULD FIND $y'(x)$ BY DIFFERENTIATING THOSE SQUARE ROOTS, BUT THAT'S MESSY—SO INSTEAD, WE **IMPLICITLY** DIFFERENTIATE THE ORIGINAL EQUATION WITH RESPECT TO x:

$$x^2 + y^2 = 1$$

$$2x + 2yy' = 0 \quad \text{AND SO}$$

$$y' = -\frac{x}{y} \quad \text{WHENEVER } y \neq 0$$

$$= \frac{x}{\sqrt{1 - x^2}} \quad \text{OR} \quad \frac{-x}{\sqrt{1 - x^2}} \qquad x \neq \pm 1$$

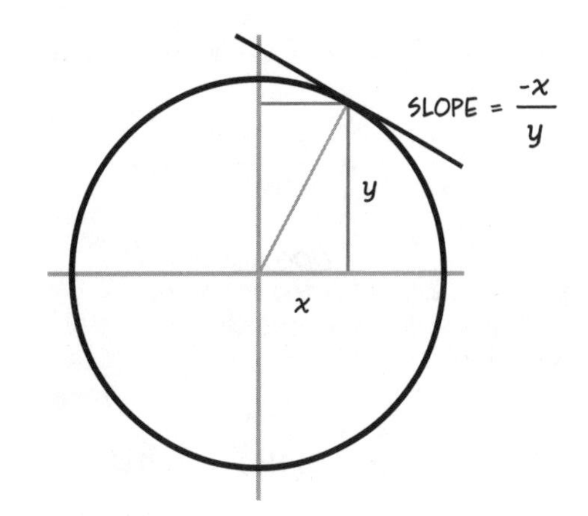
SLOPE = $\frac{-x}{y}$
y
x

DEPENDING ON WHICH SEMICIRCLE YOU CHOOSE. COMPARE THIS WITH THE EXAMPLE 4 ON PAGE 108.

More Related-Rate Examples

3. A PETROLEUM STORAGE TANK ON THE SHORELINE LEAKS OIL INTO THE WATER AT A STEADY RATE OF **2** BARRELS PER MINUTE. A CLEANUP CREW, INTENDING TO CONTAIN THE SPILL WITH A STRING OF FLOATS, ASKS HOW FAST THE SEMICIRCULAR OIL SLICK'S **CIRCUMFERENCE** IS GROWING.

C
r

GIVEN: $V'(t) = 2$, THE RATE OF CHANGE OF VOLUME

ASKED: $C'(t)$, THE RATE OF CHANGE OF CIRCUMFERENCE

LET'S ASSUME THAT THE OIL SLICK HAS UNIFORM THICKNESS, SO THAT ITS AREA IS PROPORTIONAL TO ITS VOLUME. IF 1 BARREL (BRL) OF OIL COVERS 300 SQUARE METERS, THEN AT TIME t,

$$A(t) = (300\ \text{M}^2/\text{BRL})\cdot(2\ \text{BRL/MIN})\cdot(t\ \text{MIN}) = 600t\ \text{M}^2$$

$$A'(t) = 600\ \text{M}^2/\text{MIN}$$

THE RELATED RATES COME FROM THE SPILL'S SEMI-CIRCULAR SHAPE:

$$C = \pi r,\ A = \tfrac{1}{2}\pi r^2,\ \text{SO}$$

$$A = \frac{C^2}{2\pi}$$

DIFFERENTIATING WITH RESPECT TO t,

$$A'(t) = \frac{1}{2\pi}2C(t)C'(t) = \frac{1}{\pi}C(t)C'(t) \quad \text{SO}$$

$$C'(t) = \frac{\pi A'}{C(t)} = \frac{600\pi}{C(t)}\ \text{M/MIN}$$

WHEN THE SPILL IS 1000 METERS AROUND ($C = 1000$), FOR INSTANCE, THE CIRCUMFERENCE IS GROWING AT A RATE OF

$$\frac{600\pi}{1000} \approx (0.6)(3.1416) \approx \mathbf{1.88}\ \text{METERS PER MINUTE}$$

O.K., NOW DOES ANYBODY HAVE A **CORK?**

4. DELTA IS POURING WATER INTO A CONICAL CUP 8 CM TALL AND 6 CM ACROSS AT THE TOP. IF THE VOLUME IN THE CUP AT TIME t IS $V(t)$, HOW FAST IS THE WATER **LEVEL** RISING, IN TERMS OF $V'(t)$?

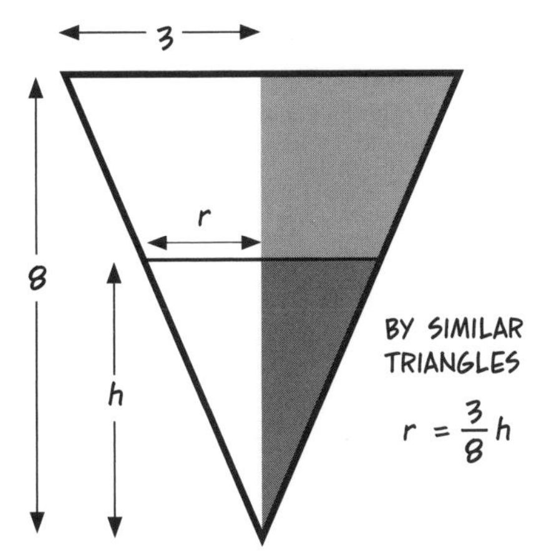

3
r
8
h

BY SIMILAR TRIANGLES

$$r = \frac{3}{8}h$$

THE WATER VOLUME IS GIVEN BY

$$(1) \quad V = \frac{1}{3}\pi r^2 h = \frac{1}{3}\pi\left(\frac{3}{8}h\right)^2 h$$

$$= \frac{1}{3}\pi\left(\frac{3}{8}\right)^2 h^3$$

NOW DIFFERENTIATE WITH RESPECT TO t:

$$V' = h'\pi\left(\frac{3}{8}\right)^2 h^2$$

WHICH GIVES

$$(2) \quad \mathbf{h' = \frac{64V'}{9\pi h^2}}$$

FOR INSTANCE, IF WATER POURS AT A CONSTANT RATE OF 10 CM^3/SEC, THEN WHEN h = 4 CM,

$$h' = \frac{(64)(10)}{9\pi(16)} \approx \frac{640}{452.4}$$

$$\approx \mathbf{1.41} \text{ CM/SEC.}$$

BY THE WAY, WHEN YOU FIRST START TO POUR AND h = 0, DO YOU SEE THAT h' IS **INFINITE?!!**

WE GLIMPSED THE INFINITE IN A CUP OF WATER... WHO'D HAVE THOUGHT...?
AND THEN... PFFT! GONE!

5. HERE'S AN ANGULAR EXAMPLE: AN AIRPLANE—AGAIN—IS FLYING AT AN ALTITUDE OF 3 KM, WITH VELOCITY $s'(t)$. THE OBSERVER IS MAKING A VIDEO RECORDING OF THE PLANE AND WOULD LIKE TO KNOW HOW FAST TO CHANGE THE **ANGLE** AT WHICH HER CAMERA IS POINTING WHEN THE ANGLE IS 60 DEGREES ($\pi/3$ RADIANS).

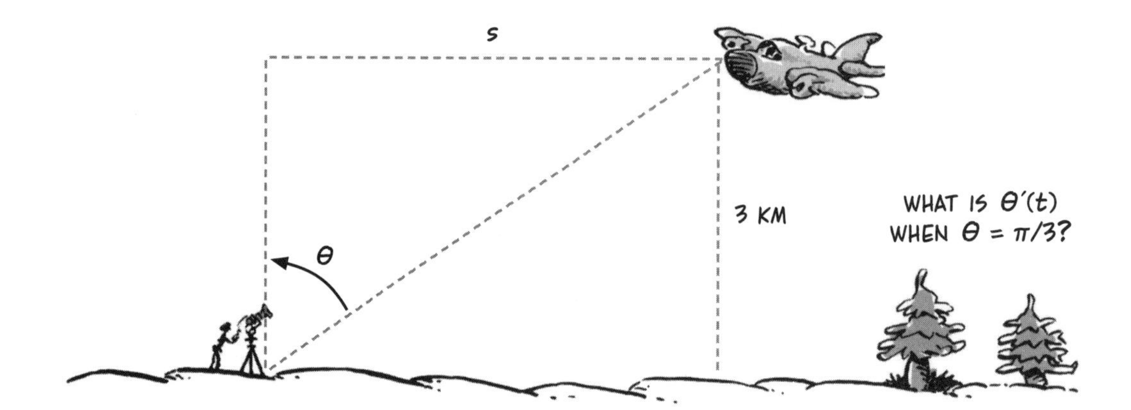
s
θ
3 KM
WHAT IS θ′(t) WHEN θ = π/3?

s IS THE HORIZONTAL DISPLACEMENT OF THE PLANE FROM THE OBSERVER. THE RELATION BETWEEN s AND θ IS

$$\tan\theta = \frac{s}{3}$$

DIFFERENTIATE WITH RESPECT TO TIME:

$$\theta' \sec^2\theta = \frac{s'}{3}$$

DIVIDING BY $\sec^2\theta$ (WHICH IS NEVER ZERO!),

$$(1) \quad \theta' = \frac{1}{3}s'\cos^2\theta$$

IF THE PLANE'S VELOCITY IS -720 KM/HR $= -12$ KM/MIN.*, AND $\theta = \pi/3$ RADIANS, THEN

$$\cos\theta = \frac{1}{2}, \quad s' = -12, \text{ AND}$$

$$\theta' = \left(\frac{1}{3}\right)(-12)\left(\frac{1}{4}\right)$$

$$= -1 \text{ RADIAN PER MINUTE}$$

$$= (1)(1/60) \approx 0.01667 \text{ RADIANS/SEC.}$$

THE ANGLE IS DECREASING AT A RATE OF 0.01667 RADIANS PER SECOND, ROUGHLY 1 DEGREE PER SECOND.

*THE VELOCITY IS NEGATIVE WHEN THE PLANE IS FLYING TOWARD THE OBSERVER.

FOR THE "B-ROLL," LET'S GET SOME REACTION SHOTS OF YOU SCRATCHING YOUR CHIN AND NODDING WISELY...

THE KEY TO THESE RELATED-RATE WORD PROBLEMS (AS TO ALL WORD PROBLEMS) IS TO EXPRESS EVERYTHING YOU KNOW FROM THE SETUP. IF A RELATIONSHIP BETWEEN TWO FUNCTIONS APPEARS, DIFFERENTIATE IT IMPLICITLY TO FIND ONE DERIVATIVE IN TERMS OF THE OTHER.

D IS THE FOURTH LETTER OF THE LATIN ALPHABET. s IS THE NINETEENTH. θ AND π ARE GREEK LETTERS, BUT I'M NOT SURE WHERE IN THE GREEK ALPHABET THEY COME, AND I'M TOO LAZY TO LOOK IT UP. THE PYTHAGOREAN THEOREM IS NAMED AFTER PYTHAGORAS, AN ANCIENT GREEK WHO LIVED IN SICILY. HE BELIEVED THAT ONLY WHOLE NUMBERS AND RATIOS OF WHOLE NUMBERS WERE REAL, SO HE WAS SHOCKED TO DISCOVER THAT $\sqrt{2}$ IS IRRATIONAL. THE "PYTHAGOREAN" THEOREM HAS BEEN PROVED IN HUNDREDS OF DIFFERENT WAYS BY MATHEMATICIANS FROM MANY CULTURES. PRESIDENT JAMES GARFIELD, AN AMATEUR MATHEMATICIAN, FOUND A PROOF THAT WAS QUITE SIMILAR TO THE TRADITIONAL CHINESE PROOF. AIRPLANES WERE INVENTED BY THE WRIGHT BROTHERS IN 1903...

I DIDN'T MEAN **ABSOLUTELY** EVERYTHING YOU KNOW!

TSK! WHY DIDN'T YOU SAY SO?

HERE ARE MORE EXAMPLES OF IMPLICIT DIFFERENTIATION WITHOUT ANY WORD PROBLEMS ATTACHED: IN THEM, WE FIND f' IN TERMS OF f, g, AND g', WHERE ALL THESE FUNCTIONS ARE ASSUMED TO DEPEND ON A VARIABLE x.

ONE MORE THING I KNOW: IT'S **WAY** EASIER CRANKING OUT FORMULAS THAN DOING ABSTRACT THOUGHT!
'SPECIALLY WHEN I'M THE ONE DOING THE CRANKING...

6. $\sin f = \ln g$

$$f'\cos f = \frac{g'}{g}$$

$$f' = \frac{g'\sec f}{g} \quad \text{WHEN } \cos f \neq 0,\ g \neq 0$$

7. $f^3 + g^2 = x$

DIFFERENTIATE WITH RESPECT TO x:

$$3f'f^2 + 2g'g = 1$$

$$f' = \frac{1 - 2g'g}{3f^2} \quad \text{WHEN } f \neq 0$$

8. $\tan^2 f + \tan f + 1 = g^2$

$$f'(2\tan f)(\sec^2 f) + f'\sec^2 f = 2g'g$$

$$f'(\sec^2 f)(1 + 2\tan f) = 2g'g$$

$$f' = \frac{2g'g\cos^2 f}{1 + 2\tan f} \quad \text{WHEN } \tan f \neq -\tfrac{1}{2}$$

Problems

1. A HEMISPHERICAL BOWL OF RADIUS R HAS VOLUME $2\pi R^3/3$. IF IT CONTAINS WATER TO A DEPTH h, THE VOLUME OF WATER IS

$$V = \pi(Rh^2 - \tfrac{1}{3}h^3)$$

(TAKE THIS ON FAITH FOR THE TIME BEING. IT WILL BE AN EXERCISE IN A LATER CHAPTER.)

IF WATER IS POURED INTO THE BOWL AT A RATE OF $V'(t)$, THEN WHAT IS $h'(t)$ IN TERMS OF V' AND h? (REMEMBER, R IS CONSTANT!)

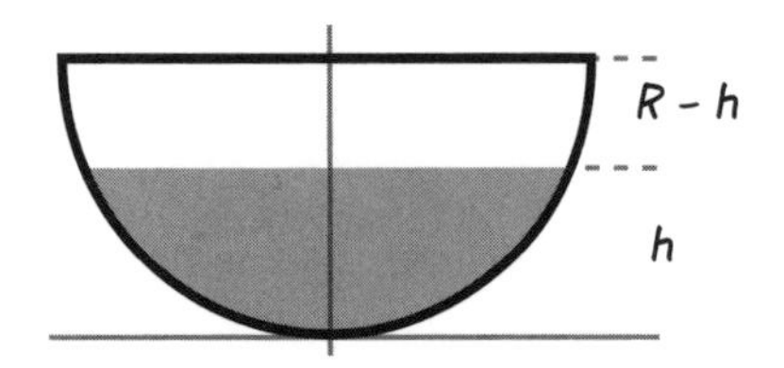

R – h
h

2. IMAGINE AN INSECT CRAWLING ON AN ELLIPTICAL WIRE. THE ELLIPSE'S EQUATION IS

$$\frac{x^2}{a^2} + \frac{y^2}{b^2} = 1$$

AT EACH INSTANT OF TIME t, THE INSECT HAS AN x-COORDINATE $x(t)$ AND A y-COORDINATE $y(t)$. REGARDLESS OF WHAT THE FUNCTIONS $x(t)$ AND $y(t)$ MAY BE, IT MUST BE TRUE THAT

$$\frac{(x(t))^2}{a^2} + \frac{(y(t))^2}{b^2} = 1$$

FIND AN EQUATION THAT RELATES x' AND y'.

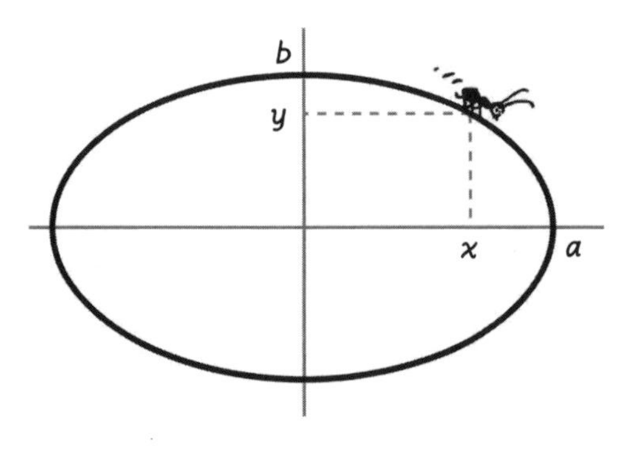

b
y
x
a

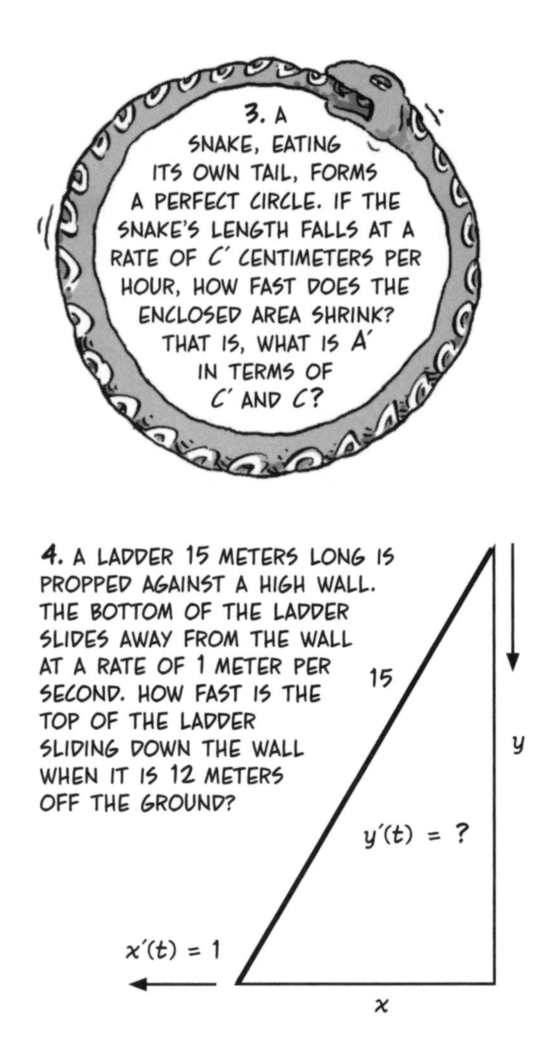

3. A SNAKE, EATING ITS OWN TAIL, FORMS A PERFECT CIRCLE. IF THE SNAKE'S LENGTH FALLS AT A RATE OF C′ CENTIMETERS PER HOUR, HOW FAST DOES THE ENCLOSED AREA SHRINK? THAT IS, WHAT IS A′ IN TERMS OF C′ AND C?
15
y
y′(t) = ?
x′(t) = 1
x

4. A LADDER 15 METERS LONG IS PROPPED AGAINST A HIGH WALL. THE BOTTOM OF THE LADDER SLIDES AWAY FROM THE WALL AT A RATE OF 1 METER PER SECOND. HOW FAST IS THE TOP OF THE LADDER SLIDING DOWN THE WALL WHEN IT IS 12 METERS OFF THE GROUND?

5. A SNAIL CREEPS ALONG THE SIDE OF A SQUARE 25 CENTIMETERS ON EACH SIDE. IF THE SNAIL MOVES FROM A TO B AT A STEADY PACE OF 1 CM/SEC, HOW FAST IS IT APPROACHING POINT C WHEN THE SNAIL HAS GONE 10 CM? HOW FAST IS IT MOVING AWAY FROM POINT D AT THE SAME MOMENT?

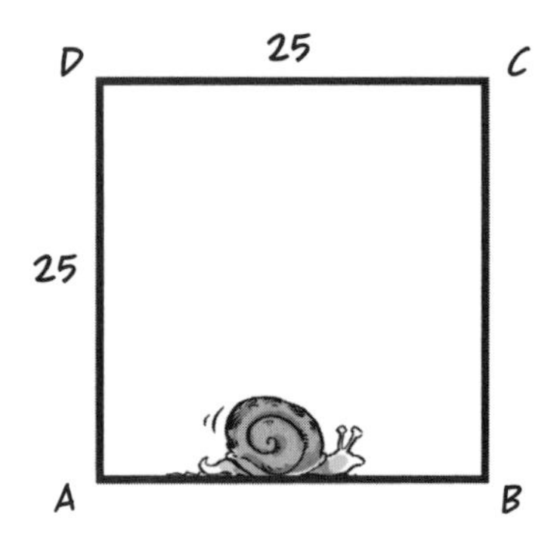

D
25
C
25
A
B

Chapter 5
Using Derivatives, Part 2: Optimization

WHEN FUNCTIONS HIT BOTTOM (OR TOP)

IN THE REAL WORLD, PEOPLE OFTEN LOOK FOR WAYS TO **OPTIMIZE** THINGS... WHICH MEANS FINDING THE **BEST** WAY TO DO SOMETHING... WE WANT TOP QUALITY—AND TOP QUANTITY!

I WANT TO LEARN THE MAXIMUM AMOUNT OF CALCULUS WITH MINIMUM HEADACHE!
I CAN HELP WITH THAT!

FOR EXAMPLE, A SHIPPING COMPANY WANTS TO MINIMIZE ITS FUEL COSTS BY SEEKING AN OPTIMAL ROUTE THAT BURNS THE SMALLEST AMOUNT OF GASOLINE. AN OIL COMPANY WANTS THE OPPOSITE!

%$##&)!!!

AN ECOLOGIST WORKING WITH A FISHING FLEET WANTS TO CALCULATE THE MAXIMUM FISH CATCH CONSISTENT WITH A SUSTAINABLE POPULATION.

A MANUFACTURER WANTS TO MAXIMIZE PROFITS.

GET ME A CALCULUS STUDENT!
revenue
cost

IN ALL OF THESE EXAMPLES, THE OPTIMAL SOLUTION IS ONE THAT **MAXIMIZES** OR **MINIMIZES** SOME FUNCTION.

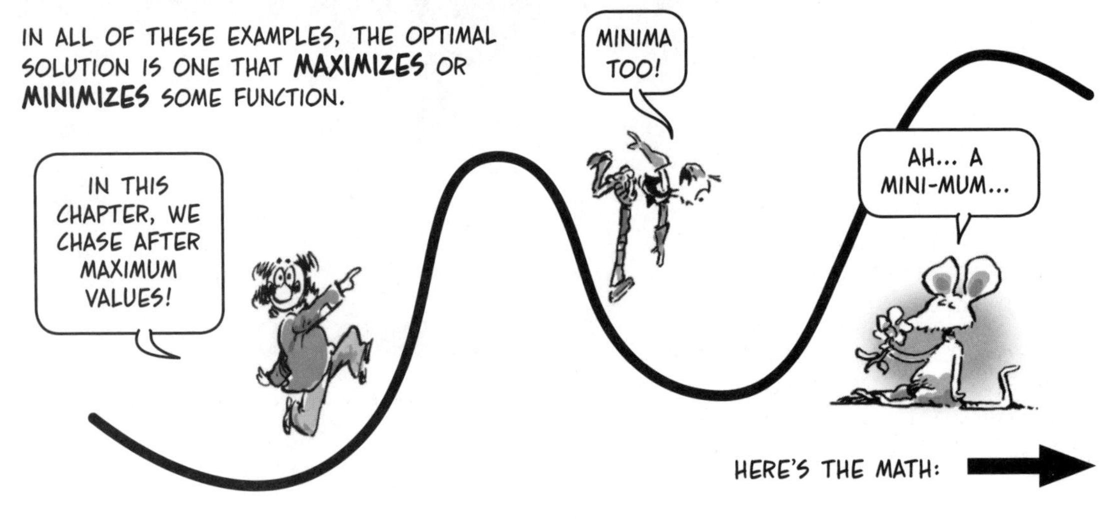
IN THIS CHAPTER, WE CHASE AFTER MAXIMUM VALUES!
MINIMA TOO!
AH... A MINI-MUM...

HERE'S THE MATH: →

A **LOCAL MAXIMUM** OF A FUNCTION IS A POINT a WHERE THE GRAPH CRESTS. AT A LOCAL MAXIMUM a OF A FUNCTION f, $f(a) \geq f(x)$ FOR ALL x IN SOME INTERVAL AROUND a. A **LOCAL MINIMUM** c IS THE BOTTOM OF A TROUGH, WHERE $f(x) \geq f(c)$ FOR NEARBY POINTS x. "LOCAL" MEANS THAT THE VALUE $f(a)$ IS COMPARED ONLY TO NEARBY POINTS. THERE MAY BE ANOTHER LOCAL MAXIMUM b WHERE f IS LARGER, I.E., $f(b) > f(a)$. WE REFER TO EITHER A LOCAL MAXIMUM OR LOCAL MINIMUM AS A LOCAL **EXTREME POINT** OR LOCAL **OPTIMUM.**

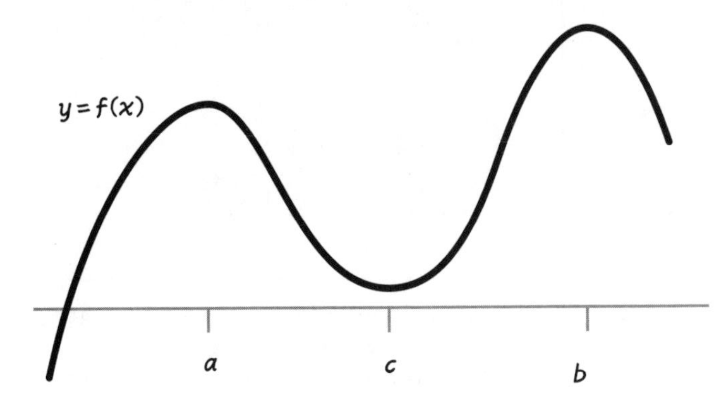

y = f(x)
a
c
b

HERE a AND b ARE BOTH LOCAL MAXIMA, AND $f(b) > f(a)$. c IS A LOCAL MINIMUM.

Extreme Fact 1: IF a IS A LOCAL EXTREME POINT OF A DIFFERENTIABLE FUNCTION f, THEN

$$f'(a) = 0$$

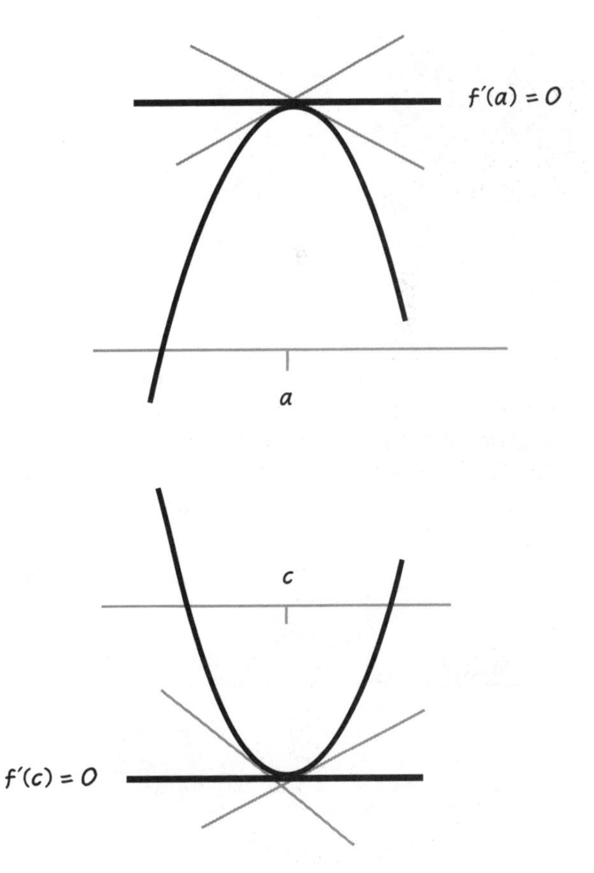

f′(a) = 0
a
c
f′(c) = 0

PROOF: SUPPOSE a IS A LOCAL MAXIMUM. THEN FOR SMALL h,

$$\frac{f(a+h) - f(a)}{h} \leq 0 \text{ WHEN } h > 0$$

$$\frac{f(a+h) - f(a)}{h} \geq 0 \text{ WHEN } h < 0$$

SO THE LIMIT AS $h \to 0$ MUST BE BOTH NON-NEGATIVE AND NON-POSITIVE, HENCE ZERO. IF a IS A LOCAL MINIMUM, THEN a IS A LOCAL MAXIMUM OF $-f$, SO AGAIN THE DERIVATIVE IS ZERO.

THE SLOPE OF THE GRAPH AT a IS FLOPPING OVER FROM POSITIVE TO NEGATIVE, OR VICE VERSA, AND SO HITS ZERO AT THE EXTREME POINT.

OUR CAR AND DRIVER CAN HELP US SEE WHY THE DERIVATIVE IS ZERO AT AN EXTREME POINT.
IF DELTA DRIVES FORWARD FOR A WHILE AND THEN REVERSES DIRECTION AT TIME $t = a$, THEN THE TURNAROUND POINT $P = s(a)$ IS A LOCAL MAXIMUM, AN EXTREME POINT: SHE GOES THAT FAR AND NO FARTHER.
$P = s(a)$
UNTIL TIME a, HER VELOCITY WAS POSITIVE; AFTER TIME a, IT WAS NEGATIVE.
$t \le a$
$v(t) \ge 0$
$t \ge a$
$v(t) \le 0$
AT THE PRECISE MOMENT $t = a$ WHEN THE CAR REACHES THE EXTREME POINT, ITS VELOCITY CHANGES FROM POSITIVE TO NEGATIVE AND SO MUST BE ZERO. $s'(a) = 0$.
$P = s(a)$
THE SAME WOULD BE TRUE IF DELTA BEGAN BY BACKING UP AND THEN REVERSED COURSE TO FORWARD MOTION. THEN THE TURNAROUND POINT WOULD BE A MINIMUM POSITION, WHERE HER VELOCITY MUST ALSO BE ZERO.
$P = s(a)$
NOTE: VELOCITY CAN ALSO BE ZERO AT TIMES THAT ARE NOT EXTREME POINTS. THE CAR COULD ROLL TO A STOP AND THEN KEEP MOVING FORWARD, AS AT A STOP SIGN. AT A TIME LIKE THAT, CALL IT b, $s'(b)$ IS 0, BUT $s(b)$ IS NOT AN EXTREME POSITION!
STOP
STOP
$P = s(b)$
SO: TO FIND EXTREMES OF A FUNCTION f, WE LOOK FOR INPUTS a FOR WHICH $f'(a) = 0$.
BUT ONCE WE FIND THEM, WE MUST CHECK WHETHER a IS REALLY AN EXTREME POINT OF THE FUNCTION, OR MERELY A "STOP SIGN."
CONFUSED?
NO, JUST TIRED OF BEING MADE AN EXAMPLE OF...
STOP

Example 1: HERE'S NEWTON ON THE TRAMPOLINE AGAIN. THE MEMBRANE IS 1 METER OFF THE GROUND, AND IT STILL FLINGS HIM UPWARD AT A VELOCITY OF 100 M/SEC. NEWTON'S ALTITUDE IN METERS, THEN, IS

$$h(t) = -4.9t^2 + 100t + 1,$$

NOW THE QUESTION IS: HOW HIGH DOES ISAAC RISE? WHAT IS HIS **MAXIMUM** ALTITUDE?

LET'S DIFFERENTIATE!

WE BEGIN BY TAKING THE DERIVATIVE OF h:

$$h'(t) = -9.8t + 100 \text{ M/SEC}$$

NEXT WE ASK: **WHEN DOES $h'(t) = 0$?** SET IT EQUAL TO ZERO, AND SOLVE FOR t:

$$h'(t) = 0$$

$$-9.8t + 100 = 0$$

$$t = \frac{100}{9.8} = \mathbf{10.20} \text{ SEC.}$$

I CAN SEE THE CURVATURE OF THE EARTH...

t = 10.2 SECONDS IS THE **TIME** WHEN NEWTON REACHES MAXIMUM HEIGHT. TO FIND THE HEIGHT ATTAINED AT THAT INSTANT, WE HAVE TO PLUG 10.2 INTO $h(t)$:

$$h(10.2) = (-4.9)(10.2)^2 + (100)(10.2) + 1$$

$$= \mathbf{1{,}125} \text{ METERS}$$

BOUNCY!

TO ASSURE OURSELVES THAT WE HAVE TRULY FOUND A MAXIMUM, LET'S RUN THROUGH THE BOUNCE IN SUPER SLO-MO:

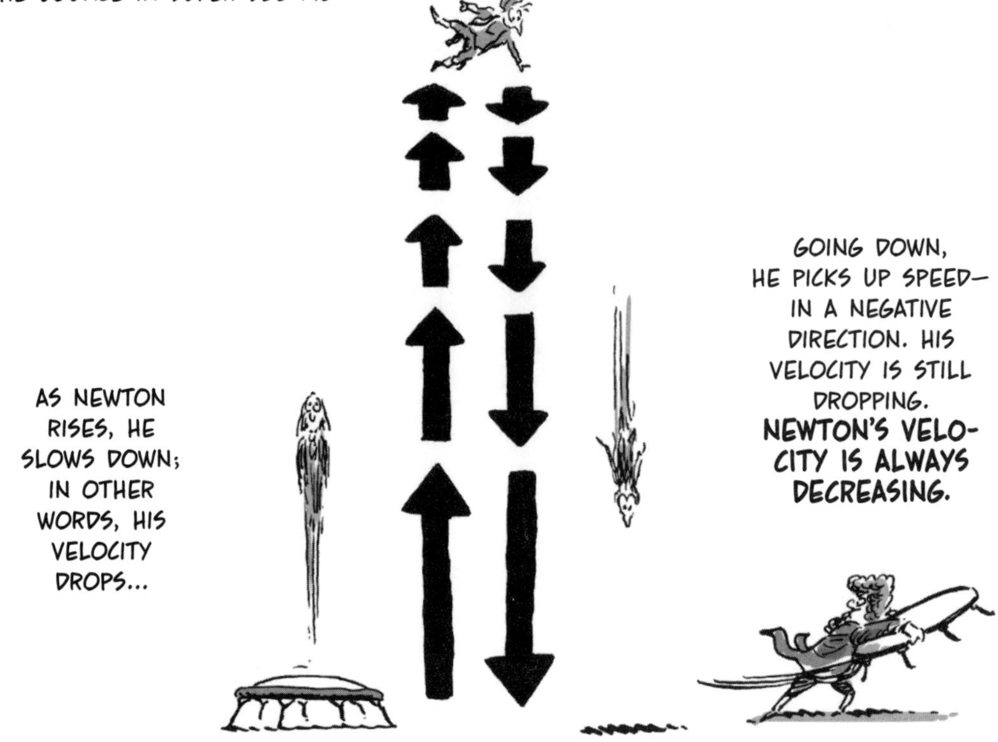
AS NEWTON RISES, HE SLOWS DOWN; IN OTHER WORDS, HIS VELOCITY DROPS...
GOING DOWN, HE PICKS UP SPEED— IN A NEGATIVE DIRECTION. HIS VELOCITY IS STILL DROPPING.
NEWTON'S VELOCITY IS ALWAYS DECREASING.

ONLY AT THE VERY TOP, AT $t = 10.20$ SECONDS, IS HIS VELOCITY PRECISELY ZERO. AT THAT ONE INSTANT, HE'S GOING NEITHER UP NOR DOWN, BUT HIS VELOCITY IS FALLING THERE TOO, CHANGING FROM POSITIVE TO NEGATIVE.

YOU CAN SEE FROM THE PARABOLIC GRAPH THAT $t = 10.2$ REALLY IS A MAXIMUM.

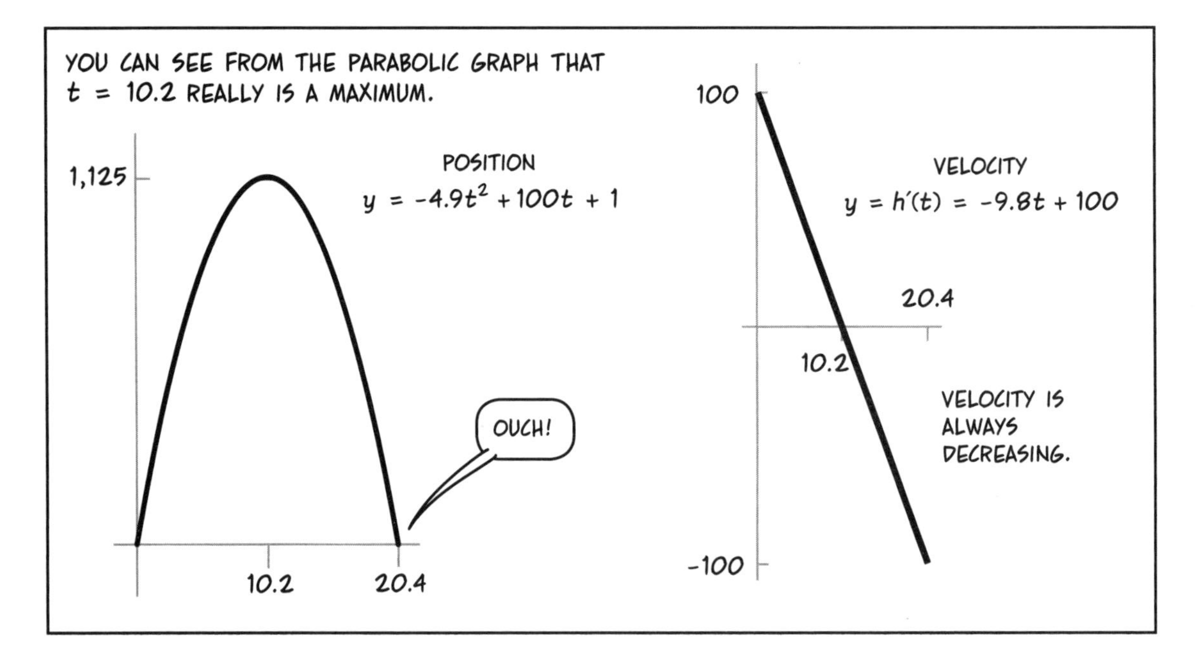
1,125
POSITION
$y = -4.9t^2 + 100t + 1$
OUCH!
10.2
20.4
100
VELOCITY
$y = h'(t) = -9.8t + 100$
20.4
10.2
VELOCITY IS ALWAYS DECREASING.
-100

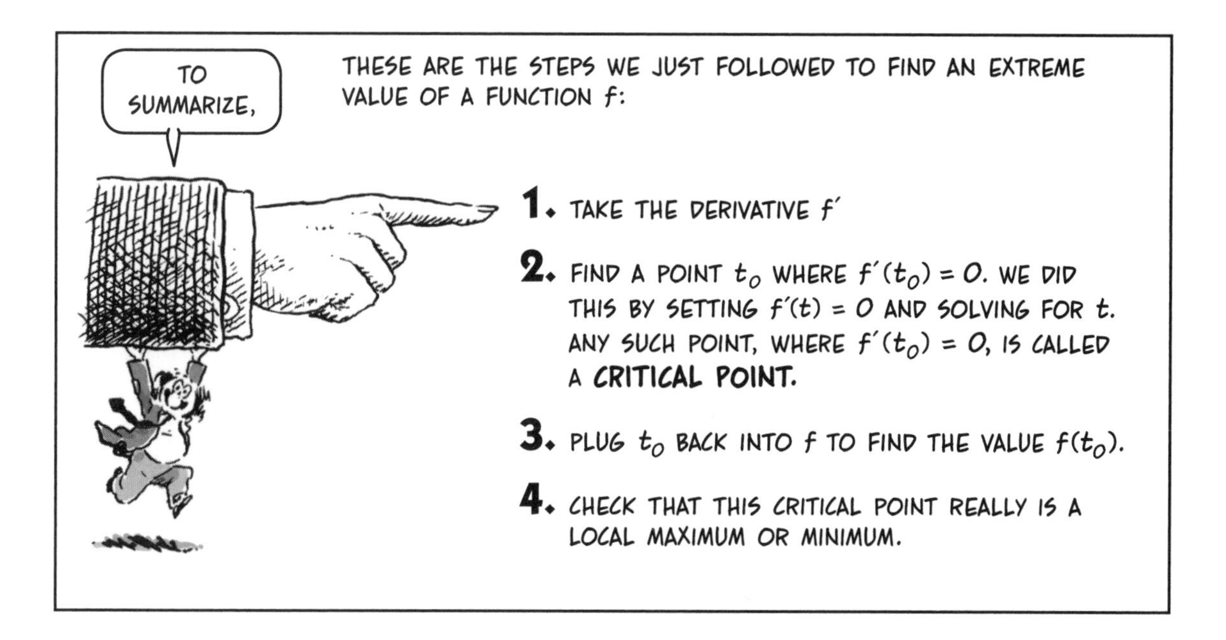
TO SUMMARIZE,

THESE ARE THE STEPS WE JUST FOLLOWED TO FIND AN EXTREME VALUE OF A FUNCTION f:

1. TAKE THE DERIVATIVE f'
2. FIND A POINT t_0 WHERE $f'(t_0) = 0$. WE DID THIS BY SETTING $f'(t) = 0$ AND SOLVING FOR t. ANY SUCH POINT, WHERE $f'(t_0) = 0$, IS CALLED A **CRITICAL POINT.**
3. PLUG t_0 BACK INTO f TO FIND THE VALUE $f(t_0)$.
4. CHECK THAT THIS CRITICAL POINT REALLY IS A LOCAL MAXIMUM OR MINIMUM.

WE FOLLOW THE SAME PROCEDURE FOR ALL OPTIMIZATION PROBLEMS. OF COURSE, IN OTHER SITUATIONS, THERE MAY BE MORE THAN ONE CRITICAL POINT; WE GOT LUCKY WITH THE TRAMPOLINE...

HOW LUCKY?
WELL, YOU DON'T HAVE TO BOUNCE AGAIN...

HERE IS ONE MORE EXAMPLE...

IN BUSINESS, PROFIT DEPENDS ON THE NUMBER OF UNITS SOLD.

Example 2: THE **SQUEEZ-U OLIVE RANCH** SELLS ITS PREMIUM OLIVE OIL FOR $100 A BOTTLE. SELLING A QUANTITY OF q BOTTLES PRODUCES A **REVENUE** $R(q)$ OF $100q$. BUT THERE ARE **COSTS,** C, WHICH ALSO DEPEND ON q ACCORDING TO THE FORMULA

$$C(q) = 800{,}000 + 4q^{\frac{5}{4}}.$$

(COSTS INCLUDE STARTUP COSTS OF $800,000 FOR LAND, PRESSES, BOTTLING EQUIPMENT, OLIVE TREES, PLUS ONGOING EXPENSES FOR WAGES, SHIPPING, WAREHOUSE FEES, BOTTLES, FERTILIZER, MAINTENANCE, WASTE DISPOSAL...)

HAVE YOU EVER THOUGHT OF SELLING YOUR OLIVE PITS TO GENERATE GREEN ENERGY?
BUT THEY'RE BLACK OLIVES...
SQUEEZ-U
PIT PIT

THE **PROFIT** P IS THE DIFFERENCE BETWEEN REVENUE AND COST. PROFIT IS A FUNCTION OF q. IT DEPENDS ON HOW MUCH IS SOLD.

$$P(q) = R(q) - C(q)$$

HOW MANY BOTTLES MUST SQUEEZ-U SELL TO **MAXIMIZE** PROFIT, AND HOW MUCH PROFIT CAN BE MADE?

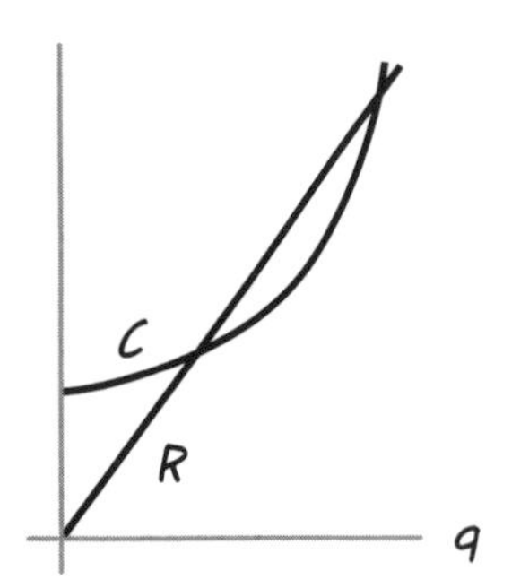
C
R
q

NOW IT'S EASY AS 1, 2, 3!
4!
JUMB

1. WE TAKE THE DERIVATIVE OF P WITH RESPECT TO q—THE RATE OF CHANGE OF PROFIT PER UNIT SOLD.

$$P(q) = 100q - 800{,}000 - 4q^{\frac{5}{4}}$$

$$P'(q) = 100 - 5q^{\frac{1}{4}}$$

2. SET $P'(q) = 0$ AND SOLVE FOR q.

$$100 - 5q^{\frac{1}{4}} = 0$$

$$q^{\frac{1}{4}} = 20$$

$$q = (20)^4 = 160{,}000 \text{ BOTTLES}$$

3. FIND THE PROFIT MADE BY SELLING 160,000 BOTTLES.

$$P(160{,}000) =$$

$$= (100)(160{,}000) - 800{,}000 - (160{,}000)^{\frac{5}{4}}$$

$$= 16{,}000{,}000 - 800{,}000 - 3{,}200{,}000$$

$$= \textbf{\$12 MILLION}$$

4. CHECK THAT $P(q)$ REACHES A MAXIMUM AT $q = 160{,}000$. IF q IS A LITTLE LESS, SAY 150,000 UNITS, THEN

$$P(150{,}000) =$$

$$(100)(150{,}000) - 800{,}000 - (150{,}000)^{\frac{5}{4}}$$

$$= 15{,}000{,}000 - 3{,}751{,}985$$

$$= 11 \text{ MILLION AND CHANGE.}$$

THIS IS LESS THAN 12 MILLION. YOU CAN TRY $q = 170{,}000$ AND OTHER NEARBY VALUES FOR YOURSELF.

BY THE WAY, ISN'T $100 A LOT TO CHARGE FOR A BOTTLE OF OLIVE OIL?
YA THINK?

A Better Test

ONE OF OUR FOUR OPTIMIZATION STEPS IS A LITTLE SQUIRRELY: THE LAST ONE. HAVING FOUND A CRITICAL POINT—A POINT WHERE THE DERIVATIVE IS ZERO—IT'S CUMBERSOME TO COMPUTE THE FUNCTION AT "NEARBY" POINTS... IT'S TIME-CONSUMING... INELEGANT!

IT'S **SO** HARD TO LOOK ELEGANT TURNING A CRANK...

IN FACT, DOING SO GUARANTEES NOTHING AT ALL. WHAT IF WE CHECKED AT POINTS THAT AREN'T "NEARBY" ENOUGH? HERE'S A GRAPH WITH A LOCAL MINIMUM AT a... BUT IF WE HAPPENED TO PICK THE POINT b FOR COMPARISON, WE WOULD FIND $f(b) < f(a)$ AND MIGHT CONCLUDE THAT $f(a)$ WAS A MAXIMUM, NOT A MINIMUM.

a
b
ER...
I HATE THOSE WIGGLY CURVES ANYWAY...

WE NEED A BETTER TEST!

THIS BEING A CALCULUS BOOK, WE WANT SOMETHING THAT USES THE **DERIVATIVE.** WE MIGHT ASK, FOR EXAMPLE, **HOW IS THE DERIVATIVE CHANGING?**

HEY, WE COULD ASK WHEN IS THE DERIVATIVE **BLUE,** TOO, BUT WHAT WOULD **THAT** TELL US?
PATIENCE... PATIENCE...

AROUND A MAXIMUM, THE DERIVATIVE $f'(x)$ GOES FROM POSITIVE TO NEGATIVE... WHILE AT A MINIMUM, f' IS GOES FROM NEGATIVE TO POSITIVE. IN PARTICULAR, AT A **MAXIMUM,** f' IS **DECREASING;** AT A **MINIMUM,** f' IS **INCREASING.**

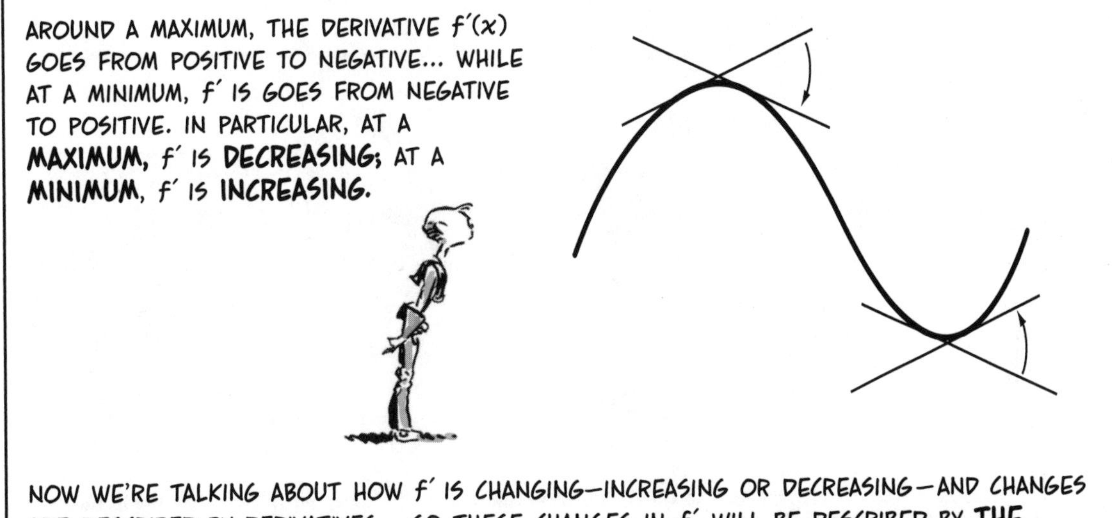

NOW WE'RE TALKING ABOUT HOW f' IS CHANGING—INCREASING OR DECREASING—AND CHANGES ARE DESCRIBED BY DERIVATIVES... SO THESE CHANGES IN f' WILL BE DESCRIBED BY **THE DERIVATIVE OF THE DERIVATIVE** $(f')'$ OR SIMPLY f'', THE **SECOND DERIVATIVE** OF f.

f'' IS ALSO WRITTEN $\frac{d^2f}{dx^2}$ OR $\frac{d^2y}{dx^2}$!!

THE ELEMENTARY FUNCTIONS CAN BE DIFFERENTIATED AGAIN AND AGAIN AS MANY TIMES AS YOU LIKE, TO GIVE FIRST, SECOND, THIRD, ... nTH DERIVATIVES:

THERE'S NO END TO 'EM!

$f(x)$	x^5	$\sin x$
$f'(x)$	$5x^4$	$\cos x$
$f''(x)$	$20x^3$	$-\sin x$
$f'''(x)$	$60x^2$	$-\cos x$
$f^{(4)}(x)$	$120x$	$\sin x$
$f^{(5)}(x)$	120	$\cos x$
$f^{(6)}(x)$	0	$-\sin x$
$f^{(7)}(x)$	0	$-\cos x$
...	...	...

BUT WHAT DO THEY MEAN?

WELL, **OBVIOUSLY** I'M THE RATE OF CHANGE OF THE RATE OF CHANGE OF THE RATE OF CHANGE OF...

WHEN IT COMES TO MOTION, THE SECOND DERIVATIVE OF POSITION, AT LEAST, IS FAMILIAR: IT'S **ACCELERATION,** THE RATE OF CHANGE OF VELOCITY.

$s(t)$ = POSITION AT TIME t
$s'(t) = v(t)$ = VELOCITY AT TIME t
$s''(t) = v'(t) = a(t)$ = ACCELERATION AT TIME t
AND THE THING ABOUT ACCELERATION IS—YOU **FEEL** IT...

WHEN A CAR SPEEDS UP, I.E., VELOCITY INCREASES, YOU FEEL PUSHED BACK INTO YOUR SEAT.*

WHEN IT SLOWS DOWN (VELOCITY FALLS), YOU'RE THROWN FORWARD.

ISAAC NEWTON (HIM AGAIN!) ANNOUNCED IT AS A NATURAL LAW, HIS SECOND: FORCE IS DIRECTLY PROPORTIONAL TO MASS AND ACCELERATION.

F = ma
DON'T YOU KNOW!

THE FACT THAT ACCELERATION ACCOMPANIES FORCE MEANS THAT WE CAN BUILD METERS TO MEASURE ACCELERATION: **ACCELEROMETERS.** THEN WE PUT THEM IN SMART-PHONES, TABLETS, AND DIGITAL CAMERAS, SO THEY RESPOND TO SHAKING AND ROTATION.

A GAME BASED ON THEFT?

*ACTUALLY, YOU FEEL THE SEAT PUSHING FORWARD AGAINST YOU. FOR MORE, SEE *THE CARTOON GUIDE TO PHYSICS!*

GRAPHICALLY, f'' DESCRIBES THE **CONCAVITY** OF f: WHEN THE SLOPE $f'(x)$ IS INCREASING, $f''(x) \geq 0$. THIS PART OF THE GRAPH IS **CONCAVE UPWARD.** WHEN f' IS DECREASING, $f'' \leq 0$, AND THE GRAPH IS **CONCAVE DOWNWARD.** A POINT c WHERE THE GRAPH CHANGES CONCAVITY IS CALLED AN **INFLECTION POINT,** AND THERE $f''(c) = 0$.

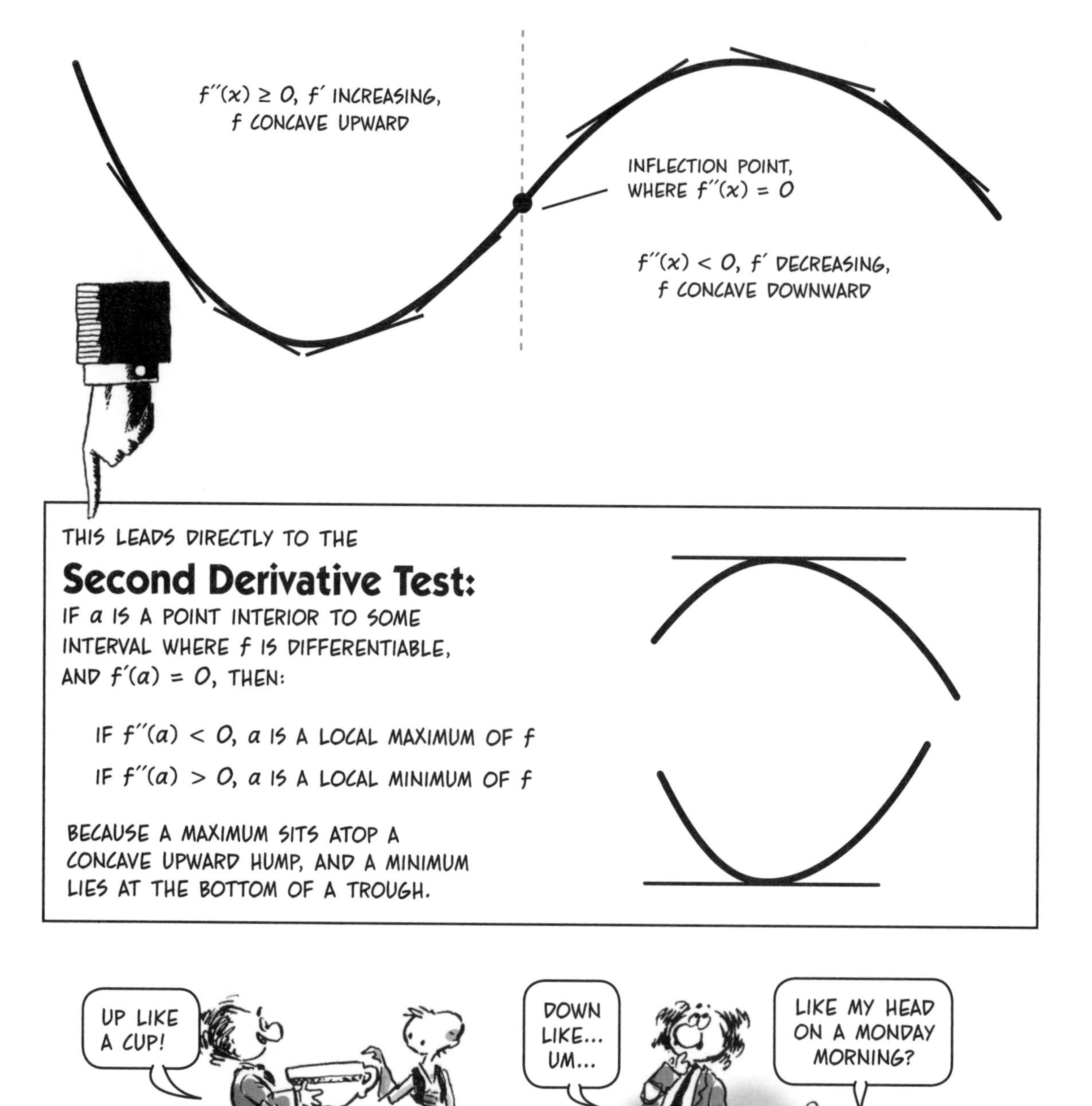
$f''(x) \geq 0$, f' INCREASING,
f CONCAVE UPWARD
INFLECTION POINT,
WHERE $f''(x) = 0$
$f''(x) < 0$, f' DECREASING,
f CONCAVE DOWNWARD
UP LIKE A CUP!
DOWN LIKE... UM...
LIKE MY HEAD ON A MONDAY MORNING?

THIS LEADS DIRECTLY TO THE

Second Derivative Test:

IF a IS A POINT INTERIOR TO SOME INTERVAL WHERE f IS DIFFERENTIABLE, AND $f'(a) = 0$, THEN:

IF $f''(a) < 0$, a IS A LOCAL MAXIMUM OF f

IF $f''(a) > 0$, a IS A LOCAL MINIMUM OF f

BECAUSE A MAXIMUM SITS ATOP A CONCAVE UPWARD HUMP, AND A MINIMUM LIES AT THE BOTTOM OF A TROUGH.

Example 3: FARMER FREDI WANTS TO PUT A RECTANGULAR SHEEP PEN AGAINST THE SIDE OF HER BARN. SHE HAS 80 METERS OF BOARDS WITH WHICH TO BUILD THE OTHER THREE SIDES. WHAT IS THE **MAXIMUM AREA** SHE CAN ENCLOSE?

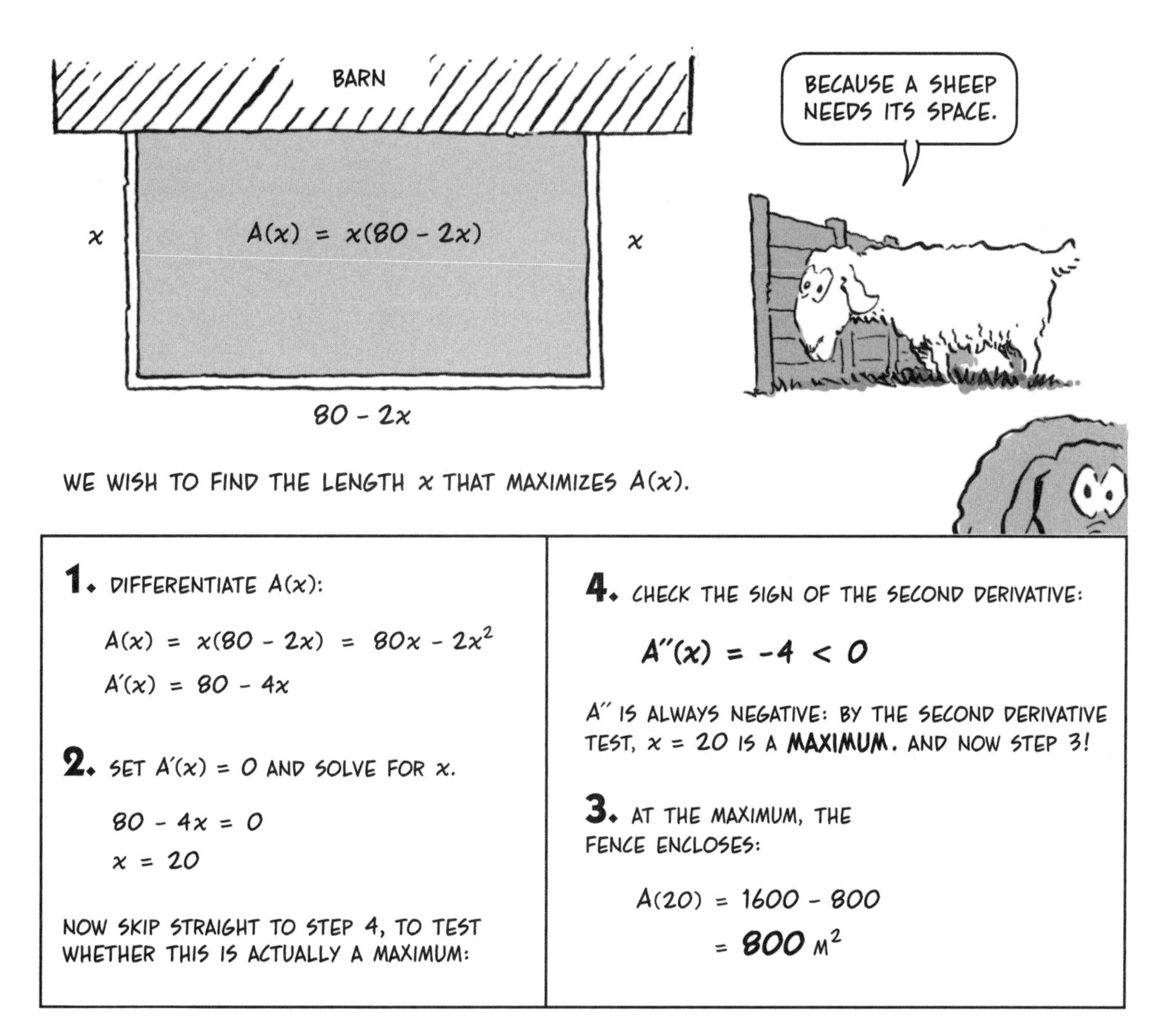
BARN
x
A(x) = x(80 - 2x)
x
80 - 2x
BECAUSE A SHEEP NEEDS ITS SPACE.

WE WISH TO FIND THE LENGTH x THAT MAXIMIZES $A(x)$.

1. DIFFERENTIATE $A(x)$:

$$A(x) = x(80 - 2x) = 80x - 2x^2$$
$$A'(x) = 80 - 4x$$

2. SET $A'(x) = 0$ AND SOLVE FOR x.

$$80 - 4x = 0$$
$$x = 20$$

NOW SKIP STRAIGHT TO STEP 4, TO TEST WHETHER THIS IS ACTUALLY A MAXIMUM:

4. CHECK THE SIGN OF THE SECOND DERIVATIVE:

$$A''(x) = -4 < 0$$

A'' IS ALWAYS NEGATIVE: BY THE SECOND DERIVATIVE TEST, $x = 20$ IS A **MAXIMUM**. AND NOW STEP 3!

3. AT THE MAXIMUM, THE FENCE ENCLOSES:

$$A(20) = 1600 - 800$$
$$= \mathbf{800}\ \text{M}^2$$

HAPPY...

Example 4:

THE **BRUTISH PETROLEUM CORP.** WANTS TO LAY PIPE FROM ONE OF ITS TANKS TO A STATION ACROSS THE RIVER. THE RIVER IS 2 KM ACROSS, AND THE DESTINATION IS 9 KM DOWNSTREAM. UNFORTUNATELY, IT COSTS MORE TO LAY PIPE ACROSS WATER THAN ON LAND: \$4 PER METER ON LAND, VS. \$8 PER METER OVER WATER. WHAT IS THE **CHEAPEST ROUTE** FOR THE PIPE?

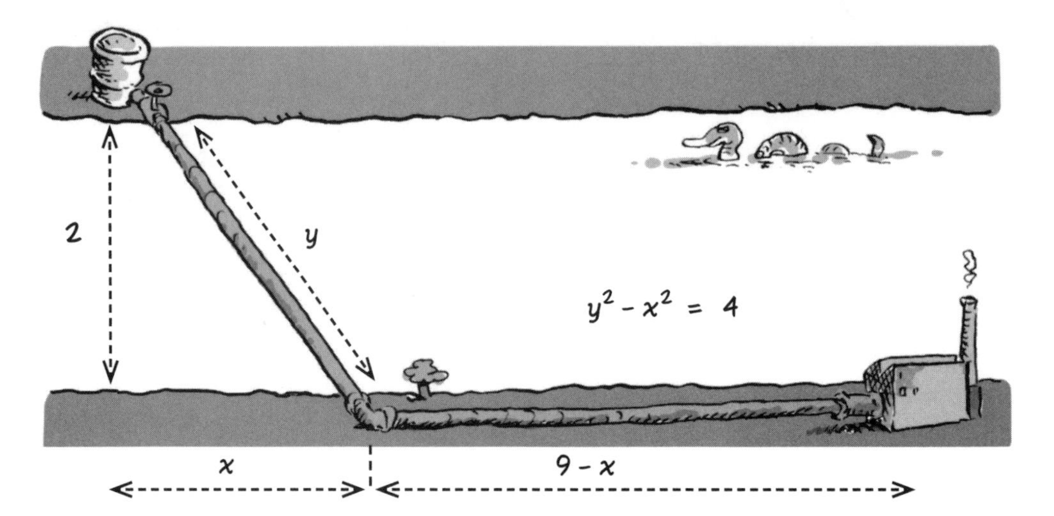
2
y
$y^2 - x^2 = 4$
x
9 − x

WE CAN ASSUME THAT THE PIPE CONSISTS OF TWO STRAIGHT SEGMENTS, BECAUSE ANYTHING CURVED WOULD BE EVEN LONGER. AS LABELED IN THE DRAWING, x AND y ARE RELATED BY:

$$(1) \quad y^2 - x^2 = 4$$

THE COST, IN THOUSANDS OF DOLLARS, IS

$$(2) \quad C(x) = 4(9 - x) + 8y$$
$$= 36 - 4x + 8y$$

WE ARE TRYING TO OPTIMIZE THE COST C WITH RESPECT TO x, THAT IS, FIND THE LENGTH x THAT MINIMIZES COST. FIRST, THEREFORE, WE HAVE TO FIND $C'(x)$.

EQUATION (1) SUGGESTS USING **IMPLICIT DIFFERENTIATION.** (THIS AVOIDS DEALING WITH MESSY SQUARE ROOTS.) DIFFERENTIATING (1) AND (2) WITH RESPECT TO x:

$$(3) \quad 2yy' - 2x = 0 \text{ SO } y' = \frac{x}{y}$$

$$(4) \quad C' = -4 + 8y'$$

TO OPTIMIZE THE COST C, SET $C' = 0$.

$$8y' - 4 = 0, \text{ SO } y' = \tfrac{1}{2}$$

BUT FROM (3), $y' = x/y$, SO WE GET

$$(5) \quad \frac{x}{y} = \frac{1}{2} \text{ OR } \mathbf{y = 2x}$$

PLUGGING THIS INTO (1) GIVES $3x^2 = 4$, SO $C'(x) = 0$ WHEN

$$\boxed{x = \frac{2}{\sqrt{3}}}$$

NOW APPLY THE SECOND-DERIVATIVE TEST BY FINDING THE SIGN OF C''. FROM (4),

$$(6) \quad C'' = 8y''$$

WHILE FROM (3), USING THE QUOTIENT RULE,

$$y'' = \frac{y - xy'}{y^2}$$

SUBSTITUTING $y' = x/y$ (AGAIN FROM (3)),

$$y'' = \frac{y^2 - x^2}{y^3} = \frac{4}{y^3} \quad \text{SO FROM (6)}$$

$$C'' = \frac{32}{y^3} > 0 \text{ BECAUSE } y > 0.$$

THE SECOND DERIVATIVE C'' IS ALWAYS POSITIVE, SO **OUR SOLUTION REALLY IS A MINIMUM.**

SAY... IF $3x^2 = 4$ AT THE CRITICAL POINT, COULDN'T x BE A **NEGATIVE** ROOT? $x = -2/\sqrt{3}$?
NO. WHEN $C' = 0$, $x = y/2$, AND y IS ALWAYS POSITIVE.

AND WHAT IS THE MINIMUM COST? WE MAY AS WELL EXPRESS C ENTIRELY IN TERMS OF x, BY SUBSTITUTING $y = \sqrt{x^2 + 4}$ IN (2):

$$C(x) = 36 - 4x + 8\sqrt{x^2 + 4}$$

AT THE CRITICAL POINT $x = 2/\sqrt{3}$, THEN,

$$C\left(\frac{2}{\sqrt{3}}\right) = 36 - 4\left(\frac{2}{\sqrt{3}}\right) + 8\sqrt{\frac{4}{3} + 4}$$

$$\approx \mathbf{49.86...}$$

SO THE TOTAL COST WILL BE **\$49,860.**

x	$C(x)$, THOUSANDS OF DOLLARS
0	52
1	49.90
$2/\sqrt{3}$	49.86
2	50.62
3	52.84
...	...
9	73.76

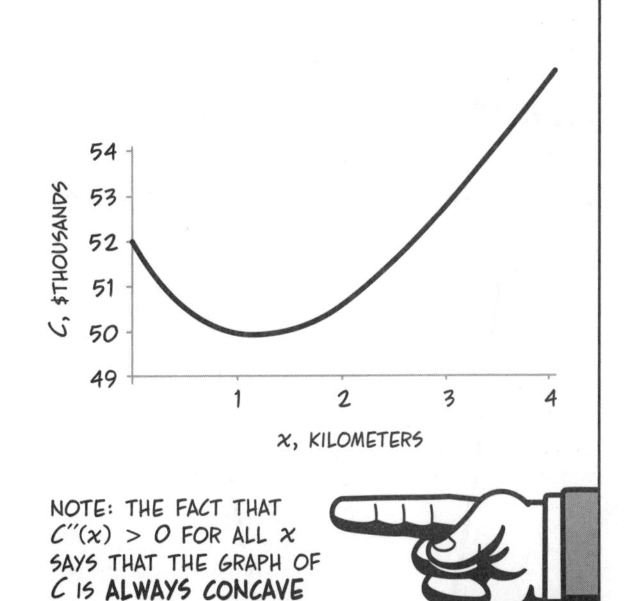
54
53
52
51
50
49
C, $THOUSANDS
1
2
3
4
x, KILOMETERS

NOTE: THE FACT THAT $C''(x) > 0$ FOR ALL x SAYS THAT THE GRAPH OF C IS **ALWAYS CONCAVE UP.** IT HAS NO INFLECTION POINTS.

Major Caution:

THE SECOND DERIVATIVE TEST IS A WONDERFUL THING WHEN IT WORKS, BUT IT DOESN'T ALWAYS WORK! WHAT HAPPENS AT A CRITICAL POINT a WHERE $f''(a) = 0$? IN THAT CASE, THE SECOND DERIVATIVE TEST FAILS; IT PROVIDES **NO INFORMATION** ABOUT WHETHER THE POINT a IS EXTREME OR NOT. TWO EXAMPLES SHOW WHAT CAN HAPPEN.

Example 5:

THE POWER FUNCTION $f(x) = x^3$ IS AN INCREASING FUNCTION WITHOUT ANY LOCAL MAXIMUM OR MINIMUM POINTS. ITS FIRST AND SECOND DERIVATIVES ARE

$$f'(x) = 3x^2 \text{ AND } f''(x) = 6x,$$

SO WHEN $x = 0$,

$$f'(0) = f''(0) = 0$$

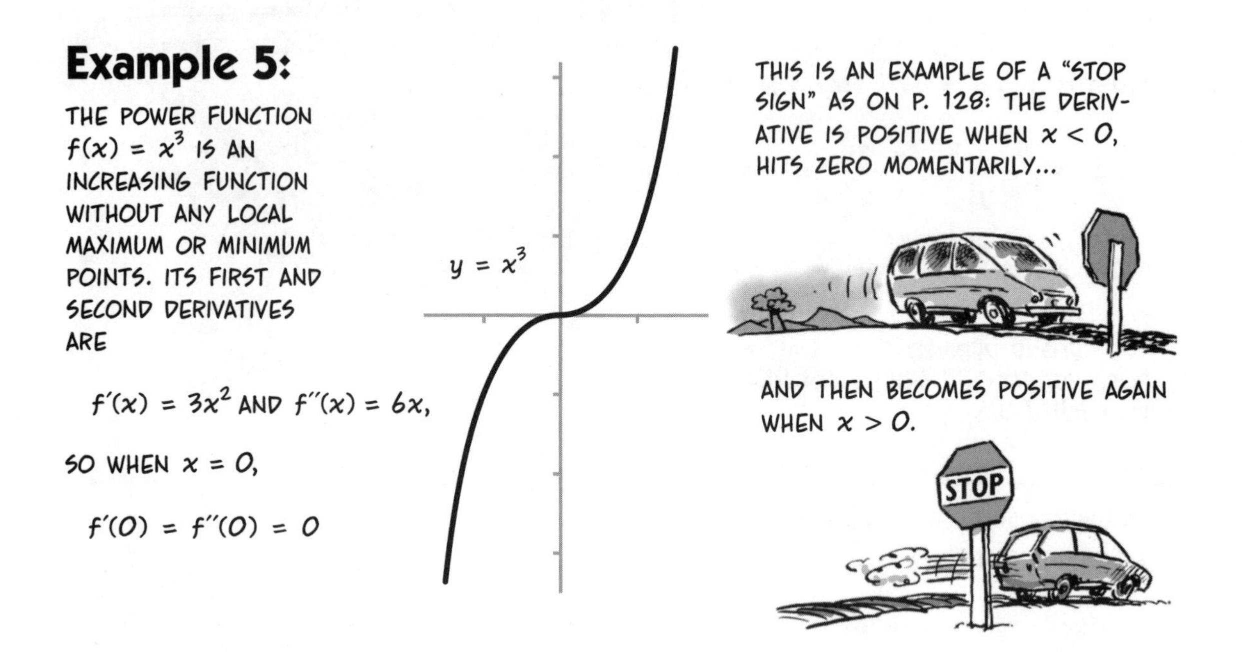
$y = x^3$
STOP

THIS IS AN EXAMPLE OF A "STOP SIGN" AS ON P. 128: THE DERIVATIVE IS POSITIVE WHEN $x < 0$, HITS ZERO MOMENTARILY...

AND THEN BECOMES POSITIVE AGAIN WHEN $x > 0$.

Example 6:

ON THE OTHER HAND, $g(x) = x^4$ DOES SOMETHING DIFFERENT AT $x = 0$. THE FIRST TWO DERIVATIVES ARE $g'(x) = 4x^3$ AND $g''(x) = 12x^2$. AGAIN $g'(0) = g''(0) = 0$, BUT HERE THE POINT $x = 0$ IS CLEARLY A MINIMUM.

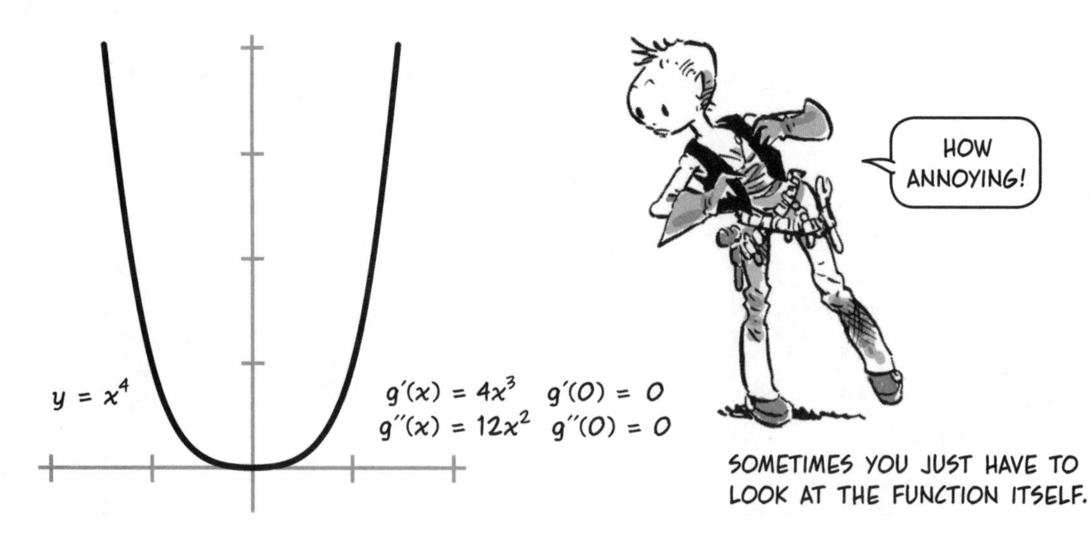
$y = x^4$
$g'(x) = 4x^3 \quad g'(0) = 0$
$g''(x) = 12x^2 \quad g''(0) = 0$
HOW ANNOYING!

SOMETIMES YOU JUST HAVE TO LOOK AT THE FUNCTION ITSELF.

THE SECOND DERIVATIVE IS GOOD FOR MORE THAN JUST TESTING FOR MAXIMA: IT TELLS YOU SOMETHING ABOUT THE SHAPE OF A FUNCTION'S GRAPH.

IN A GROWING ECONOMY, FOR INSTANCE, A NEGATIVE SECOND DERIVATIVE (OF TOTAL PRODUCTION, SAY) WOULD MEAN THAT THE BOOM IS LEVELING OFF AND COULD BE ABOUT TO TOP OUT...

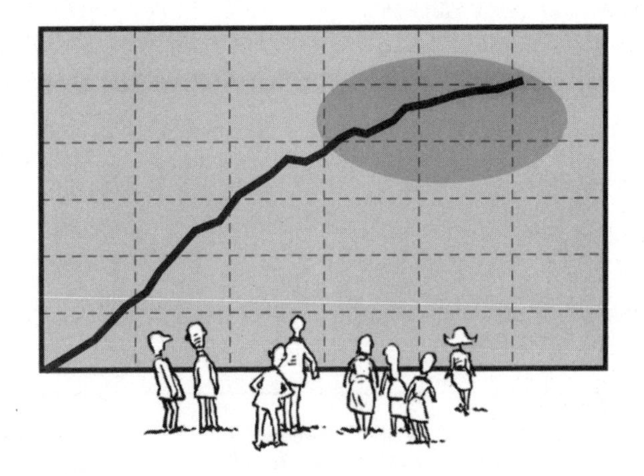

LIKEWISE, A POSITIVE f'' DURING A SLUMP MIGHT BE A SIGN THAT THE WORST IS OVER, AND THAT THINGS WILL SOON TURN AROUND.

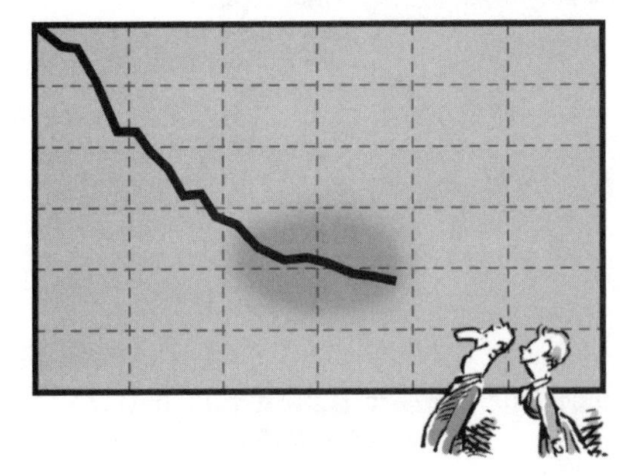

NOT NECESSARILY, THOUGH!

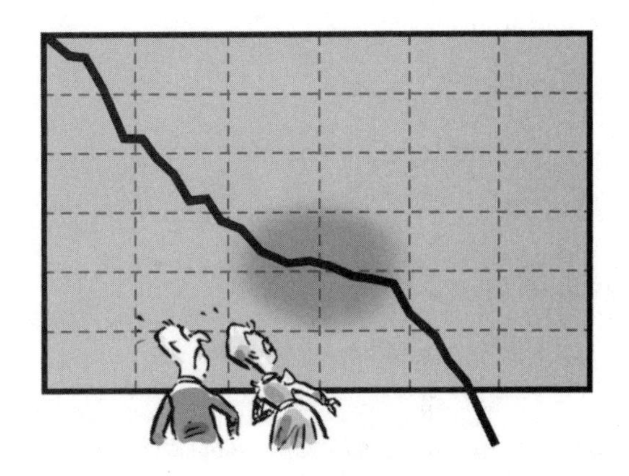

AND ONE OTHER THING: THE DERIVATIVE TESTS HELP LOCATE **LOCAL** EXTREME POINTS, BUT SOMETIMES WE WANT TO KNOW THE **"GLOBAL"** OR OVERALL MAXIMUM OR MINIMUM OF A FUNCTION. IF f IS DEFINED ON A CLOSED INTERVAL $[a, b]$, THE MOST EXTREME VALUE OF f MAY OCCUR AT ONE OF THE ENDPOINTS. YOU HAVE TO COMPARE THE VALUES $f(a)$ AND $f(b)$ WITH THE VALUE OF f AT THE LOCAL HIGHS OR LOWS.

HERE THE GLOBAL MAXIMUM IS AT THE INTERIOR POINT c, AND THE GLOBAL MINIMUM OCCURS AT THE ENDPOINT b.

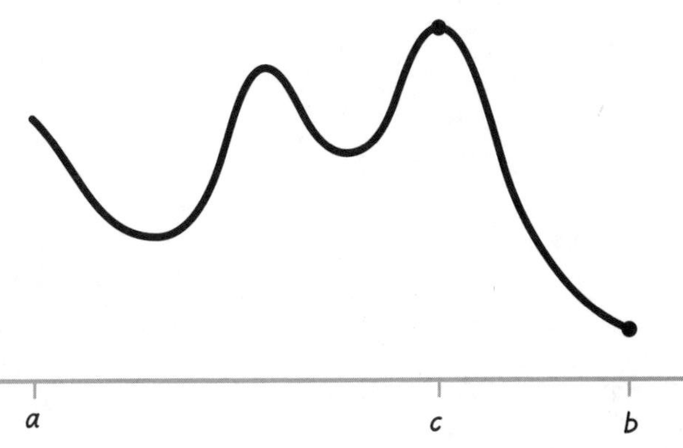
a
c
b

Problems

1. FIND ALL LOCAL EXTREME POINTS OF THESE FUNCTIONS. IDENTIFY WHICH ARE MAXIMA AND WHICH ARE MINIMA, AND DRAW GRAPHS.

a. $f(x) = x^2 + x - 1$

b. $g(x) = x^3 - 3x + 8$

c. $h(t) = 2t^3 - 3t^2 - 36t - 1$

d. $S(x) = \sin^2 x$

e. $F(\theta) = \cos\theta + \sin\theta$

f. $A(x) = \sqrt{4 - x^2}$

g. $Q(x) = x \ln x$

h. $s(t) = e^{-t}\cos t$

2. WHAT IS THE TENTH DERIVATIVE OF $f(x) = \sin x$? WHAT IS THE 110TH?

3. SHOW THAT OF ALL RECTANGLES WITH PERIMETER P, THE ONE ENCLOSING THE LARGEST AREA IS A SQUARE OF SIDE $P/4$.

4. A CATAPULT FLINGS A COW INTO THE AIR AT AN ANGLE θ WITH THE GROUND WITH AN INITIAL VELOCITY v_0. THIS VELOCITY HAS A HORIZONTAL COMPONENT $v_0 \cos\theta$ AND A VERTICAL COMPONENT $v_0 \sin\theta$.

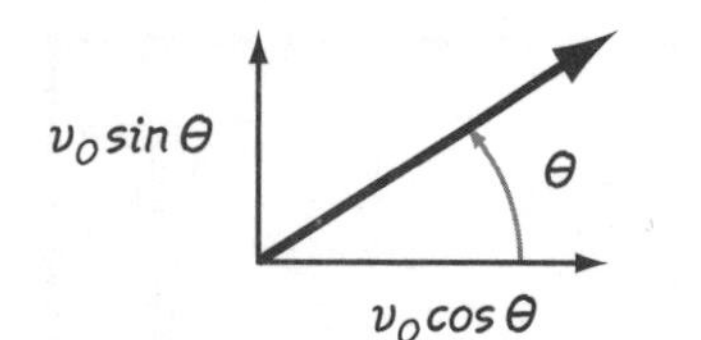
$v_0 \sin\theta$
θ
$v_0 \cos\theta$

THE COW'S HEIGHT ABOVE THE GROUND AT TIME t IS GIVEN BY

$$h(t) = -4.9t^2 + (v_0 \sin\theta)t$$

a. FIND THE TIME T WHEN THE COW REACHES MAXIMUM HEIGHT. (THIS WILL DEPEND ON θ.)

THE HORIZONTAL DISTANCE TRAVELED DURING THAT TIME WILL BE $D(\theta) = (v_0 \cos\theta)T$, AND THE TOTAL DISTANCE TRAVELED WHEN THE COW HITS THE EARTH WILL BE TWICE THAT, OR

$$D(\theta) = (2v_0 \cos\theta)T$$

θ
D

b. FIND THE ANGLE θ THAT MAXIMIZES D. (DON'T FORGET THAT T IS A FUNCTION OF θ!)

5. THE PAVE-ALL COMPANY WANTS TO BUILD A ROAD FROM A POINT ON A CIRCULAR POND TO THE POINT DIAMETRICALLY OPPOSITE, 2 KM AWAY. IT COSTS \$5 PER METER TO BUILD OVER WATER AND \$4 PER METER ON DRY LAND. DESCRIBE THE FINAL ROUTE.

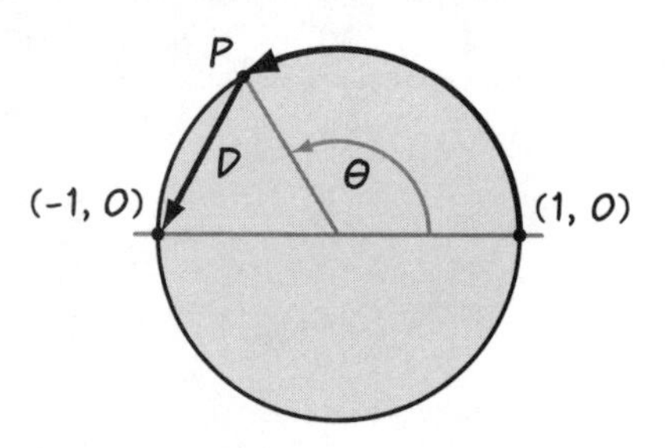
P
D
θ
(-1, 0)
(1, 0)

HINT: THE DISTANCE D FROM THE TURNING POINT P TO THE DESTINATION SATISFIES

$$D^2 = (\cos\theta + 1)^2 + \sin^2\theta$$

6. TWO CONSTRUCTION WORKERS ARE CARRYING A PIECE OF WALLBOARD DOWN A HALL WITH A RIGHT-ANGLE TURN. THE HALL IS 3 METERS WIDE IN ONE DIRECTION AND 4 METERS WIDE IN THE OTHER. FIND THE LENGTH OF THE LONGEST PIECE OF WALLBOARD THAT CAN MAKE THE TURN. (HINT: FIND THE SHORTEST PIECE THAT JUST FITS. ANYTHING SHORTER WILL WORK.)

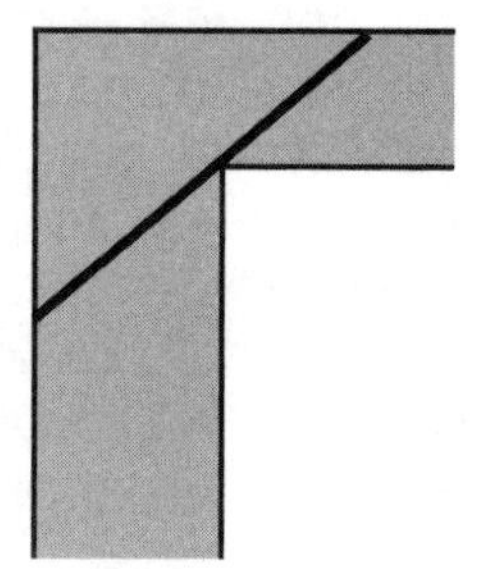

Chapter 6
Acting Locally

IN WHICH WE FOLLOW A LINE

NOW LET'S SHIFT OUR PERSPECTIVE SLIGHTLY. INSTEAD OF WATCHING THE DERIVATIVE ROAM AROUND ITS DOMAIN, LET'S CONFINE OUR ATTENTION TO A **SINGLE POINT.** YOU MAY BE SURPRISED AT HOW MUCH WE'LL FIND THERE...

ON PAGE 113 WE DESCRIBED SMALL CHANGES OF A FUNCTION f AROUND A POINT a BY SOMETHING I CALLED THE **FUNDAMENTAL EQUATION** OF CALCULUS:

$$f(a+h) - f(a) = hf'(a) + \text{FLEA}$$

THIS EQUATION SAYS THAT THE DISCREPANCY BETWEEN $f(a+h) - f(a)$, OR Δf, ON THE ONE HAND, AND $hf'(a)$ ON THE OTHER IS SMALL COMPARED WITH h. THIS MAKES IT EASY TO CALCULATE APPROXIMATE VALUES OF f.

AND I WANT TO DO THIS WHY?
WIN BAR BETS?

IN MATH, SOMETIMES A LITTLE CHANGE OF NOTATION CAN MASSIVELY SHIFT YOUR ENTIRE PERSPECTIVE...
THAT'S WHAT I'VE ALWAYS SAID!
KICK ME

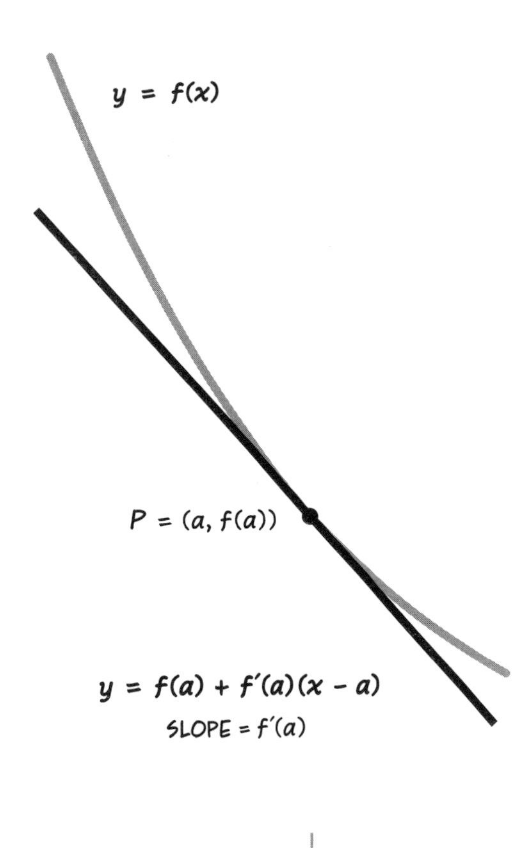
y = f(x)
P = (a, f(a))
y = f(a) + f′(a)(x − a)
SLOPE = f′(a)
a

LET'S WRITE $x = a + h$, SO $h = x - a$. THEN THE FUNDAMENTAL EQUATION BECOMES

$$f(x) - f(a) = f'(a)(x - a) + \text{FLEA}$$

OR

$$f(x) = f(a) + f'(a)(x - a) + \text{FLEA}$$

THAT IS ONE WAY, THEN, OF DESCRIBING THE ORIGINAL FUNCTION f NEAR a. NOW SUBTRACT THE FLEA TO GET A SIMPLER FUNCTION.

$$T_a(x) = f(a) + f'(a)(x - a)$$

ITS GRAPH IS A STRAIGHT LINE—THE ONE AND ONLY STRAIGHT LINE, IN FACT, **PASSING THROUGH a AND HAVING SLOPE $f'(a)$.**

THIS LINE, THE **TANGENT LINE** TO THE GRAPH $y = f(x)$ AT a, TOUCHES THE CURVE AT THE POINT $P = (a, f(a))$ AND HAS SLOPE EQUAL TO THE DERIVATIVE OF f THERE. IT IS A STRAIGHT-LINE FUNCTION WITH THE SAME VALUE AND DERIVATIVE AS f AT a.

AND T_a DIFFERS FROM f BY A FLEA—WHICH MEANS, YOU RECALL, THAT NOT ONLY DOES

$$\lim_{x \to a} (T_a(x) - f(x)) = 0$$

BUT ALSO

$$\lim_{x \to a} \frac{1}{(x - a)} (T_a(x) - f(x)) = 0$$

x
a
$f(x)$
$T_a(x)$
$f(a) = T_a(a)$
THAT IS, NEAR THE POINT a, THE DIFFERENCE BETWEEN $T_a(x)$ AND $f(x)$ IS SMALL **EVEN COMPARED TO $x - a$.**

WE CAN EXPRESS THIS BY SAYING **THE CLOSER WE ZOOM IN ON THE POINT P, THE MORE THE GRAPH $y = f(x)$ LOOKS LIKE A STRAIGHT LINE.**

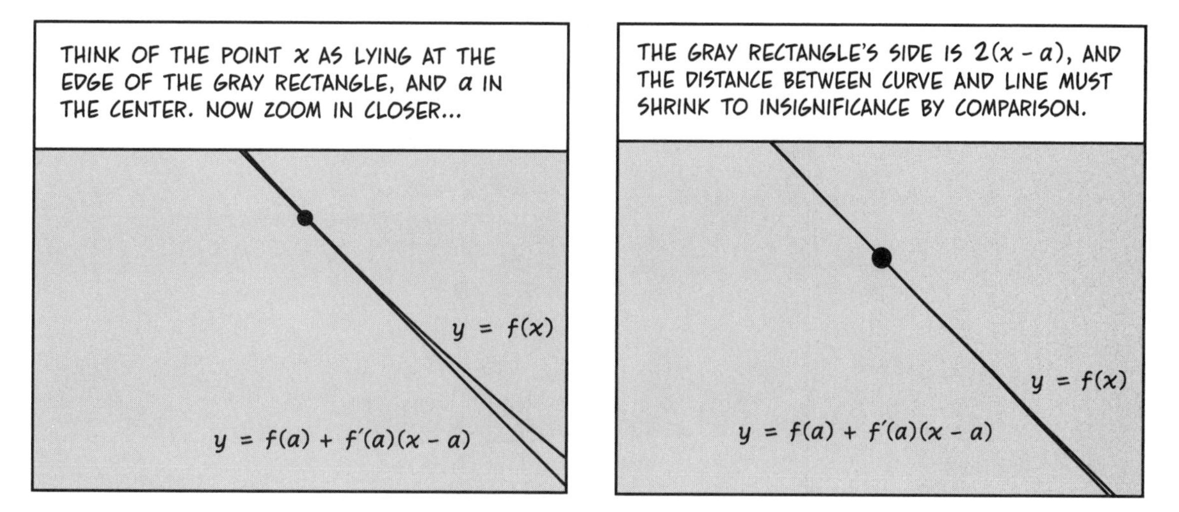

THINK OF THE POINT x AS LYING AT THE EDGE OF THE GRAY RECTANGLE, AND a IN THE CENTER. NOW ZOOM IN CLOSER...
$y = f(x)$
$y = f(a) + f'(a)(x - a)$
THE GRAY RECTANGLE'S SIDE IS $2(x - a)$, AND THE DISTANCE BETWEEN CURVE AND LINE MUST SHRINK TO INSIGNIFICANCE BY COMPARISON.
$y = f(x)$
$y = f(a) + f'(a)(x - a)$

ANOTHER WAY OF SAYING THE SAME THING: **FOR x NEAR a, THE NUMBER $f(a) + f'(a)(x - a)$ IS A GOOD APPROXIMATION FOR $f(x)$.** THIS GIVES US A WAY TO CALCULATE **APPROXIMATE** VALUES FOR FUNCTIONS.

TEN BUCKS SAYS THAT THE SQUARE ROOT OF 70 IS 8.375 TO WITHIN ONE PART IN A THOUSAND!!

Examples: LET $f(x) = \sqrt{x}$ AND $a = 1$. WE CAN APPROXIMATE SQUARE ROOTS OF NUMBERS NEAR 1, BECAUSE WE KNOW $f(a)$ AND $f'(a)$. $f(1) = \sqrt{1} = 1$, OF COURSE, AND

$$f'(x) = \frac{1}{2\sqrt{x}} \quad \text{SO} \quad f'(1) = \tfrac{1}{2}$$

IF x IS NEAR 1, THEN,

$$f(x) \approx f(1) + f'(1)(x - 1) = 1 + \tfrac{1}{2}(x - 1)$$

FOR INSTANCE:

$$\sqrt{1.3} \approx 1 + (\tfrac{1}{2})(1.3 - 1) = \mathbf{1.15}$$

THE ACTUAL VALUE IS 1.1402... SO THE APPROXIMATION IS ACCURATE TO WITHIN BETTER THAN ONE PART IN A HUNDRED.

SIMILARLY, WE CAN APPROXIMATE THE NATURAL LOGARITHM $\ln x$ FOR x NEAR e:

$$f(x) = \ln x, \quad f(e) = 1,$$

$$f'(x) = \frac{1}{x}, \quad f'(e) = \frac{1}{e}, \text{ SO}$$

$$\ln 3 \approx 1 + \frac{(3 - e)}{e}$$

$$\approx 1 + \frac{0.282}{2.718}$$

$$\approx \mathbf{1.104...}$$

THE ACTUAL VALUE IS 1.0986... SO THE APPROXIMATION IS GOOD TO ROUGHLY FIVE PARTS IN 1000—NOT BAD!

SLOW NIGHT?

A DIFFERENTIABLE FUNCTION'S GRAPH "FLATTENS OUT" WHEN YOU ZOOM IN... SO ANY FUNCTION WHOSE GRAPH DOES **NOT** FLATTEN NEAR A POINT a MUST NOT HAVE A DERIVATIVE AT a!
OW!!!
CAREFUL, THERE...

THE ABSOLUTE VALUE FUNCTION $g(x) = |x|$ IS AN EXAMPLE. AT $a=0$, g HAS NO DERIVATIVE: ITS GRAPH TURNS A SHARP CORNER, AND NO AMOUNT OF MAGNIFICATION WILL MAKE IT LOOK LIKE ANYTHING **OTHER** THAN A SHARP CORNER. THE DIFFERENCE QUOTIENTS CAN'T APPROACH A LIMIT AT 0.

$$\lim_{h \to 0} \frac{|h|}{h} = \begin{cases} -1 & \text{WHEN } h<0 \\ 1 & \text{WHEN } h>0 \end{cases}$$

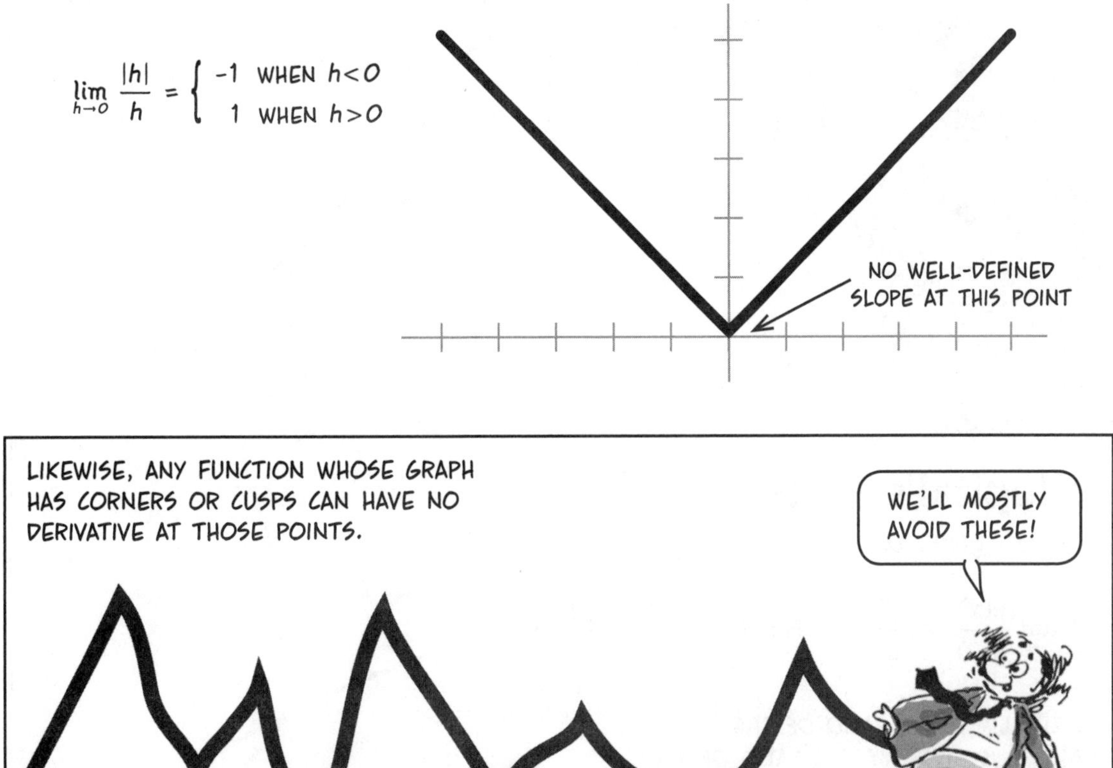
NO WELL-DEFINED SLOPE AT THIS POINT
LIKEWISE, ANY FUNCTION WHOSE GRAPH HAS CORNERS OR CUSPS CAN HAVE NO DERIVATIVE AT THOSE POINTS.
WE'LL MOSTLY AVOID THESE!

NOW BACK TO STRAIGHT LINES...

HERE'S SOMETHING ABOUT LINES YOU MAY NEVER HAVE NOTICED: SUPPOSE TWO NON-VERTICAL STRAIGHT LINES, $y = L_1(x)$ AND $y = L_2(x)$ CROSS ON THE x-AXIS AT A POINT a. IF THE TWO SLOPES ARE m AND p, THEN THE LINES HAVE THESE EQUATIONS:

$$y = L_1(x) = m(x - a)$$
$$y = L_2(x) = p(x - a)$$

ASSUME $p \neq 0$. THEN WHEN $x \neq a$,

$$\frac{L_1(x)}{L_2(x)} = \frac{m(x-a)}{p(x-a)} = \frac{m}{p}$$

ALTHOUGH THE FUNCTIONS L_1 AND L_2 APPROACH 0, THEIR RATIO IS ALWAYS THE **RATIO OF THE SLOPES.**

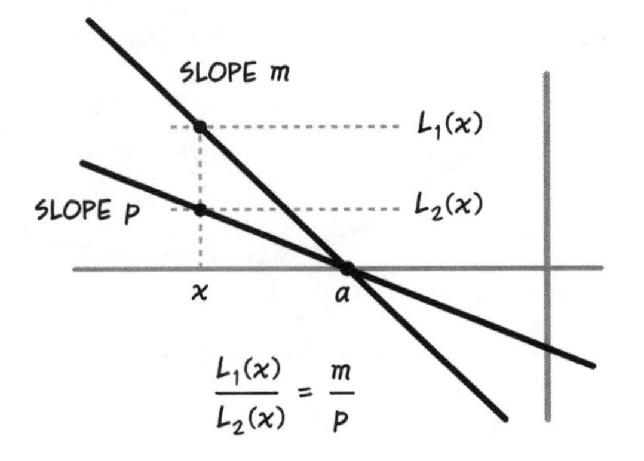
SLOPE m
L1(x)
SLOPE p
L2(x)
x
a
L1(x)/L2(x) = m/p

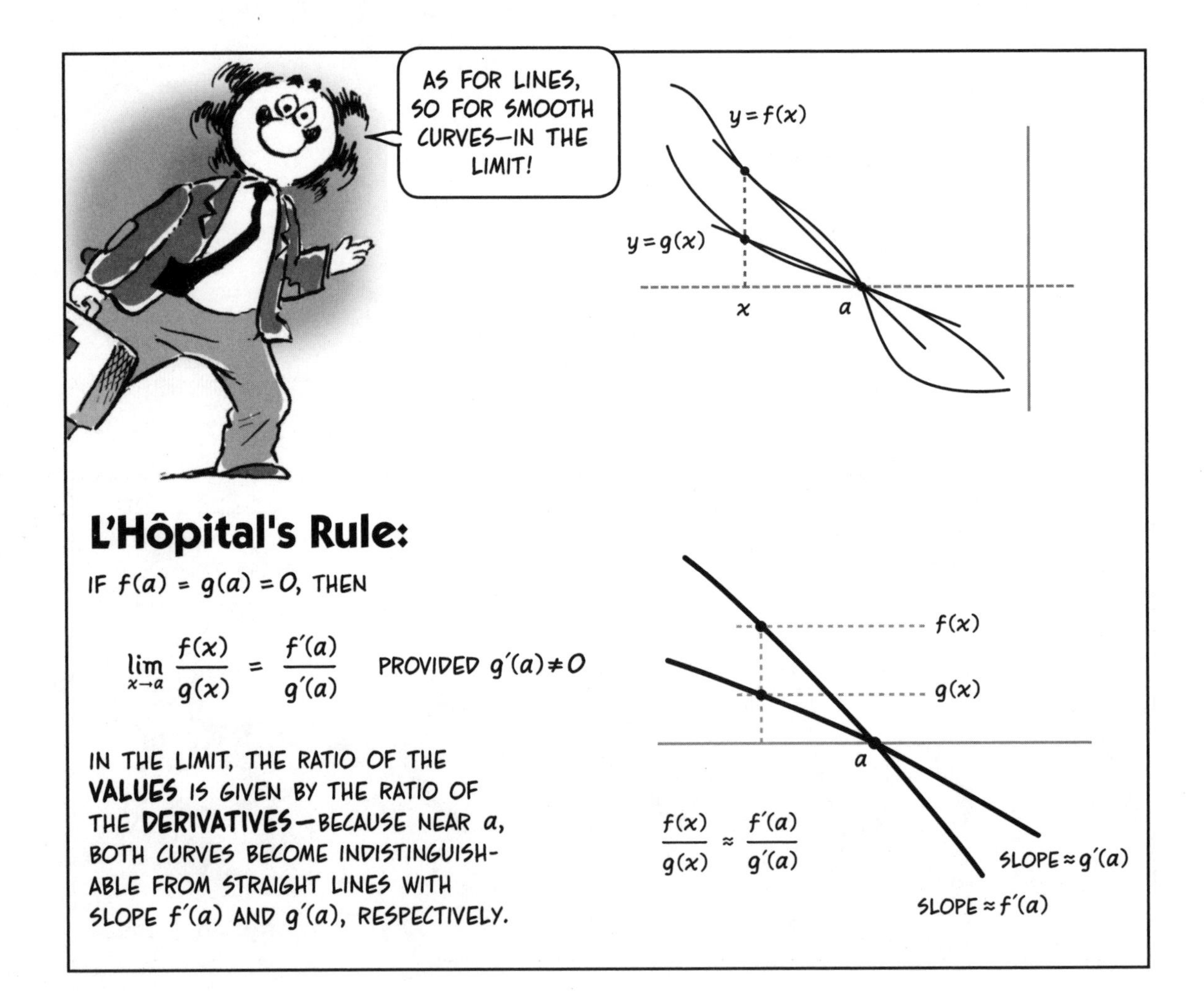
AS FOR LINES, SO FOR SMOOTH CURVES—IN THE LIMIT!
y=f(x)
y=g(x)
x
a
f(x)
g(x)
a
f(x)/g(x) ≈ f′(a)/g′(a)
SLOPE ≈ g′(a)
SLOPE ≈ f′(a)

L'Hôpital's Rule:

IF $f(a) = g(a) = 0$, THEN

$$\lim_{x \to a} \frac{f(x)}{g(x)} = \frac{f'(a)}{g'(a)} \quad \text{PROVIDED } g'(a) \neq 0$$

IN THE LIMIT, THE RATIO OF THE **VALUES** IS GIVEN BY THE RATIO OF THE **DERIVATIVES**—BECAUSE NEAR a, BOTH CURVES BECOME INDISTINGUISHABLE FROM STRAIGHT LINES WITH SLOPE $f'(a)$ AND $g'(a)$, RESPECTIVELY.

Example: FIND $\lim_{x\to 0} \frac{e^x - 1}{\sin 2x}$

FIRST NOTE THAT BOTH NUMERATOR AND DENOMINATOR ARE 0 WHEN $x = 0$.

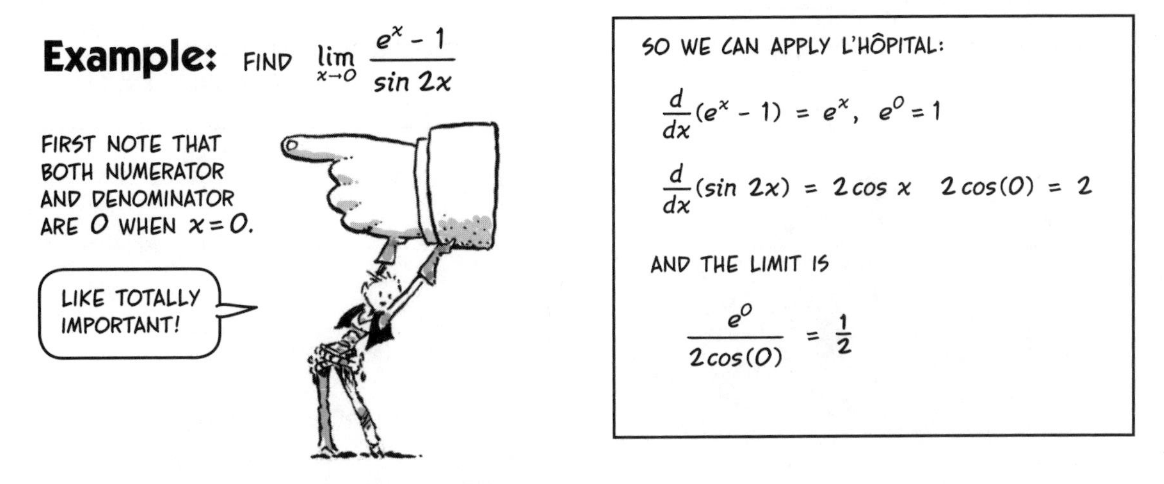
LIKE TOTALLY IMPORTANT!

SO WE CAN APPLY L'HÔPITAL:

$$\frac{d}{dx}(e^x - 1) = e^x, \quad e^0 = 1$$

$$\frac{d}{dx}(\sin 2x) = 2\cos x \quad 2\cos(0) = 2$$

AND THE LIMIT IS

$$\frac{e^0}{2\cos(0)} = \frac{1}{2}$$

WHAT HAPPENS IF $f(a)$, $g(a)$, $f'(a)$, AND $g'(a)$ ARE **ALL** ZERO? THEN WE GO TO THE SECOND DERIVATIVE, AND IF $f''(a) = g''(a) = 0$, THEN WE GO TO THE THIRD, ETC.! THIS MORE GENERAL FORM OF L'HÔPITAL'S RULE SAYS:

IF $f(a) = g(a) = 0$, AND $\lim_{x\to a} \frac{f'(x)}{g'(x)}$ EXISTS, THEN

$$\lim_{x\to a} \frac{f(x)}{g(x)} = \lim_{x\to a} \frac{f'(x)}{g'(x)}$$

Example: FIND $\lim_{x\to 0} \frac{e^{3x} - 1 - 3x}{1 - \cos x}$

REMEMBER: TO APPLY L'HÔPITAL'S RULE, WE **MUST** CHECK THAT NUMERATOR AND DENOMINATOR ARE BOTH ZERO AT THE LIMIT POINT! CALL THE NUMERATOR f AND THE DENOMINATOR g. WE SEE THAT $f(0) = g(0) = 0$.

UNFORTUNATELY THEIR **DERIVATIVES** ARE ALSO BOTH ZERO AT $x = 0$.

$$f'(x) = 3e^{3x} - 3 \quad f'(0) = 0$$

$$g'(x) = \sin x \quad g'(0) = 0$$

MOST UNFOR-TUNATE...

TERRIBLY, **TERRIBLY** SAD...

NO PROBLEM! WE LOOK AT THE **SECOND** DERIVATIVES:

$$f''(x) = 9e^{3x} \quad f''(0) = 9$$

$$g''(x) = \cos x \quad g''(0) = 1$$

AND CONCLUDE

$$\lim_{x\to 0} \frac{e^{3x} - 1 - 3x}{1 - \cos x} = \lim_{x\to 0} \frac{f'(x)}{g'(x)}$$

$$= \frac{f''(0)}{g''(0)} = \frac{9}{1} = \mathbf{9}$$

WHY IS THIS RULE NAMED AFTER A FRENCH HOSPITAL?
'CAUSE IT'S SO **SICK!**
WHERE YOU GOING?
MY HEAD HURTS. I'M CHECKING IN TO THE HOSPITAL AT INFINITY...

L'HÔPITAL'S RULE ALSO WORKS FOR LIMITS AT INFINITY, INCLUDING INFINITE LIMITS:

IF $\lim\limits_{x \to \infty} f(x) = \lim\limits_{x \to \infty} g(x) = \infty$, OR

$\lim\limits_{x \to \infty} f(x) = \lim\limits_{x \to \infty} g(x) = 0$, THEN

$$\lim_{x \to \infty} \frac{f(x)}{g(x)} = \lim_{x \to \infty} \frac{f'(x)}{g'(x)}$$

IF THE LATTER LIMIT EXISTS.

Example at infinity:

FIND

$$\lim_{x \to \infty} \frac{x^p}{\ln x},\ p > 0$$

BOTH NUMERATOR AND DENOMINATOR GO TO INFINITY AS $x \to \infty$. TO APPLY L'HÔPITAL WE TAKE THE DERIVATIVE OF EACH FUNCTION:

$$\frac{d}{dx}(x^p) = px^{p-1} \qquad \frac{d}{dx}(\ln x) = \frac{1}{x} \quad \text{SO}$$

$$\lim_{x \to \infty} \frac{x^p}{\ln x} = \lim_{x \to \infty} \frac{px^{p-1}}{\frac{1}{x}} = \lim_{x \to \infty} px^p = \infty$$

THIS SAYS THAT $\ln x$ GOES TO INFINITY SLOWER THAN **ANY POSITIVE POWER FUNCTION.** x^p BECOMES INFINITELY GREATER THAN $\ln x$ AS $x \to \infty$. THE LOGARITHM IS A VERY SLOW GROWER!

NOTE THAT YOU DON'T SEE IT IN THIS GRAPH, WHERE x IS SMALL... BUT FOR LARGER x, $\ln x$ REALLY STRUGGLES TO GET OFF THE GROUND!

x	$\ln x$	$x^{\frac{1}{3}}$
$e^{10} \approx 220{,}026$	10	28.02
$e^{15} \approx 3{,}269{,}017$	15	148.3
$e^{20} \approx 485{,}000{,}000$	20	785.2
...	...	...
e^N	N	$e^{N/3}$
...	...	...

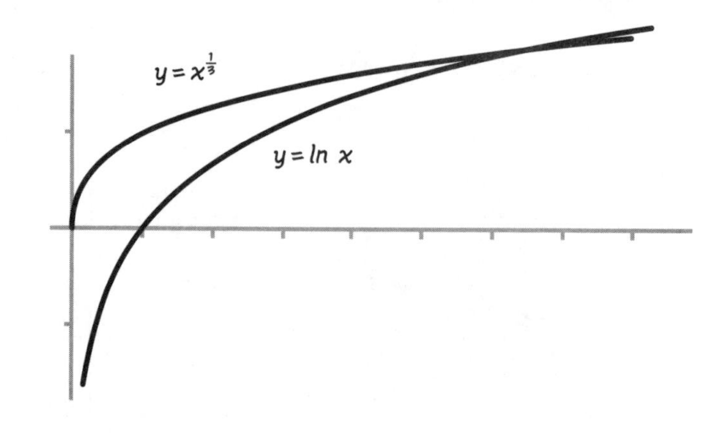
$y = x^{\frac{1}{3}}$
$y = \ln x$

IT'S TAKING **FOREVER** TO GET THERE...
YES... BUT LOOK HOW FAR WE'VE COME!

THE LAST SIX CHAPTERS HAVE EXPLORED THE FIRST BIG TOPIC OF CALCULUS, THE **DERIVATIVE.** BEFORE GOING ON TO TOPIC 2, THE INTEGRAL, LET'S REVIEW WHAT USES WE'VE FOUND FOR NEWTON AND LEIBNIZ'S GREAT INVENTION, A FUNCTION'S INSTANTANEOUS RATE OF CHANGE.

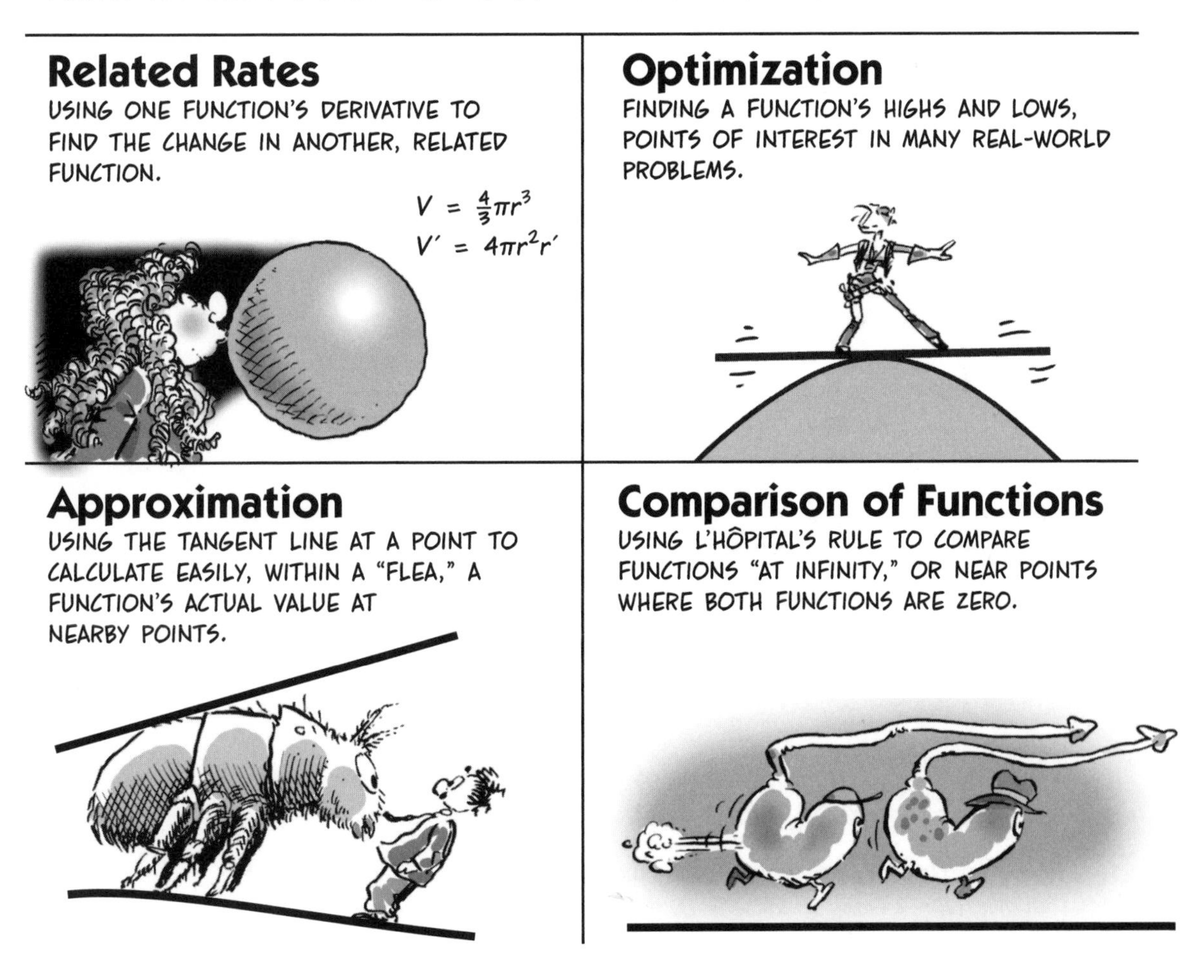
$V = \frac{4}{3}\pi r^3$
$V' = 4\pi r^2 r'$

Related Rates

USING ONE FUNCTION'S DERIVATIVE TO FIND THE CHANGE IN ANOTHER, RELATED FUNCTION.

Optimization

FINDING A FUNCTION'S HIGHS AND LOWS, POINTS OF INTEREST IN MANY REAL-WORLD PROBLEMS.

Approximation

USING THE TANGENT LINE AT A POINT TO CALCULATE EASILY, WITHIN A "FLEA," A FUNCTION'S ACTUAL VALUE AT NEARBY POINTS.

Comparison of Functions

USING L'HÔPITAL'S RULE TO COMPARE FUNCTIONS "AT INFINITY," OR NEAR POINTS WHERE BOTH FUNCTIONS ARE ZERO.

Problems

1. ESTIMATE $\sqrt{5}$ BY USING THE APPROXIMATION

$$f(x) \approx f(4) + f'(4)(x - 4)$$

2. ESTIMATE $\sqrt{67}$. (HINT: USE A NEARBY PERFECT SQUARE.) COMPARE YOUR ESTIMATE WITH THE VALUE OBTAINED FROM A CALCULATOR.

3. ESTIMATE $\sin 3$.

4. ESTIMATE $\arctan(1.1)$. (REMEMBER THAT $\arctan 1 = \pi/4$.)

USE L'HÔPITAL'S RULE, IF APPROPRIATE, TO EVALUATE THESE LIMITS. (REMEMBER TO CHECK THE LIMITS OF NUMERATOR AND DENOMINATOR FIRST! THERE MAY BE SOME HERE WHERE L'HÔPITAL DOES NOT APPLY...)

5. $\lim_{x \to 0} \dfrac{\sin(x^2)}{\cos x - 1}$

6. $\lim_{x \to 0} \dfrac{x}{\sin 2x}$

7. $\lim_{x \to 0} \dfrac{e^{-8x^2} - 1}{\cos 2x - 1}$

8. $\lim_{x \to 1} \dfrac{x^7 - 1}{x^3 - 1}$

9. $\lim_{x \to 0} \dfrac{6 \sin x - 6x + x^3}{2 \cos x + x^2 - 2}$

10. $\lim_{x \to \infty} x^{\frac{1}{x}}$ HINT: TAKE THE LOGARITHM.

11. $\lim_{x \to 1} \dfrac{\ln x}{x - 1}$

12. $\lim_{x \to \pi} \dfrac{\sin x}{\cos x - 1}$

13a. GIVEN A POLYNOMIAL $P(x) = a_0 + a_1 x + a_2 x^2 + \ldots + a_n x^n$, SHOW THAT $P'(0) = a_1$, $P''(0) = 2a_2$, AND $P^{(m)}(0) = m! a_m$ FOR ALL $m \leq n$.

13b. IF f IS ANY FUNCTION DIFFERENTIABLE AT a, SHOW THAT THE POLYNOMIAL

$$P_n(x) = f(0) + f'(0)x + \frac{f''(0)}{2!}x^2 + \ldots + \frac{f^{(m)}(0)}{m!}x^m + \ldots + \frac{f^{(n)}(0)}{n!}x^n$$

HAS $P(0) = f(0)$ AND $P^{(m)}(0) = f^{(m)}(0)$ FOR $m = 1, 2, \ldots, n$. THE POLYNOMIAL P_n IS CALLED THE nTH **TAYLOR POLYNOMIAL** OF f AT $x = 0$.

13c. WRITE AN 8TH-DEGREE POLYNOMIAL HAVING THE SAME VALUE AND FIRST EIGHT DERIVATIVES AS $\cos x$ AT $x = 0$.

Chapter 7
The Mean Value Theorem

SOME FINAL, FRENZIED, FEORETICAL FOUGHTS

(WHICH YOU MAY SKIP IF ALL YOU CARE ABOUT IS HOW TO USE CALCULUS, AND HAVE NO APPRECIATION OF ITS DEEP, BEAUTIFUL, AND ELEGANT FOUNDATIONS—SEE IF I CARE!)

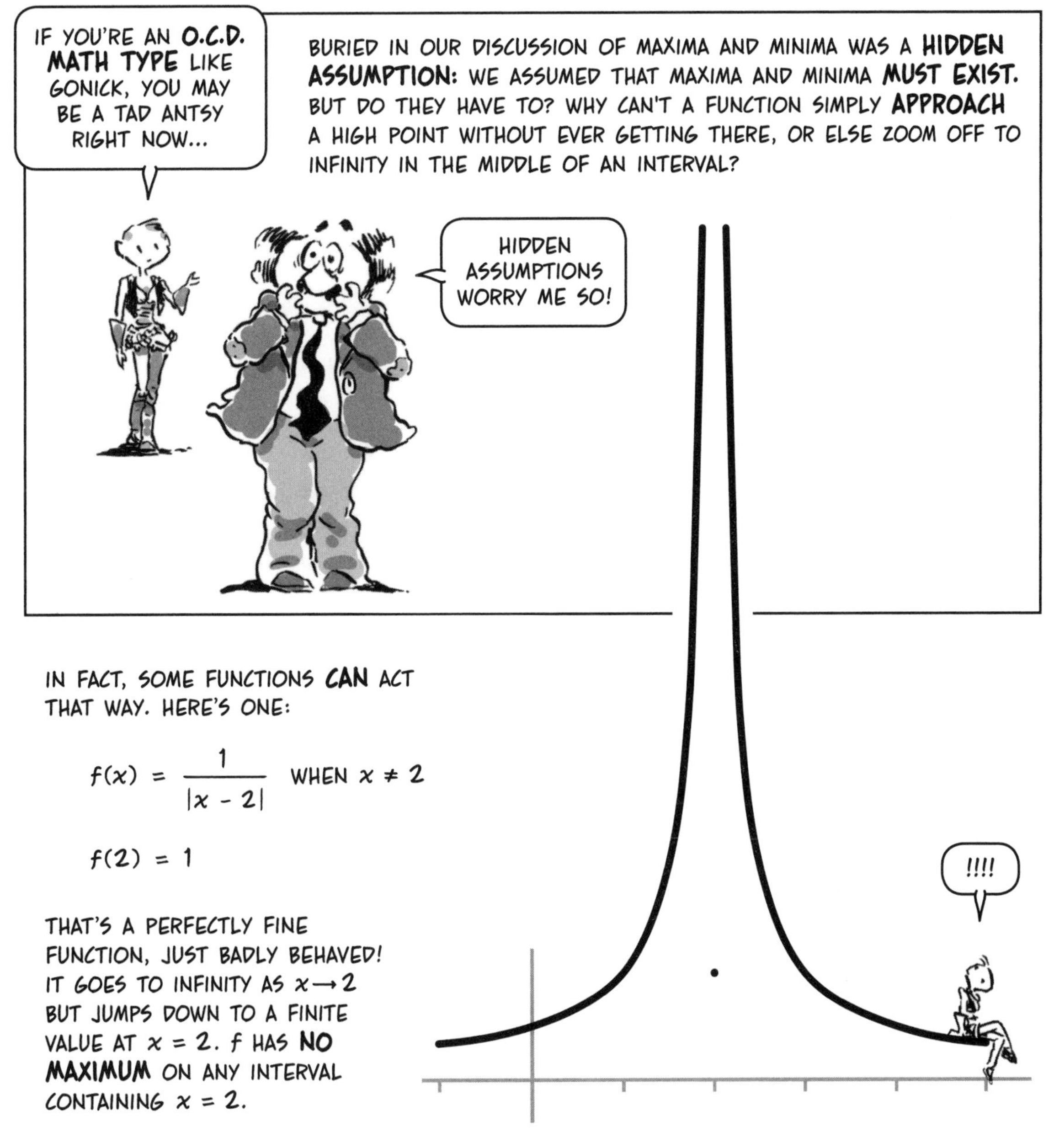
IF YOU'RE AN O.C.D. MATH TYPE LIKE GONICK, YOU MAY BE A TAD ANTSY RIGHT NOW...
BURIED IN OUR DISCUSSION OF MAXIMA AND MINIMA WAS A HIDDEN ASSUMPTION: WE ASSUMED THAT MAXIMA AND MINIMA MUST EXIST. BUT DO THEY HAVE TO? WHY CAN'T A FUNCTION SIMPLY APPROACH A HIGH POINT WITHOUT EVER GETTING THERE, OR ELSE ZOOM OFF TO INFINITY IN THE MIDDLE OF AN INTERVAL?
HIDDEN ASSUMPTIONS WORRY ME SO!
!!!!

IN FACT, SOME FUNCTIONS **CAN** ACT THAT WAY. HERE'S ONE:

$$f(x) = \frac{1}{|x-2|} \quad \text{WHEN } x \neq 2$$

$$f(2) = 1$$

THAT'S A PERFECTLY FINE FUNCTION, JUST BADLY BEHAVED! IT GOES TO INFINITY AS $x \to 2$ BUT JUMPS DOWN TO A FINITE VALUE AT $x = 2$. f HAS **NO MAXIMUM** ON ANY INTERVAL CONTAINING $x = 2$.

THE PROBLEM WITH THAT FUNCTION IS THE ISOLATED POINT (2, 1) ON ITS GRAPH... THE FUNCTION DOESN'T **APPROACH** THAT POINT, IT JUST **JUMPS** THERE, SO TO SPEAK... SO LET'S LOOK AT FUNCTIONS WITHOUT ANY JUMPS... FUNCTIONS WHOSE GRAPH CAN BE DRAWN WITHOUT LIFTING PENCIL FROM PAPER. SUCH "UNJUMPY" FUNCTIONS ARE CALLED **CONTINUOUS.**

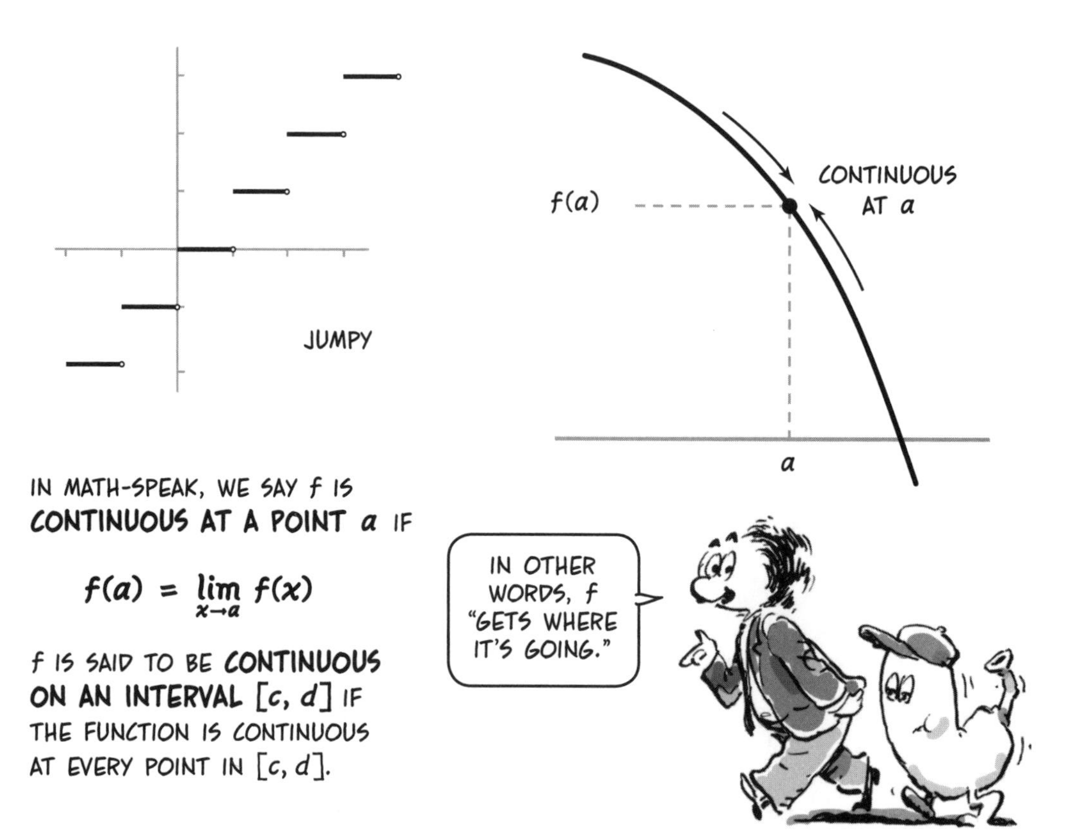
JUMPY
CONTINUOUS AT a
$f(a)$
a
IN OTHER WORDS, f "GETS WHERE IT'S GOING."

IN MATH-SPEAK, WE SAY f IS **CONTINUOUS AT A POINT** a IF

$$f(a) = \lim_{x \to a} f(x)$$

f IS SAID TO BE **CONTINUOUS ON AN INTERVAL** $[c, d]$ IF THE FUNCTION IS CONTINUOUS AT EVERY POINT IN $[c, d]$.

ALL DIFFERENTIABLE FUNCTIONS ARE CONTINUOUS, BUT NOT VICE VERSA. IF f IS DIFFERENTIABLE AT a, THEN WE KNOW $f(x) - f(a) = f'(a)(x - a)$ + FLEA, SO $\lim_{x \to a}(f(x) - f(a)) = 0$ OR $\lim_{x \to a} f(x) = f(a)$. ON THE OTHER HAND, A CONTINUOUS FUNCTION MAY HAVE SHARP CORNERS WHERE IT IS NOT DIFFERENTIABLE.

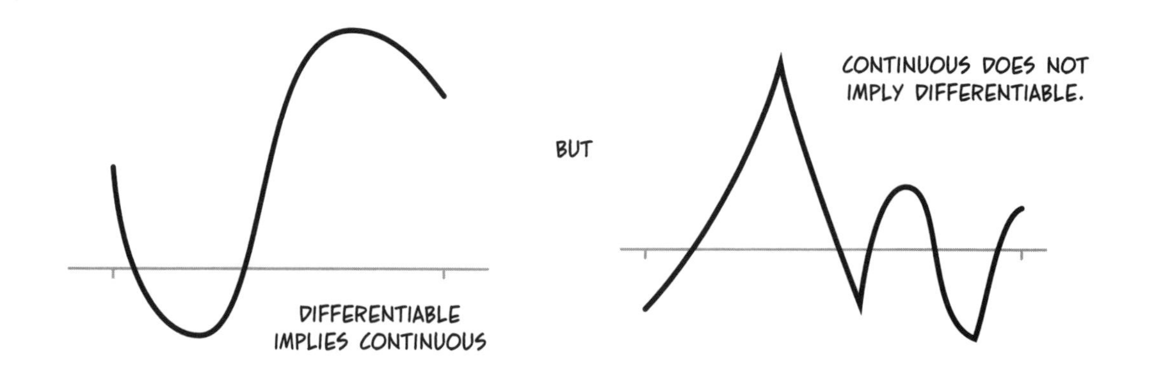
DIFFERENTIABLE IMPLIES CONTINUOUS
BUT
CONTINUOUS DOES NOT IMPLY DIFFERENTIABLE.

CONTINUOUS FUNCTIONS DO WHAT WE WANT:

Extreme Value Theorem:

A CONTINUOUS FUNCTION f DEFINED ON A **CLOSED** INTERVAL $[c, d]$ ATTAINS A MAXIMUM VALUE M ON THE INTERVAL: I.E., THERE IS A POINT a IN $[c, d]$ WHERE $f(a) = M$ AND $f(x) \leq M$ FOR ALL OTHER x IN $[c, d]$.

(NOTE THAT THIS ALSO IMPLIES THE EXISTENCE OF A MINIMUM, BECAUSE $-f$ MUST HAVE A MAXIMUM!)

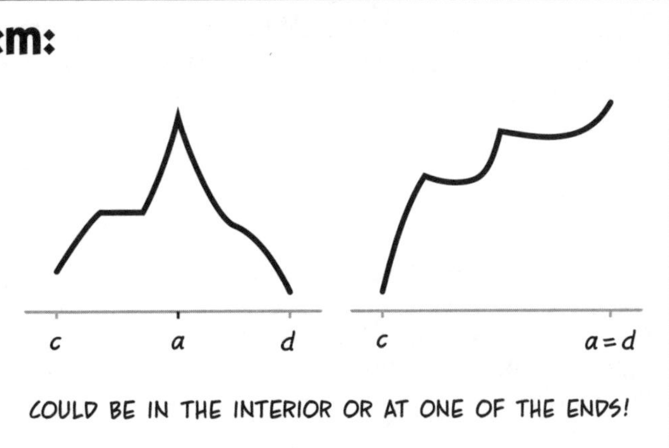
c
a
d
c
a=d

COULD BE IN THE INTERIOR OR AT ONE OF THE ENDS!

WE MUST OMIT THE PROOF, WHICH RELIES ON DEEP AND SUBTLE PROPERTIES OF THE REAL NUMBERS.

HOW CAN SOMETHING SO ONE-DIMENSIONAL BE SO DEEP?

THE EXTREME VALUE THEOREM HAS THIS CONSEQUENCE FOR CALCULUS:

Rolle's Theorem:

IF f IS CONTINUOUS ON A CLOSED INTERVAL $[c, d]$ AND DIFFERENTIABLE ON (c, d), AND $f(c) = f(d) = 0$, THEN THERE IS AT LEAST ONE POINT a IN THE OPEN INTERVAL (c, d) WHERE $f'(a) = 0$.

PROOF: IF f IS THE CONSTANT FUNCTION $f = 0$, THEN THE RESULT IS TRIVIAL: ANY POINT BETWEEN c AND d WILL DO.

IF f IS NOT CONSTANT, THEN IT HAS NON-ZERO VALUES. THEREFORE, IT ATTAINS EITHER A MAXIMUM $M > 0$ OR A MINIMUM $m < 0$ AT SOME POINT a, BY THE EXTREME VALUE THEOREM. a IS NOT ONE OF THE ENDPOINTS BECAUSE $f(c) = f(d) = 0$, SO $f'(a) = 0$.

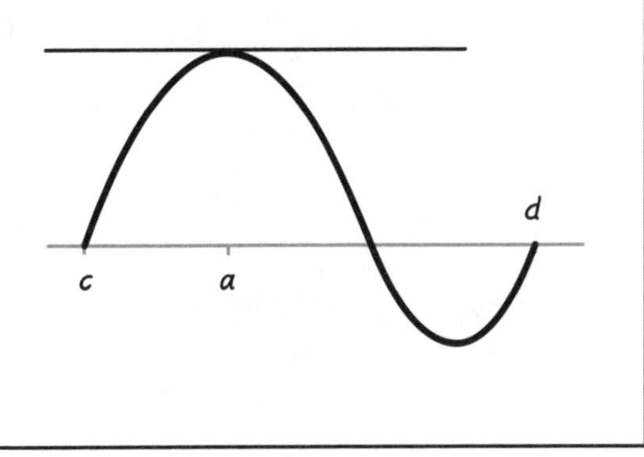
d
c
a

ROLLE, IN TURN, IMPLIES THIS SURPRISING, IMPORTANT, SKEWED VERSION:

Mean Value Theorem: IF f IS CONTINUOUS ON A CLOSED INTERVAL $[c, d]$ AND DIFFERENTIABLE ON THE OPEN INTERVAL (c, d), THEN THERE IS AN INTERIOR POINT a IN (c, d) WHERE

$$f'(a) = \frac{f(d) - f(c)}{d - c}$$

THAT IS, THERE MUST BE AT LEAST ONE INTERIOR POINT WHERE THE TANGENT LINE **PARALLELS** THE CHORD JOINING THE GRAPH'S ENDPOINTS.

c a d

NOTE THAT ALL THREE OF THESE THEOREMS MERELY ALLEGE **EXISTENCE.** THEY PROVE THAT THERE ARE POINTS WITH THE REQUIRED PROPERTIES—WITHOUT OFFERING ANY WAY OF FINDING THOSE POINTS! THE PROOFS ARE NOT "CONSTRUCTIVE."

LIKE SOME MATHEMATICIANS I KNOW...

PROOF OF MEAN VALUE THEOREM: GIVEN f AS DESCRIBED, DEFINE A NEW FUNCTION g BY SUBTRACTING THE CHORD FROM f:

$$g(x) = f(x) - \frac{f(d) - f(c)}{d - c}(x - c) - f(c)$$

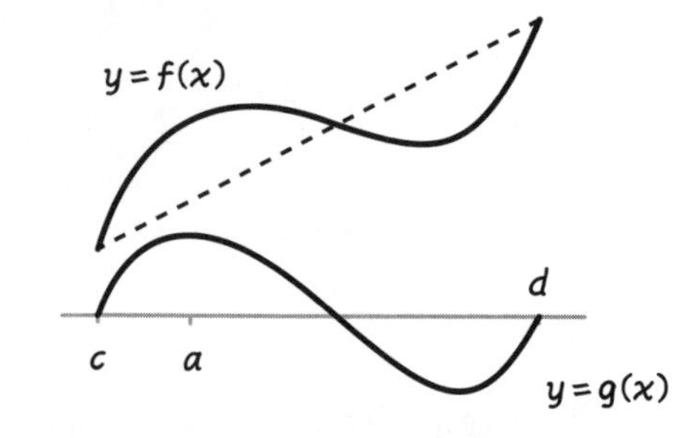
y = f(x)
d
c
a
y = g(x)

g SATISFIES THE HYPOTHESIS OF ROLLE'S THEOREM: $g(c) = g(d) = 0$. THEREFORE, THERE IS AN INTERIOR POINT a WHERE $g'(a) = 0$. BUT

$$g'(x) = f'(x) - \frac{f(d) - f(c)}{d - c}$$

SINCE $g'(a) = 0$, IT FOLLOWS THAT

$$f'(a) = \frac{f(d) - f(c)}{d - c}$$

THE MEAN VALUE THEOREM HAS POWERFUL CONSEQUENCES:

ASSUME THE FUNCTION f IS CONTINUOUS ON A CLOSED INTERVAL $[c, d]$ AND DIFFERENTIABLE ON THE OPEN INTERVAL (c, d).

DO YOU EVER SEE FUNCTIONS THAT **AREN'T** LIKE THAT?
IN MY NIGHTMARES!

1. A POSITIVE DERIVATIVE IMPLIES A STRICTLY INCREASING FUNCTION: SUPPOSE $f'(x) > 0$ (STRICTLY!) FOR EVERY x IN AN INTERVAL (c, d). THEN f IS STRICTLY INCREASING ON THE INTERVAL.

PROOF: TAKE ANY TWO POINTS $a < b$ IN THE INTERVAL. BY THE MEAN VALUE THEOREM, THERE IS A POINT x_0 BETWEEN a AND b SUCH THAT

$$f'(x_0) = \frac{f(b) - f(a)}{b - a}$$

WE ASSUMED THAT $f'(x_0) > 0$, SO $f(b) - f(a) > 0$, I.E., f IS STRICTLY INCREASING.

IF A FUNCTION ALWAYS TRENDS UPWARD, HOW CAN IT EVER GET SMALLER?
a
x_0
b

2. ONLY CONSTANT FUNCTIONS HAVE A CONSTANT ZERO DERIVATIVE: IF $f'(x) = 0$ FOR EVERY x IN AN INTERVAL (c, d), THEN f IS CONSTANT ON THE INTERVAL.

PROOF: TAKE ANY TWO POINTS $a < b$ IN THE INTERVAL. THE MEAN VALUE THEOREM SAYS THAT THERE IS A POINT x_0 SUCH THAT

$$f'(x_0) = \frac{f(b) - f(a)}{b - a}$$

BUT $f'(x_0)$ IS ASSUMED TO BE ZERO, SO $f(a) = f(b)$ AND THE FUNCTION IS CONSTANT.

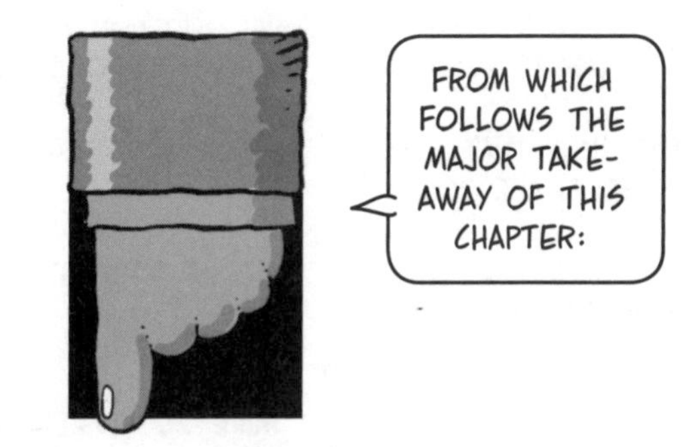
FROM WHICH FOLLOWS THE MAJOR TAKE-AWAY OF THIS CHAPTER:

3. COROLLARY: IF f AND g ARE ANY TWO FUNCTIONS WITH $f' = g'$, THEN f AND g DIFFER BY A CONSTANT. THIS FOLLOWS FROM THE PREVIOUS RESULT, APPLIED TO THE FUNCTION $f - g$.

AND NOW, ON TO THE INTEGRAL!

Problems

FOR EACH FUNCTION f, FIND THE SLOPE $m = (f(b) - f(a))/(b - a)$ OF THE SECANT LINE JOINING THE ENDPOINTS OF THE GRAPH ON THE GIVEN INTERVAL. THEN FIND ALL POINTS c ON THE INTERVAL WHERE $f'(c) = m$. USE A CALCULATOR WHEN NECESSARY.

1. $f(x) = x^3 + 2x + 3$ ON $[0, 2]$

2. $f(x) = e^{-x}$ ON $[-1, 3]$

3. $f(x) = \dfrac{4 + x}{4 - x}$ ON $[0, 2]$

4. $f(x) = \cos x$ ON $[0, 3\pi]$

5. $f(x) = 2x^4 - x^2$ ON $[-50, 50]$

6. $f(x) = \tan x$ ON $[-a, a]$, FOR ANY a WITH $a < 0 < \pi/2$

NOTE THAT ROLLE'S THEOREM IMPLIES THAT IF THE DERIVATIVE $f'(x)$ OF A CONTINUOUS, DIFFERENTIABLE FUNCTION f IS **NEVER ZERO** ON AN INTERVAL, THEN THERE CANNOT BE TWO POINTS a AND b IN THE INTERVAL WHERE $f(a) = f(b)$.

7. SHOW THAT THE EQUATION $y = 3x - \sin x + 7$ HAS AT MOST ONE ROOT. DOES IT HAVE ANY ROOTS? WHY OR WHY NOT??

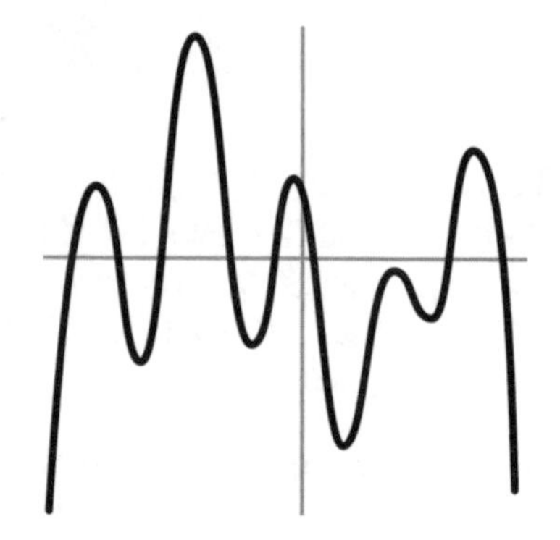

8a. SHOW THAT A POLYNOMIAL $P(x) = x^2 + bx + c$ OF DEGREE TWO HAS AT MOST TWO ROOTS.

8b. SHOW THAT A POLYNOMIAL OF DEGREE 3 HAS AT MOST THREE ROOTS.

8c. SHOW THAT A POLYNOMIAL OF DEGREE n HAS AT MOST n ROOTS.

9. A RACECAR DRIVER IS AT MILE 20. IF HER SPEED NEVER EXCEEDS 150 MI/HR, WHAT IS THE MAXIMUM MILEPOST SHE CAN REACH IN THE NEXT TWO HOURS?

10. A FUNCTION f, CONTINUOUS ON AN INTERVAL $[a, b]$ AND DIFFERENTIABLE ON (a, b), HAS $f(a) = 2$. IF $f'(x) \leq 7$ FOR EVERY x IN (a, b), WHAT IS THE LARGEST VALUE $f(x)$ CAN POSSIBLY ATTAIN ON THE INTERVAL? (HINT: COMPARE WITH PROBLEM 9.)

11. LET $f(x) = (x - 2)^{-2}$. SHOW THAT THERE IS NO VALUE OF c IN THE INTERVAL $(1, 3)$ SUCH THAT $f(3) - f(1) = f'(c)(3 - 1)$. WHY DOES THIS NOT VIOLATE THE MEAN VALUE THEOREM?

12. SUPPOSE f AND g SATISFY THE HYPOTHESES OF THE MEAN VALUE THEOREM ON THE INTERVAL $[a, b]$, AND THAT $f(a) = g(a)$. SHOW THAT IF $f'(x) > g'(x)$ FOR EVERY x IN (a, b), THEN $f(b) > g(b)$.

13. SHOW THAT ANY FUNCTION WHOSE DERIVATIVE IS ITSELF MUST HAVE THE FORM $f(x) = Ce^x$ FOR SOME CONSTANT C. (HINT: SUPPOSE $f'(x) = f(x)$, DIFFERENTIATE THE FUNCTION

$$g(x) = \frac{f(x)}{e^x},$$

AND APPLY COROLLARY 2.)

Chapter 8
Introducing Integration

PUTTING TWO AND TWO AND TWO AND TWO TOGETHER

CALCULUS, AS WE'VE SEEN, SLICES QUANTITIES INTO SMALL SLIVERS, MINUTE MOUSY THINGS WITH NAMES LIKE h, Δx, Δy, Δt, AND Δf. IF P IS A PIE, THEN ΔP IS A THIN SLICE OF PIE.

MAN! EDGEWISE, YOU CAN HARDLY SEE IT...

UP TO NOW, WE HAVE LOOKED AT WHAT HAPPENS WHEN WE **DIVIDE** ONE OF THESE THINGS BY ANOTHER TO MAKE RATIOS LIKE $\Delta f/h$... BUT NOW WE WANT TO DO SOMETHING ELSE WITH OUR NUMBER-CRUMBS: **ADD THEM TOGETHER.**

WELL, HOW MANY PIECES DO YOU WANT?
ALL OF THEM, FROM HERE TO THERE...
ADDITION IS EASIER THAN MULTIPLICATION... THAT'S WHY WE LEARN IT FIRST IN SCHOOL... AND IN FACT, MATHEMATICIANS USED SUMMING-UP PROCESSES FOR THOUSANDS OF YEARS BEFORE NEWTON AND LEIBNIZ INVENTED THE DERIVATIVE.
HAVE YOU CONSIDERED CUTTING IT CROSSWISE?

THERE IS STANDARD NOTATION FOR SUMMING MANY ITEMS. IT USES A CAPITAL **SIGMA,** THE GREEK LETTER S, STANDING FOR "SUM."

DON'T WORRY... IT'S NEVER BITTEN ANYONE...

IF WE HAVE A SEQUENCE OF n NUMBERS

$$a_1,\ a_2,\ a_3,\ \ldots\ a_i,\ \ldots\ a_n$$

a_i ("AE-EYE") IS CALLED THE ***i*TH TERM** OF THE SEQUENCE, AND THE SUM OF ALL TERMS IS WRITTEN

$$\sum_{i=1}^{n} a_i$$

READ "THE SUM, AS i GOES FROM 1 TO n, OF a_i." THE LETTER i IS CALLED THE **INDEX** OF THE SEQUENCE.

THE SUM OF THE CONSECUTIVE TERMS FROM a_p TO a_q, INCLUSIVE, IS

$$\sum_{i=p}^{q} a_i = a_p + a_{p+1} + \ldots + a_q$$

FOR EXAMPLE, CONSIDER THE FIVE-TERM SEQUENCE $\{2, 4, 8, 16, 32\}$. HERE $a_i = 2^i$ AND $n = 5$.

i	a_i
1	2
2	4
3	8
4	16
5	32

IN THIS CASE,

$$\sum_{i=1}^{5} a_i = 2 + 4 + 8 + 16 + 32 = 62$$

$$\sum_{i=2}^{4} a_i = 4 + 8 + 16 = 28$$

O.K.... I THINK IT'S UNDER CONTROL...

IF WE WERE TO DIVIDE A PIE P INTO n (POSSIBLY UNEQUAL) SLICES, CALLED ΔP_1, ΔP_2, ΔP_3 ..., ΔP_n, THEN THE WHOLE PIE WOULD BE THE SUM:

$$P = \sum_{i=1}^{n} \Delta P_i$$

THEN, AS WE LIKE TO DO IN CALCULUS, WE SHRINK THE SIZE OF THESE SLICES (TO AN INFINITESIMAL dP, AS LEIBNIZ LIKED TO SAY). AT THAT POINT WE'LL WRITE THE THING WITH A DIFFERENT SORT OF "S," A STRETCHED ONE CALLED AN **INTEGRAL SIGN.**

$$P = \int dP$$

WHY ANOTHER "S"?
BECAUSE IT'S SSORTA SSIMILAR TO A SSUM...
THAT SYMBOL'S ANOTHER ONE OF MINE, BY THE WAY...

O.K.... THAT'S OUR NOTATION... IT'S ALL JUST ADDING FROM HERE ON OUT...

HEY! WAIT A MINUTE...

A good question:

NOW YOU MIGHT WONDER, IF ADDING IS SIMPLER THAN DIVIDING, AND THE ANCIENTS DID INTEGRALS LONG BEFORE NEWTON DID DERIVATIVES, WHY DIDN'T WE START THE BOOK WITH **THIS** SECTION?

SURELY YOU DON'T THINK I DID IT THIS WAY OUT OF SOME SORT OF PERVERSE DESIRE TO MESS WITH YOUR MIND?
NEVER OCCURRED TO ME UNTIL JUST NOW.

THE SURPRISING ANSWER: ALTHOUGH SUMS MAY BE EASIER TO **IMAGINE,** THEY CAN BEST BE **CALCULATED** BY USING **DERIVATIVES!!** AS NEWTON AND LEIBNIZ DISCOVERED, THERE IS A SURPRISING RELATIONSHIP BETWEEN SUMS AND DERIVATIVES!

AS WE ARE ABOUT TO SEE...

SUPPOSE DELTA IS DRIVING HER CAR ALONG A STRAIGHT COURSE AGAIN, EXCEPT THAT NOW HER WINDOWS ARE BLACKED OUT.

WHAT??!!

ALL SHE CAN SEE ARE THE VELOCIMETER AND THE TIME. CAN SHE FIGURE OUT WHERE SHE IS AFTER, SAY, 10 UNITS OF TIME?

IF I SURVIVE THAT LONG....
10:00

BY CHECKING t AND $v(t)$ OFTEN, DELTA GETS A SERIES OF READINGS $v(t_0)$, $v(t_1)$, $v(t_2)$, ... $v(t_i)$, ETC., AT TIMES t_0, t_1, t_2, ... t_i, ..., t_n, WHERE $t_0 = 0$ AND $t_n = 10$.

WHAT A WEIRD EXPERIMENT!

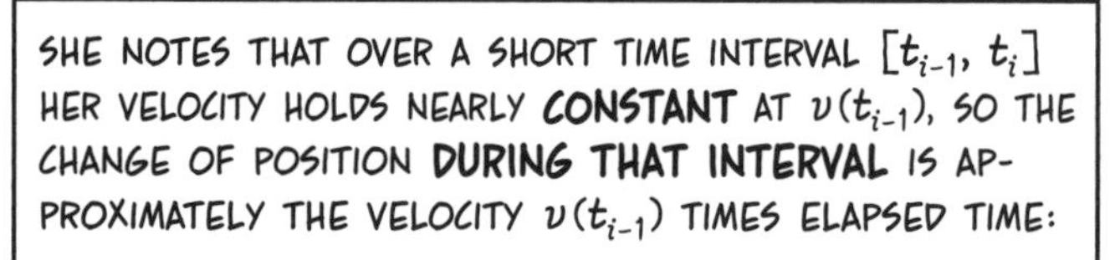

SHE NOTES THAT OVER A SHORT TIME INTERVAL $[t_{i-1}, t_i]$ HER VELOCITY HOLDS NEARLY **CONSTANT** AT $v(t_{i-1})$, SO THE CHANGE OF POSITION **DURING THAT INTERVAL** IS APPROXIMATELY THE VELOCITY $v(t_{i-1})$ TIMES ELAPSED TIME:

$$s(t_i) - s(t_{i-1}) \approx v(t_{i-1})(t_i - t_{i-1})$$

$$= v(t_{i-1})\Delta t_i$$

WHERE $\Delta t_i = t_i - t_{i-1}$. THE CHANGE OF POSITION OVER THE iTH INTERVAL IS VERY NEARLY $v(t_{i-1})\Delta t_i$.

ADDING ALL THESE UP GIVES—APPROXIMATELY—THE **TOTAL** CHANGE OF POSITION BETWEEN $t_0 = 0$ AND 10:

$s(10) - s(0) \approx \sum_{i=1}^{n} v(t_{i-1})\Delta t_i$
YES!

ON A GRAPH, EACH TERM IS THE AREA OF A SLENDER RECTANGLE OF HEIGHT $v(t_{i-1})$ AND BASE Δt_i.* THE TOTAL CHANGE OF POSITION IS THE **SUM OF THESE AREAS.**

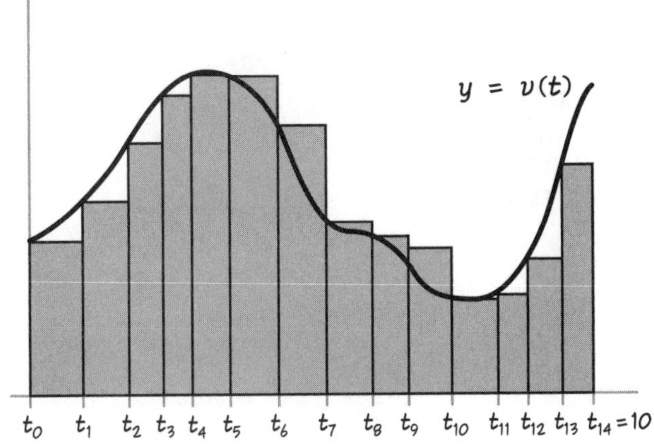

$y = v(t)$
t_0 t_1 t_2 t_3 t_4 t_5 t_6 t_7 t_8 t_9 t_{10} t_{11} t_{12} t_{13} $t_{14} = 10$

THE FIRST RECTANGLE HAS AREA $v(t_0)\Delta t_1$, AND THE iTH RECTANGLE HAS AREA $v(t_{i-1})\Delta t_i$.

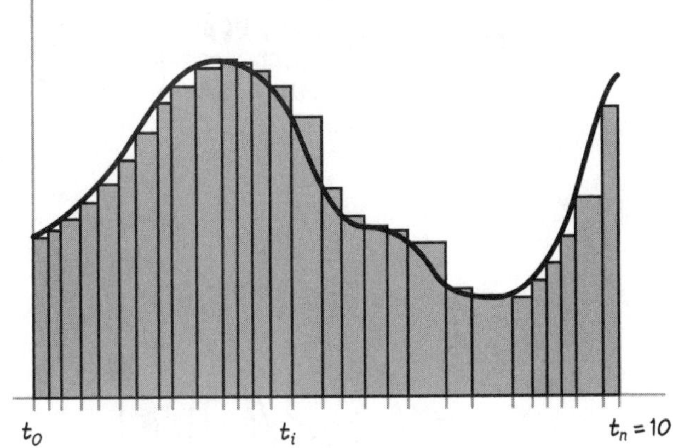

t_0
t_i
$t_n = 10$

IF THE VELOCIMETER IS READ MORE OFTEN, SO THAT THE WIDEST Δt_i GETS SMALLER, THE SUM GIVES A BETTER APPROXIMATION OF THE ACTUAL CHANGE OF POSITION, AND THE RECTANGLES SQUEEZE IN TOWARD THE GRAPH.

NEXT TIME TAKE A READING EVERY MILLISECOND!
YOU DON'T ASK MUCH, DO YOU?

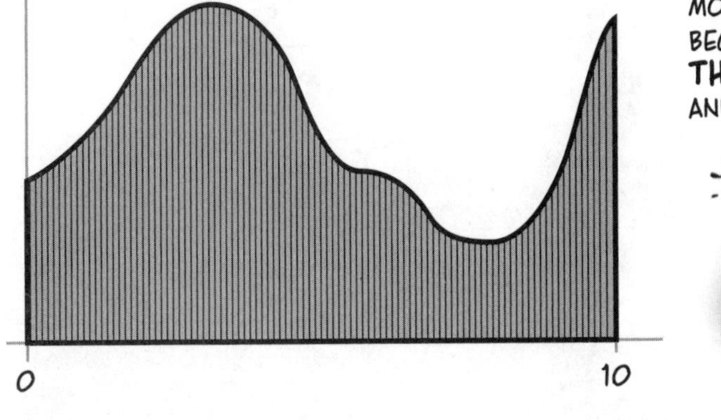

0
10

AS $\Delta t \rightarrow 0$, THE APPROXIMATION BECOMES MORE PERFECT, AND THE RECTANGLES BEGIN TO LOOK LIKE THE **AREA UNDER THE CURVE** $y = v(t)$ BETWEEN $t = 0$ AND $t = 10$.

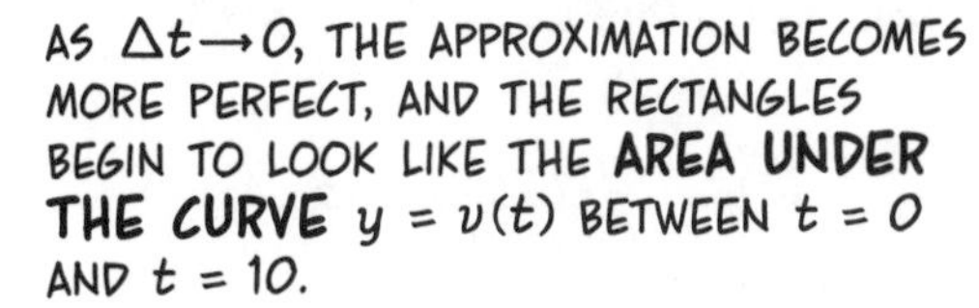

$s(10) - s(0)$ IS THE AREA UNDER THE GRAPH OF v!

*ASSUMING, FOR THE TIME BEING, THAT THE VELOCITY IS NON-NEGATIVE

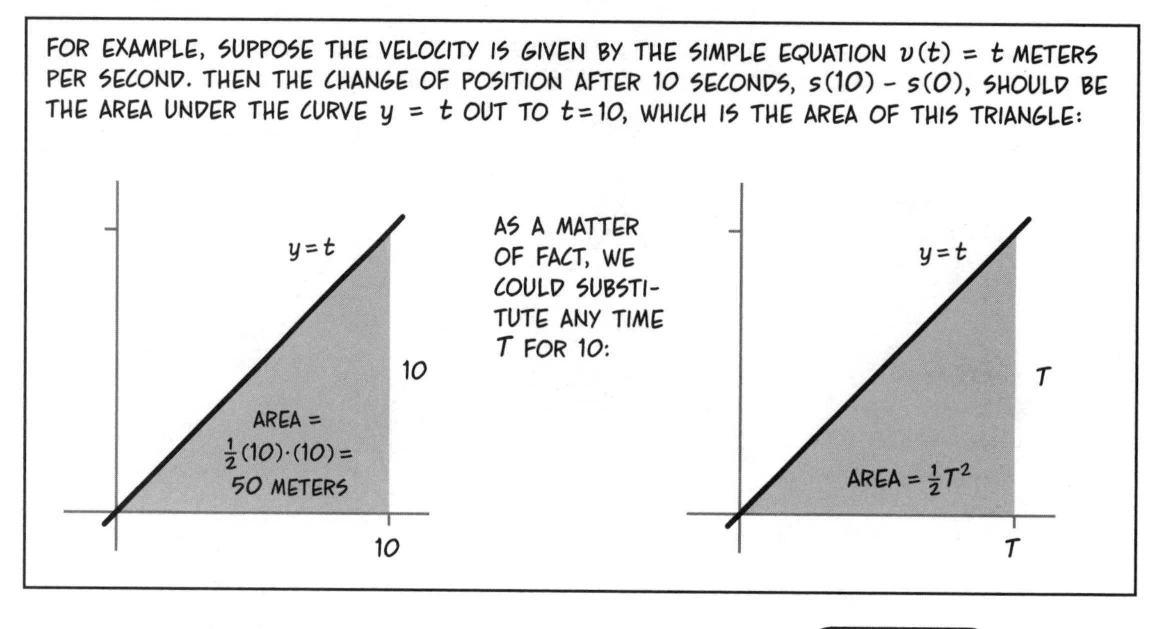
FOR EXAMPLE, SUPPOSE THE VELOCITY IS GIVEN BY THE SIMPLE EQUATION $v(t) = t$ METERS PER SECOND. THEN THE CHANGE OF POSITION AFTER 10 SECONDS, $s(10) - s(0)$, SHOULD BE THE AREA UNDER THE CURVE $y = t$ OUT TO $t = 10$, WHICH IS THE AREA OF THIS TRIANGLE:
$y = t$
10
AREA = $\frac{1}{2}(10)\cdot(10)$ = 50 METERS
10
AS A MATTER OF FACT, WE COULD SUBSTITUTE ANY TIME T FOR 10:
$y = t$
T
AREA = $\frac{1}{2}T^2$
T

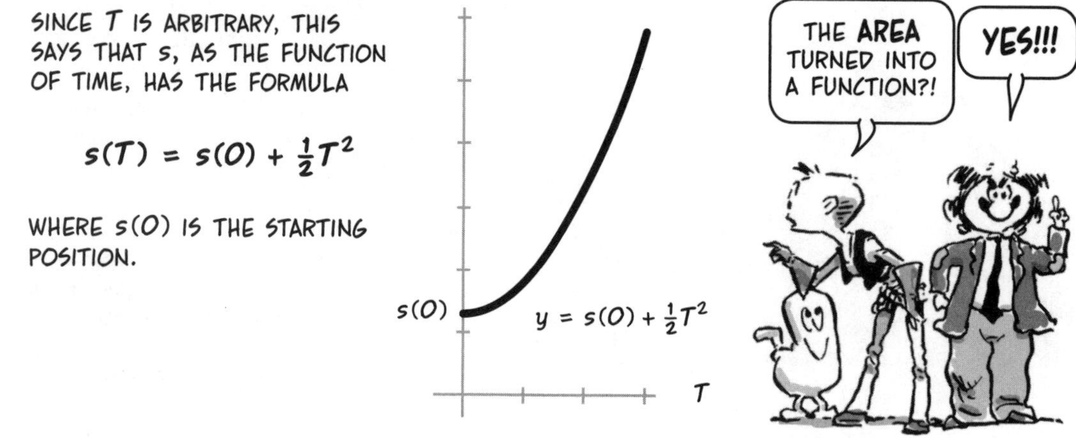
SINCE T IS ARBITRARY, THIS SAYS THAT s, AS THE FUNCTION OF TIME, HAS THE FORMULA
$s(T) = s(0) + \frac{1}{2}T^2$
WHERE $s(0)$ IS THE STARTING POSITION.
$s(0)$
$y = s(0) + \frac{1}{2}T^2$
T
THE AREA TURNED INTO A FUNCTION?!
YES!!!

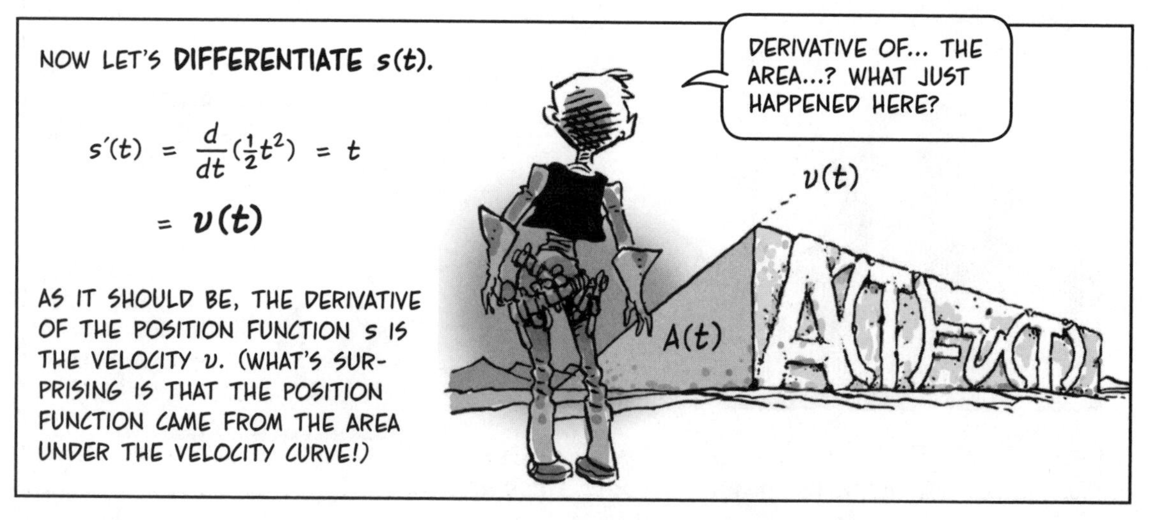
NOW LET'S DIFFERENTIATE $s(t)$.
$s'(t) = \frac{d}{dt}(\frac{1}{2}t^2) = t$
$= v(t)$
AS IT SHOULD BE, THE DERIVATIVE OF THE POSITION FUNCTION s IS THE VELOCITY v. (WHAT'S SURPRISING IS THAT THE POSITION FUNCTION CAME FROM THE AREA UNDER THE VELOCITY CURVE!)
DERIVATIVE OF... THE AREA...? WHAT JUST HAPPENED HERE?
$v(t)$
$A(t)$

BY FINDING POSITION FROM VELOCITY, WE ARE **DIFFERENTIATING IN REVERSE.** GIVEN A FUNCTION v, WE SOUGHT A FUNCTION s WHOSE DERIVATIVE IS v.

UP TO THIS POINT, WE'VE ALWAYS GONE FROM A FUNCTION f TO ITS DERIVATIVE f'. NOW WE WANT TO GO THE OTHER WAY, FROM f TO SOME FUNCTION F, WHERE $F' = f$.

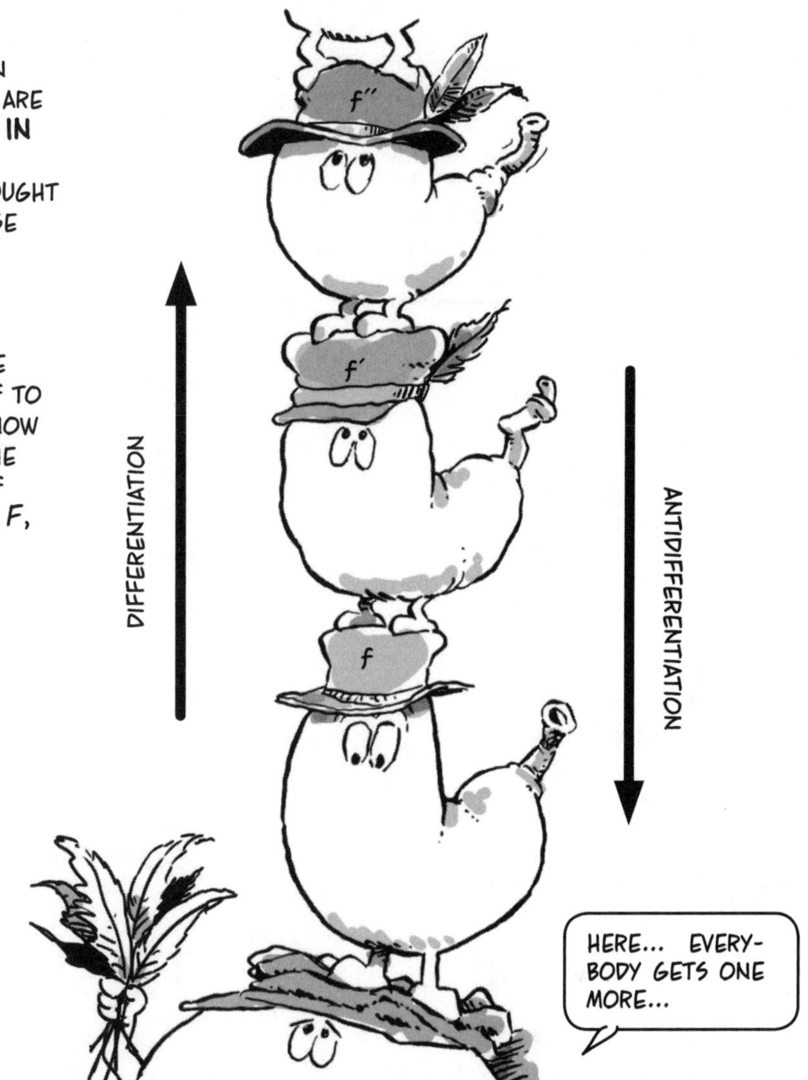
DIFFERENTIATION
ANTIDIFFERENTIATION
f''
f'
f
HERE... EVERY-BODY GETS ONE MORE...

THIS FUNCTION f IS CALLED AN **ANTIDERIVATIVE** OF f. FOR EXAMPLE, POSITION s IS AN ANTIDERIVATIVE OF VELOCITY v.

f
F
FUNNY TO THINK YOU WERE THERE ALL ALONG...

IF OUR VELOCITY EXAMPLE IS ANY GUIDE (AND IT IS!), THIS REVERSAL INVOLVES A PROCESS OF SUMMING UP... AND THAT, IN TURN, UNLOCKS THE PROBLEM OF FINDING AREAS.

Problems

SUPPOSE A CAR'S VELOCITY AT TIME t IS $v(t) = 3t^2$ METERS PER SECOND. MAKE AN ESTIMATE OF THE DISTANCE TRAVELED BETWEEN $t = 0$ AND $t = 4$ SECONDS BY ADDING UP RECTANGLES: BEGIN BY DIVIDING THE INTERVAL $[0, 4]$ INTO FOUR EQUAL SEGMENTS. LET $t_i = i$ FOR $i = 0, 1, 2, 3, 4$. EACH SEGMENT HAS LENGTH $\Delta t_i = 1$.

1. GET A LOW ESTIMATE BY ADDING THE RECTANGLES **UNDER** THE CURVE. FIND:

$$E_{LOW} = \sum_{i=0}^{3} f(t_i)\Delta t_i = \sum_{i=0}^{3} 3i^2$$

2. GET A HIGH ESTIMATE BY ADDING THE RECTANGLES **ABOVE** THE CURVE. FIND:

$$E_{HIGH} = \sum_{i=1}^{4} f(t_i)\Delta t_i = \sum_{i=1}^{4} 3i^2$$

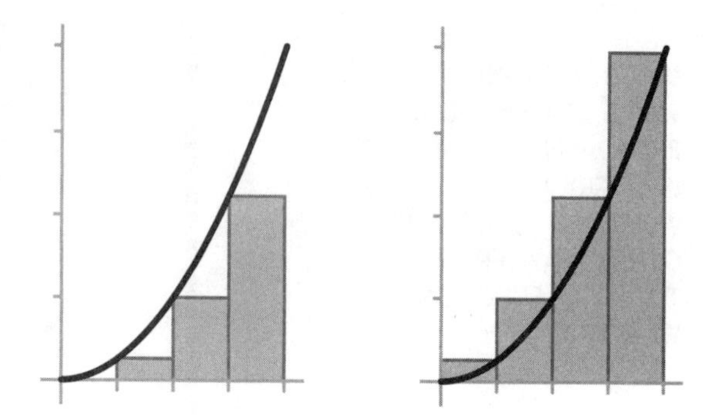

3. WHAT DO YOU GET WHEN YOU SPLIT THE DIFFERENCE? FIND:

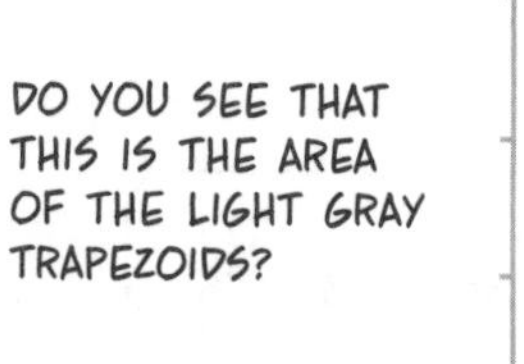

$$\tfrac{1}{2}(E_{HIGH} + E_{LOW})$$

DO YOU SEE THAT THIS IS THE AREA OF THE LIGHT GRAY TRAPEZOIDS?

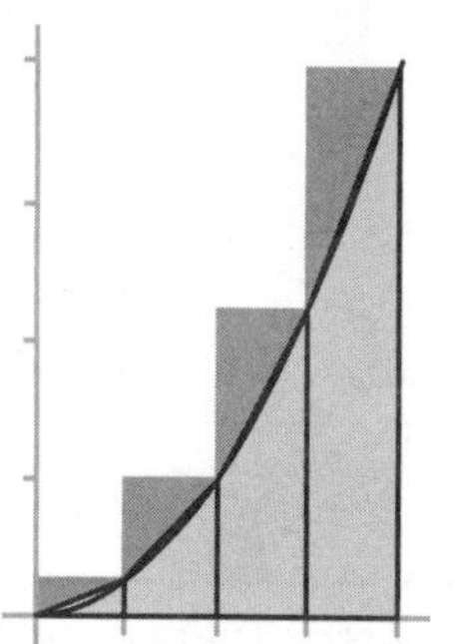

4. TRY ONE MORE ESTIMATE: TAKE t_i TO BE THE **MIDPOINT** OF THE SEGMENT $[i, i+1]$, THAT IS, $t_i = (2i + 1)/2$. FIND

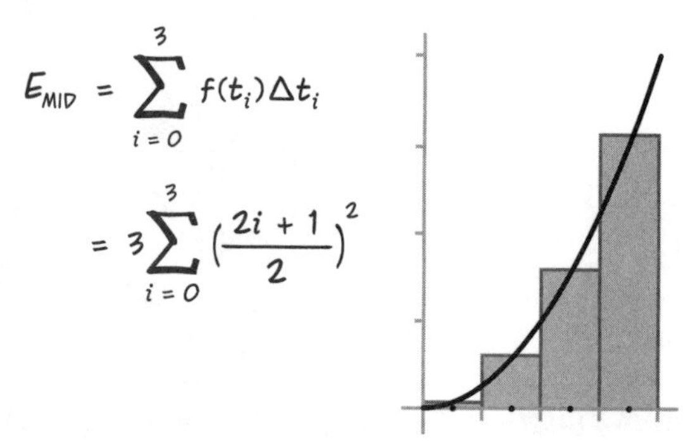

$$E_{MID} = \sum_{i=0}^{3} f(t_i)\Delta t_i$$

$$= 3\sum_{i=0}^{3} \left(\frac{2i+1}{2}\right)^2$$

5. CAN YOU THINK OF A FUNCTION $s(t)$ WITH $s'(t) = 3t^2$? WHAT IS $s(4) - s(0)$? IS IT CLOSE TO ANY OF YOUR ESTIMATES? WHICH ESTIMATE IS CLOSEST TO $s(4) - s(0)$?

6. DO THE SAME THING AS IN PROBLEMS 1–5 WITH THE FUNCTION $v(t) = 1/t$ BETWEEN THE POINTS $t = 1$ AND $t = e^2$. USE RECTANGLES WITH THEIR BOTTOM CORNERS AT THE POINTS 1, 2, ..., 7, e^2. (SO YOU'LL HAVE SIX RECTANGLES OF BASE $\Delta t_i = 1$ AND ONE THINNER RECTANGLE OF BASE $\Delta t_7 = e^2 - 7 \approx 0.39$.)

7. MAKE AN ESTIMATE OF THE AREA UNDER BOTH GRAPHS BY USING TWICE AS MANY RECTANGLES HALF AS WIDE.

1 2 3 4 5 6 7 e^2

Chapter 9
Antiderivatives

PLUS A CONSTANT!

UNFORTUNATELY, THE PROCESS OF FINDING ANTIDERIVATIVES IS SLIGHTLY **MESSIER** THAN THE REVERSE PROCESS OF DIFFERENTIATION.
POSSIBLY THE BIGGEST UNDERSTATEMENT OF THE LAST 400 YEARS...
FOR EXAMPLE, IF $f(x) = x^3$, THEN $F(x) = \frac{1}{4}x^4$ IS AN ANTIDERIVATIVE:
$F'(x) = \frac{1}{4}(4x^3) = x^3$
IN GENERAL, $g(x) = x^n$ HAS AN ANTIDERIVATIVE
$G(x) = \frac{1}{n+1}x^{n+1}$
x^n
$\frac{x^n}{n+1}$
THAT'S **AN** ANTIDERIVATIVE OF g, NOT **THE** ANTIDERIVATIVE, BECAUSE THERE ARE PLENTY OF OTHERS. ALL THESE HAVE DERIVATIVE x^n:
$G(x) = \frac{x^{n+1}}{n+1} + 3$
$H(x) = \frac{x^{n+1}}{n+1} + 7$
$P(x) = \frac{x^{n+1}}{n+1} + C$
BECAUSE THE DERIVATIVE OF A CONSTANT IS ZERO.
WHERE C IS ANY CONSTANT.

IF F IS AN ANTIDERIVATIVE OF A FUNCTION f, THEN SO IS $F + C$, FOR ANY CONSTANT C. $(F + C)' = F' = f$. SLIDING THE GRAPH $y = F(x)$ STRAIGHT UP AND DOWN DOESN'T AFFECT THE SLOPE AT ANY POINT x.

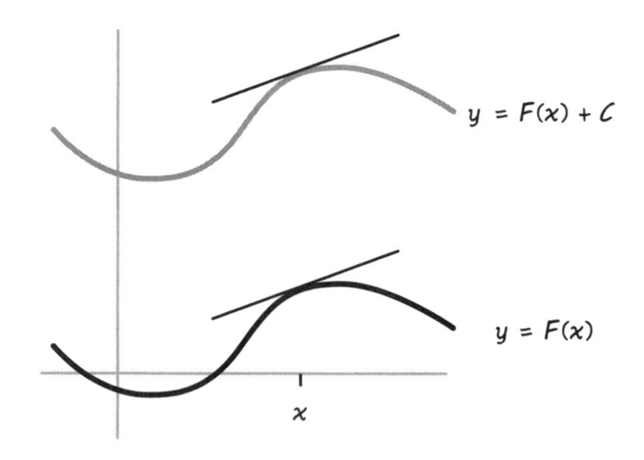
y = F(x) + C
y = F(x)
x

CONVERSELY, IF $F' = f$, THEN **ANY ANTIDERIVATIVE** OF f DIFFERS FROM F BY A CONSTANT. **PROOF:** IF G IS ANY OTHER ANTIDERIVATIVE, THEN $(F - G)'(x) = f(x) - f(x) = 0$ FOR ALL x. BUT BY CONSEQUENCE (3) OF THE MEAN VALUE THEOREM (P. 158), THE ONLY FUNCTIONS WITH ZERO DERIVATIVE ARE CONSTANTS, SO $F - G = C$ FOR SOME CONSTANT C.

THAT WAS ONE MEAN VALUE THEOREM!
ALL POSSIBLE FUNCTIONS WITH ZERO DERIVATIVE

HERE IS HOW TO WRITE THE FORMULA THAT SAYS "THE ANTIDERIVATIVE OF f IS $F + C$":

$$\int f = F + C \quad \text{OR} \quad \int f(x)\,dx = F(x) + C$$

THE TALL SYMBOL IS CALLED AN **INTEGRAL SIGN...** THE FUNCTION f IS CALLED THE **INTEGRAND.** THE SYMBOL dx IS THERE ONLY TO IDENTIFY THE VARIABLE, AS IT IS IN df/dx, AND IS NOT A SEPARATE FACTOR IN THE EQUATION. AND AS USUAL, THE NAME OF THE VARIABLE DOESN'T MATTER: ALL THESE EXPRESSIONS MEAN THE SAME THING, NAMELY THE ANTIDERIVATIVE OF f:

$$\int f(x)\,dx, \quad \int f(t)\,dt, \quad \text{AND} \quad \int f(y)\,dy$$

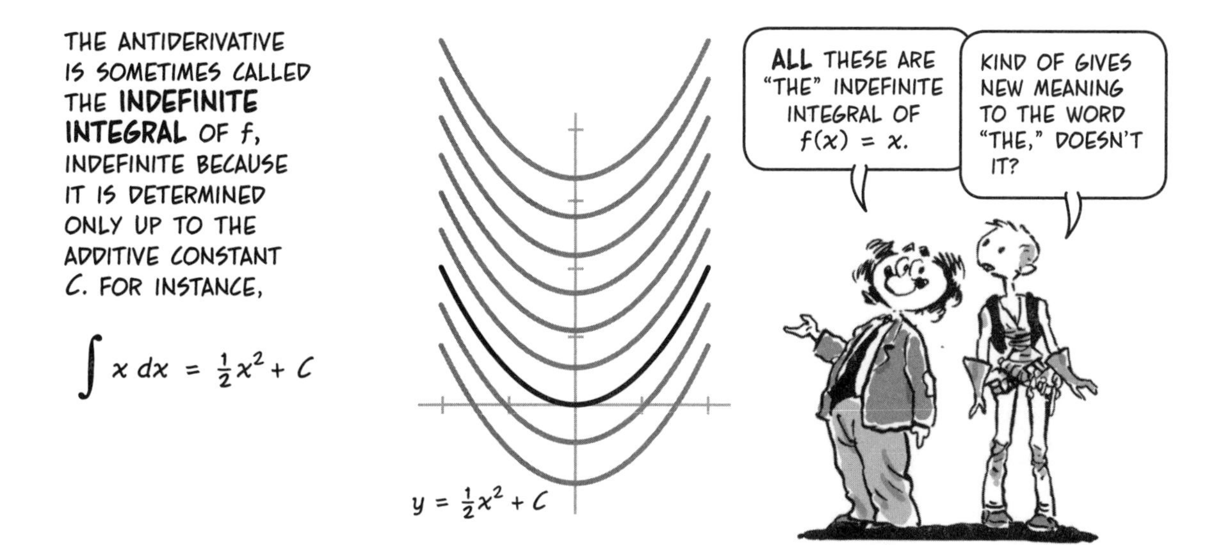
THE ANTIDERIVATIVE IS SOMETIMES CALLED THE **INDEFINITE INTEGRAL** OF f, INDEFINITE BECAUSE IT IS DETERMINED ONLY UP TO THE ADDITIVE CONSTANT C. FOR INSTANCE,
$\int x\,dx = \frac{1}{2}x^2 + C$
$y = \frac{1}{2}x^2 + C$
ALL THESE ARE "THE" INDEFINITE INTEGRAL OF $f(x) = x$.
KIND OF GIVES NEW MEANING TO THE WORD "THE," DOESN'T IT?

HAVING FOUND MANY DERIVATIVES ALREADY, WE ALREADY KNOW THESE FORMULAS:

$$\int dx = x + C$$

(THERE'S AN UNWRITTEN NUMBER 1 AFTER THE INTEGRAL SIGN.)

$$\int x^p\,dx = \frac{1}{p+1}x^{p+1} + C$$

$$\int e^x\,dx = e^x + C$$

$$\int \sin x\,dx = -\cos x + C$$

$$\int \cos x\,dx = \sin x + C$$

$$\int \frac{dx}{1+x^2} = \arctan x + C$$

$$\int \frac{dx}{\sqrt{1-x^2}} = \arcsin x + C$$

$$\int \frac{1}{x}\,dx = \ln|x| + C$$

NOTE: THE ABSOLUTE VALUE SIGN IN THE LAST EQUATION IS JUSTIFIED, BECAUSE IF $x < 0$, THEN

$$\frac{d}{dx}\ln(-x) = \frac{-1}{(-x)} = \frac{1}{x}$$

IF $x > 0$, THEN $\frac{d}{dx}(\ln x) = \frac{1}{x}$ ALSO.

TOGETHER THESE IMPLY $\frac{d}{dx}\ln|x| = \frac{1}{x}$, $x \neq 0$

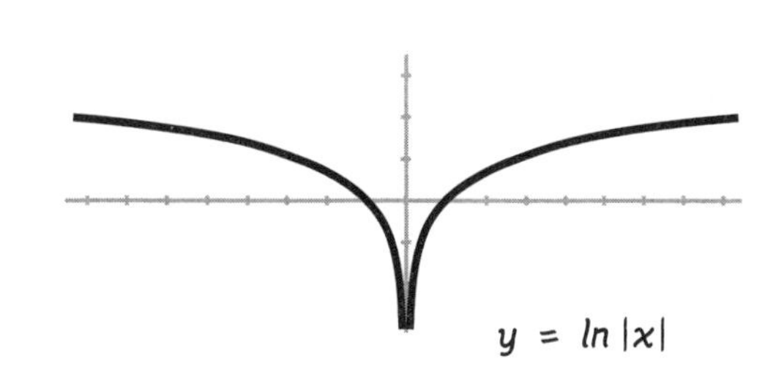
$y = \ln|x|$

AND $\int \ln x \, dx$ IS... UM... AHHH... AHEM! DOES THIS LOOK FAMILIAR?

NO... I THINK I'D REMEMBER IF WE'D SEEN **THIS** BEFORE...

UNFORTUNATELY, TO INTEGRATE A FUNCTION, WE HAVE TO RECOGNIZE IT AS THE DERIVATIVE OF SOMETHING ELSE WE'VE ALREADY SEEN. AND SO FAR, NOTHING HAS TURNED UP WITH $\ln x$ AS ITS DERIVATIVE.

MUST BE HERE SOMEWHERE...

UNLIKE DIFFERENTIATION, WHICH WE DO BY APPLYING SIMPLE RULES, INTEGRATION REQUIRES SOME EXPERIENCE. THE MORE DERIVATIVES YOU'VE SEEN, THE BETTER ABLE YOU'LL BE TO FIND ANTIDERIVATIVES...

WHO WAS THAT?
A CALCULUS STUDENT WHO DIDN'T DO THE PROBLEM SETS.

IF THE FUNCTION UNDER THE INTEGRAL SIGN (KNOWN AS THE **INTEGRAND)** IS "SOMETHING LIKE" A KNOWN DERIVATIVE, WE CAN OFTEN FIND ITS ANTIDERIVATIVE SIMPLY BY GUESSING AND THEN MAKING SOME LITTLE ADJUSTMENT.

Example 1: $\int e^{2x}\,dx$

WE KNOW THAT $f(x) = e^{2x}$ IS SOMETHING LIKE THE DERIVATIVE OF $G(x) = e^{2x}$. IN FACT, $G'(x) = 2e^{2x}$, WHICH IS OFF BY A FACTOR OF TWO. NOW WE TRY $F(x) = \frac{1}{2}e^{2x}$ AND FIND

$$F'(x) = \tfrac{1}{2}(2)e^{2x} = e^{2x} = f(x)$$

F IS AN ANTIDERIVATIVE, AND WE CONCLUDE THAT

$$\int e^{2x}\,dx = \tfrac{1}{2}e^{2x} + C$$

WE FOLLOWED THESE STEPS:

1. SEE IF THE INTEGRAND f LOOKS LIKE A CONSTANT MULTIPLE OF A KNOWN DERIVATIVE.
2. GUESS A LIKELY ANTIDERIVATIVE G.
3. DIFFERENTIATE G.
4. IF G' IS A CONSTANT MULTIPLE OF f, MULTIPLY G BY AN APPROPRIATE FACTOR TO MAKE A BETTER GUESS, F.
5. CHECK THAT $F' = f$.
6. REPEAT IF NECESSARY.

THIS PROCEDURE HAS A NAME: THE

Guess-and-Check Method.

IT SOUNDS BETTER IN THE ORIGINAL LATIN.

Example 2: $\int \frac{1}{4+x^2}\,dx$

1. NOTE THAT THE INTEGRAND f LOOKS SOMETHING LIKE

$$\frac{1}{1+x^2}$$

WHICH IS THE DERIVATIVE OF THE ARCTANGENT. LET'S WRITE IT AS

$$f(x) = \frac{1}{4}\,\frac{1}{\left(1+\left(\frac{x}{2}\right)^2\right)}$$

2. SO WE GUESS $G(x) = \arctan \frac{x}{2}$

3. DIFFERENTIATING GIVES

$$G'(x) = \frac{1}{2}\,\frac{1}{\left(1+\left(\frac{x}{2}\right)^2\right)} = 2f(x)$$

TOO HIGH BY A FACTOR OF 2.

4. TAKE $F(x) = \frac{1}{2}\arctan\left(\frac{x}{2}\right)$.

5. THE LAST STEP, CHECKING THAT $F'(x) = f(x)$, IS LEFT TO YOU, LUCKY READER! AND WE CONCLUDE THAT

$$\int \frac{1}{4+x^2}\,dx = \frac{1}{2}\arctan\left(\frac{x}{2}\right) + C$$

THE ONLY STEP REQUIRING THOUGHT WAS #1... THE REST WAS CRANKING...

SOMETIMES THE CHAIN RULE HELPS US IDENTIFY A FUNCTION AS A DERIVATIVE. THE CHAIN RULE SAYS:

$$\frac{d}{dx}(u(v(x))) = v'(x)u'(v(x))$$

IF AN INTEGRAND LOOKS LIKE THE RIGHT-HAND SIDE—I.E., IT CONTAINS AN INSIDE FUNCTION WHOSE DERIVATIVE APPEARS AS A FACTOR—THIS IDENTIFIES THE INTEGRAND AS A DERIVATIVE, AND WE CAN "UNCHAIN" THE FUNCTION TO GET THE ANTIDERIVATIVE $F(x) = u(v(x))$.

PLUS A CONSTANT!
u
v

Example 3: $\int 2xe^{x^2}\,dx$

1. IN THE INTEGRAND, THE FACTOR $2x$ IS THE DERIVATIVE OF THE EXPONENTIAL'S INSIDE FUNCTION x^2, SO WE MIGHT TRY:

2. $F(x) = e^{x^2}$.

3. TEST:

$$F'(x) = 2xe^{x^2} = f(x)$$

WE'RE IN LUCK: WE GOT IT RIGHT THE FIRST TIME! SO WE CAN WRITE:

$$\int 2xe^{x^2}\,dx = e^{x^2} + C$$

Example 4: $\int \frac{x}{\sqrt{1+x^2}}\,dx$

AGAIN, WE'RE LOOKING FOR AN INSIDE FUNCTION AND ITS DERIVATIVE AS A FACTOR!

1. THE x IN THE NUMERATOR IS, WITHIN A CONSTANT FACTOR, THE DERIVATIVE OF THE INSIDE FUNCTION $1 + x^2$.

2. WE GUESS $G(x) = (1 + x^2)^{\frac{1}{2}}$.

3. $G'(x) = (2x)\frac{1}{2}(1 + x^2)^{-\frac{1}{2}} = x(1 + x^2)^{-\frac{1}{2}}$

= THE INTEGRAND.

NO CORRECTION IS NECESSARY, SO WE SKIP STEPS 4 AND 5, AND CAN IMMEDIATELY WRITE:

$$\int \frac{x}{\sqrt{1+x^2}}\,dx = \sqrt{1+x^2} + C$$

CALL IT THE "UNCHAIN" RULE!

Example 5a: $\int \sin^n\theta \cos\, d\theta$

1. REMEMBER, IF f IS ANY FUNCTION, THEN f^n HAS DERIVATIVE $nf^{n-1}f'$. IN THE INTEGRAND, WE SEE A POWER OF THE SINE TIMES ITS DERIVATIVE, THE COSINE. IS THIS $\frac{d}{d\theta}(\sin^{n+1}\theta)$?

AND IT GETS EVEN MESSIER...

2. TRY $G(\theta) = \sin^{n+1}\theta$

3. TEST. $G'(\theta) = (n + 1)\sin^n\theta\cos\theta$. THIS IS OFF BY A FACTOR OF $n + 1$.

4. $F(\theta) = \frac{\sin^{n+1}\theta}{n+1}$, THEN, HAS THE RIGHT DERIVATIVE (**5.** YOU CHECK!), AND

$$\int \sin^n\theta\cos\, d\theta = \frac{\sin^{n+1}\theta}{n+1} + C$$

MORE OF THESE INTEGRATION TRICKS, ER, **TECHNIQUES,** WILL APPEAR IN MORE DETAIL IN A LATER CHAPTER... BUT FIRST—

Problems

FIND THE ANTIDERIVATIVES. DON'T FORGET THE ADDED CONSTANT!

1. $\int 6\, dx$

2. $\int \frac{2}{3}x^4\, dx$

3. $\int (x - 2)^{50}\, dx$

4. $\int (1 - x)^{-2}\, dx$

5. $\int (a - x)^n\, dx$

6. $\int \frac{2x}{9 + x^2}\, dx$

7. $\int \frac{1}{\sqrt{4 - x^2}}\, dx$

8. $\int \sin 2x\, dx$

9. $\int 2 \sin x \cos x\, dx$

10. RECALL FROM TRIG THAT $\sin 2x = 2 \sin x \cos x$. CONCLUDE THAT

$$\cos 2x = -2\sin^2 x + C$$

FOR SOME CONSTANT C.

11. WHAT IS C IN PROBLEM 10?

12. $\int \frac{3}{2}x^2 e^{(x^3 + 1)}\, dx$

13. $\int \sin x\, e^{\cos x}\, dx$

14. $\int \frac{x^2 - 4x}{\sqrt{x^3 - 6x^2}}\, dx$

15. $\int \frac{1}{x + 1}\, dx$

16. $\int \frac{1}{x^2 - 1}\, dx$

(HINT: DECOMPOSE THE INTEGRAND INTO PARTIAL FRACTIONS, AS SHOWN ON PP. 27-28.)

17. SHOW THAT IF F IS AN ANTIDERIVATIVE OF f, AND G IS AN ANTIDERIVATIVE OF g, AND C AND D ARE ANY TWO CONSTANTS, THEN $CF + DG$ IS AN ANTIDERIVATIVE OF $Cf + Dg$. (HINT: DIFFERENTIATE $CF + DG$.)

FIND THE ANTIDERIVATIVES:

18. $\int 2x^3 + 15x^2 - \frac{1}{2}x - 7\, dx$

19. $\int \sin^2\theta \cos\theta + \cos^2\theta \sin\theta\, d\theta$

20. $\int \frac{e^x + e^{-x}}{2}\, dx$

21. $\int \frac{3t^2}{t^3 - t^2 + 1}\, dt - \int \frac{2t}{t^3 - t^2 + 1}\, dt$

22. $\int t^{3/2} + t^{5/2} - 4t^{-7/2}\, dt$

23. $\int |x|\, dx$

(HINT: CONSIDER POSITIVE AND NEGATIVE VALUES OF x SEPARATELY.) DRAW THE GRAPH OF THE ANTIDERIVATIVE.

24. IF $F'(x) = f(x)$, WHAT IS

$$\int f(x - a)\, dx\ ?$$

25. IF f IS A DIFFERENTIABLE FUNCTION, WHAT IS

$$\int \frac{f'(x)}{f(x)}\, dx\ ?$$

Chapter 10
The Definite Integral

AREAS, OVER AND UNDER

WHAT DO WE MEAN BY THE AREA INSIDE A FIGURE? IF THE REGION IS RECTANGULAR OR TRIANGULAR, OR A BUNCH OF RECTANGLES AND TRIANGLES PASTED TOGETHER, WE HAVE A PRETTY GOOD IDEA: JUST ADD THE AREA OF THE TRIANGLES OR RECTANGLES.

NOW IF ONLY I COULD REMEMBER HOW TO FIND THE AREA OF A TRIANGLE...

BUT WHAT IF THE FIGURE HAS A **CURVED BOUNDARY?** WHAT IS THE AREA THEN?

EH-EH-EH! NO STRAIGHTENING ALLOWED!

IF YOU'VE GOTTEN THIS FAR, YOU PROBABLY SUSPECT THAT THE ANSWER MAY HAVE SOMETHING TO DO WITH A **LIMITING PROCESS...**

FOR SIMPLICITY'S SAKE, WE CONSIDER A SPECIAL TYPE OF REGION, BOUNDED ON THREE SIDES BY STRAIGHT LINES: LEFT AND RIGHT BY THE VERTICAL LINES $x = a$, $x = b$, BELOW BY THE x-AXIS, AND ABOVE BY THE GRAPH OF SOME FUNCTION $y = f(x)$, WHICH WE ASSUME, FOR THE TIME BEING, TO BE NON-NEGATIVE. THIS REGION HAS ONLY ONE CURVY SIDE.

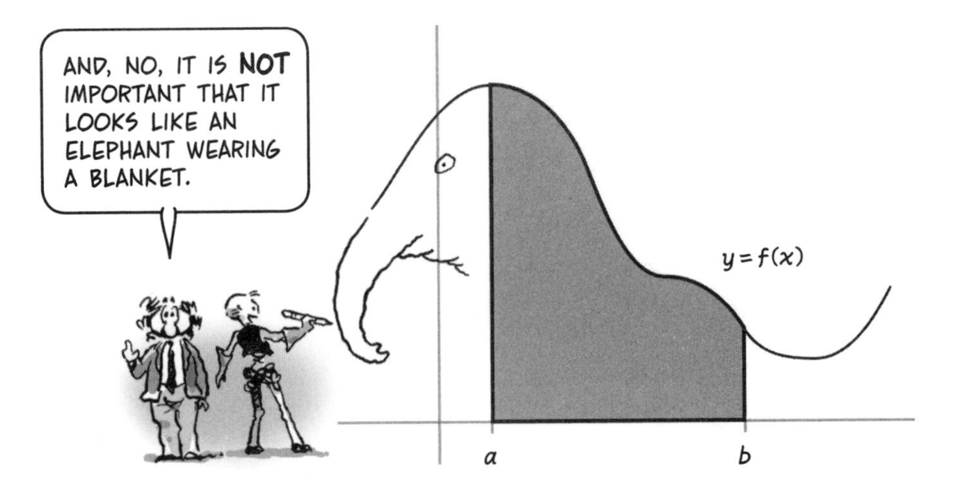
AND, NO, IT IS **NOT** IMPORTANT THAT IT LOOKS LIKE AN ELEPHANT WEARING A BLANKET.
y = f(x)
a
b

WE PROCEED MORE OR LESS AS WE DID ON PAGE 163: SUBDIVIDE THE INTERVAL $[a, b]$ INTO n SUBINTERVALS BY A SPRINKLING OF POINTS $x_0, x_1, x_2, \ldots x_i, \ldots x_n$, WHERE $x_0 = a$ AND $x_n = b$. FOR EACH $i \geq 1$, PICK ANY POINT x_i^* IN THE iTH INTERVAL $[x_{i-1}, x_i]$, AND RAISE A RECTANGLE ON THAT INTERVAL OF HEIGHT $f(x_i^*)$, ITS BASE BEING $\Delta x_i = x_i - x_{i-1}$. FINALLY, SUM THE AREAS OF THE RECTANGLES TO GET AN APPROXIMATE VALUE FOR THE DESIRED AREA.

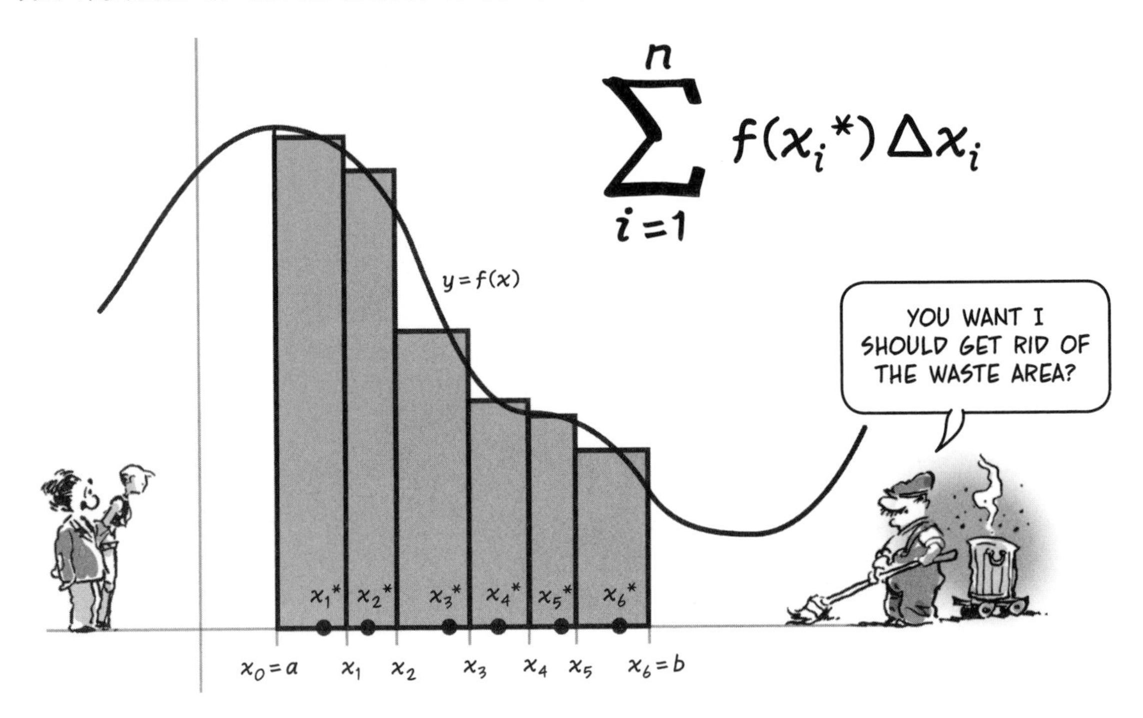
$\sum_{i=1}^{n} f(x_i^*) \Delta x_i$
y = f(x)
YOU WANT I SHOULD GET RID OF THE WASTE AREA?
x_1^* x_2^* x_3^* x_4^* x_5^* x_6^*
$x_0 = a$ x_1 x_2 x_3 x_4 x_5 $x_6 = b$

THIS EXPRESSION IS CALLED A **RIEMANN SUM,** AFTER BERNHARD RIEMANN, A 19TH-CENTURY MATHEMATICIAN SO ORIGINAL AND BRILLIANT THAT HE WON PRAISE EVEN FROM THE GREAT GAUSS, WHO PRAISED NO ONE.

UM... IF GAUSS PRAISES NO ONE, AND GAUSS PRAISES ME... THEN... I'M NO ONE?
BEST STICK TO CALCULUS, BERNIE, AND LEAVE LOGIC ALONE...
Riemann

THE PLAN, THEN, IS TO LET THE SUBDIVISION GET FINER AND FINER, MEANING THAT THE LARGEST $\Delta x_i \to 0$, AND SEE IF THE SUM OF THE RECTANGULAR AREAS APPROACHES A LIMIT.

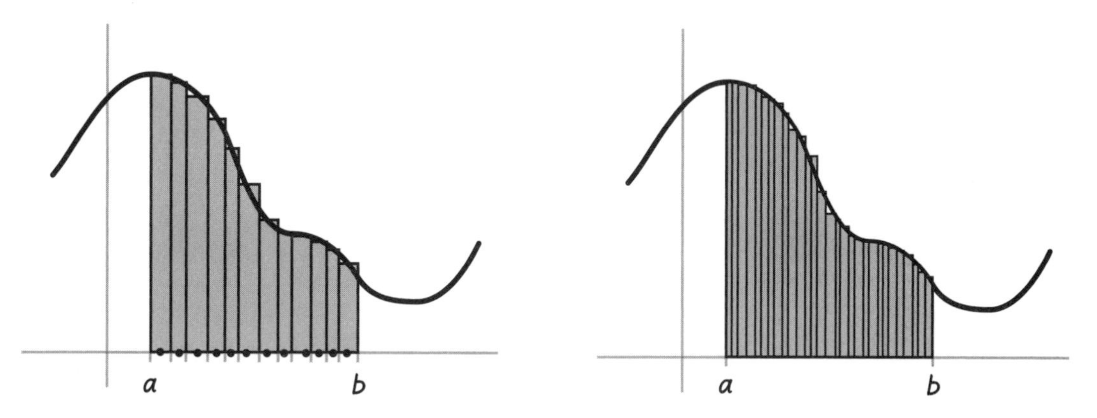
a
b
a
b

THE ANSWER (WHY WAIT?) IS **YES,** PROVIDED THE FUNCTION f IS CONTINUOUS ON THE INTERVAL $[a, b]$ (SEE P. 156). IN THAT CASE, THE LIMITING VALUE IS CALLED THE **DEFINITE INTEGRAL,** INTERPRETED AS THE AREA UNDER THE CURVE AND WRITTEN LIKE THIS:

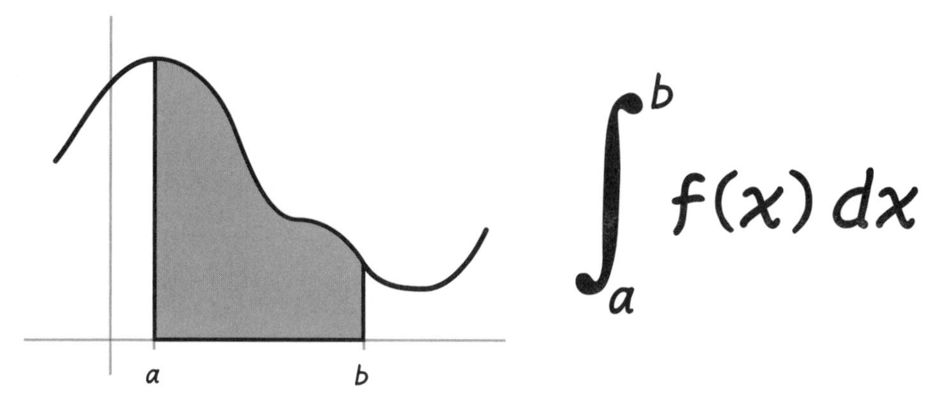
a
b

$$\int_a^b f(x)\,dx$$

WARNING! THEORY ALERT! THESE TWO PAGES SKETCH THE PROOF THAT RIEMANN SUMS CONVERGE ON A UNIQUE NUMBER, THE DEFINITE INTEGRAL, WHEN f IS CONTINUOUS. PURELY PRACTICAL-MINDED READERS CAN SKIP AHEAD TO PAGE 182 AND STILL LEAD HEALTHY, PRODUCTIVE LIVES...

IF YOU **PROMISE** ME I'LL STILL GET INTO MEDICAL SCHOOL...

SKETCH OF PROOF: ASSUME f IS CONTINUOUS ON $[a, b]$. LET $\{a = x_0, x_1, \ldots x_n = b\}$ BE A SUBDIVISION OF THE INTERVAL. ON EACH SUBINTERVAL $[x_{i-1}, x_i]$, BY THE EXTREME VALUE THEOREM, f ATTAINS A MAXIMUM M_i AND A MINIMUM m_i.

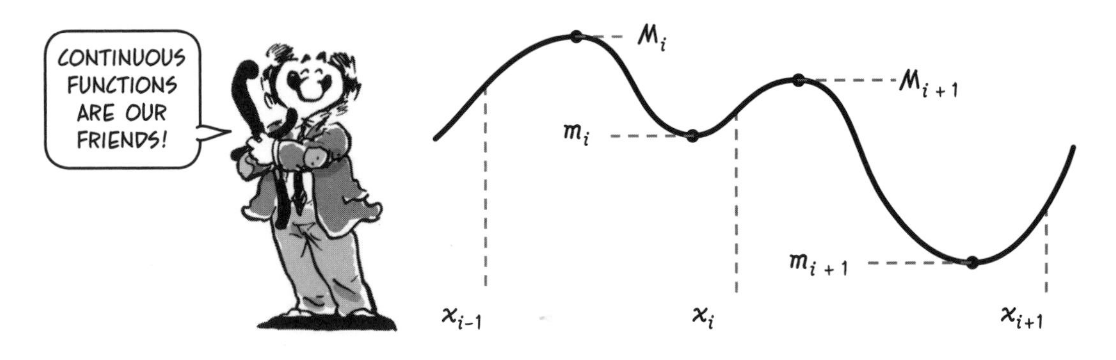
CONTINUOUS FUNCTIONS ARE OUR FRIENDS!
M_i
M_{i+1}
m_i
m_{i+1}
x_{i-1}
x_i
x_{i+1}

NOW WE MAKE SPECIAL RIEMANN SUMS THAT COME AT THE GRAPH FROM ABOVE AND BELOW.

THE **LOWER SUM** OF THIS SUBDIVISION IS DEFINED BY

$$s = \sum_{i=1}^{n} m_i \Delta x_i$$

THE **UPPER SUM** IS DEFINED BY

$$S = \sum_{i=1}^{n} M_i \Delta x_i$$

CLEARLY, $S > s$... AND IT'S NOT TOO HARD TO SHOW THAT **EVERY** UPPER SUM IS GREATER THAN **EVERY** LOWER SUM, REGARDLESS OF WHAT SUBDIVISION THEY'RE BASED ON...

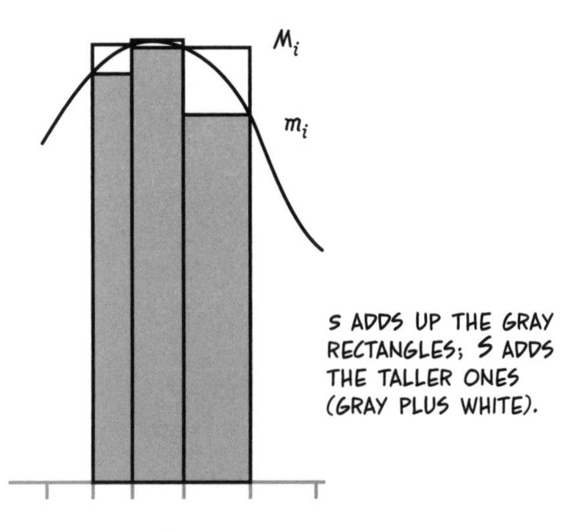
M_i
m_i

s ADDS UP THE GRAY RECTANGLES; S ADDS THE TALLER ONES (GRAY PLUS WHITE).

NEXT WE INVOKE THIS FACT, GIVEN WITHOUT PROOF: GIVEN ANY $\varepsilon > 0$, IT IS POSSIBLE TO SUBDIVIDE $[a, b]$ SO THAT $|f(c) - f(d)| < \varepsilon$ **WHENEVER c AND d ARE IN THE SAME SUBINTERVAL.** FOR THAT SUBDIVISION, THEN, $M_i - m_i < \varepsilon$ FOR **EVERY** i.

WARNING! BOTTOMLESS PIT AHEAD!
THIS DEPENDS ON THOSE DEEP PROPERTIES OF NUMBERS, OFTEN MENTIONED BUT NEVER DESCRIBED...

THIS IMPLIES THAT THE UPPER AND LOWER SUMS SQUEEZE TOGETHER AS THE SUBDIVISIONS GET FINER. FOR, GIVEN ANY SMALL $\varepsilon > 0$, WE CAN MAKE A SUBDIVISION SO FINE THAT

$$M_i - m_i < \frac{\varepsilon}{b-a} \quad \text{FOR EVERY } i.$$

IN THAT CASE,

$$S - s = \sum_{i=1}^{n} (M_i - m_i)\Delta x_i$$

$$< \sum_{i=1}^{n} \frac{\varepsilon}{b-a}\Delta x_i = \frac{\varepsilon}{b-a}\sum_{i=1}^{n} \Delta x_i$$

$$= \frac{\varepsilon}{b-a}(b-a) = \varepsilon$$

ε

BECAUSE THE UPPER SUMS AND LOWER SUMS SQUEEZE ARBITRARILY CLOSE TO EACH OTHER, THERE MUST BE **EXACTLY ONE NUMBER** SANDWICHED BETWEEN THEM. (ANOTHER CONSEQUENCE OF DEEP, SUBTLE, ETC.). THE DEFINITE INTEGRAL OF f FROM a TO b IS **DEFINED** TO BE THIS NUMBER!

JUST ONE NUMBER FITS IN HERE...
BIGGER THAN ANY LOWER SUM! SMALLER THAN ANY UPPER SUM! I'M UNIQUE!
$\int_a^b f(x)\,dx$

NOW BACK TO STUFF YOU REALLY NEED TO KNOW.

HELLO AGAIN!

FOR THE SAKE OF ILLUSTRATION, WE BEGAN THIS CHAPTER WITH A **NON-NEGATIVE** FUNCTION... BUT ACTUALLY RIEMANN SUMS CONVERGE TO A DEFINITE INTEGRAL FOR **ANY** CONTINUOUS FUNCTION ON A CLOSED INTERVAL.

WHAT HAPPENS WHERE A FUNCTION g IS NEGATIVE? WHENEVER THE VALUE $g(x_i^*) < 0$, SO IS THE TERM $g(x_i^*)\Delta x_i$ IN THE RIEMANN SUM. (BECAUSE Δx_i IS POSITIVE.)

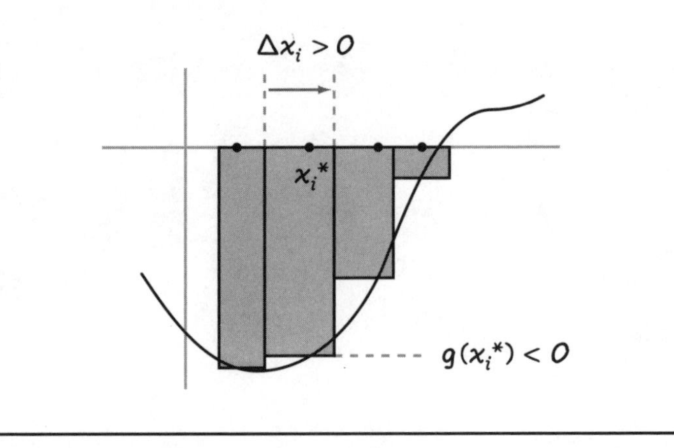
$\Delta x_i > 0$
x_i^*
$g(x_i^*) < 0$

IN OTHER WORDS, AREAS **BELOW** THE x-AXIS ARE TAKEN AS **NEGATIVE.** IN THE DEFINITE INTEGRAL, AREAS BELOW THE AXIS OFFSET AREAS ABOVE THE AXIS. JUST AS THE DERIVATIVE IS "SIGNED SPEED," SO THE INTEGRAL IS "SIGNED AREA."

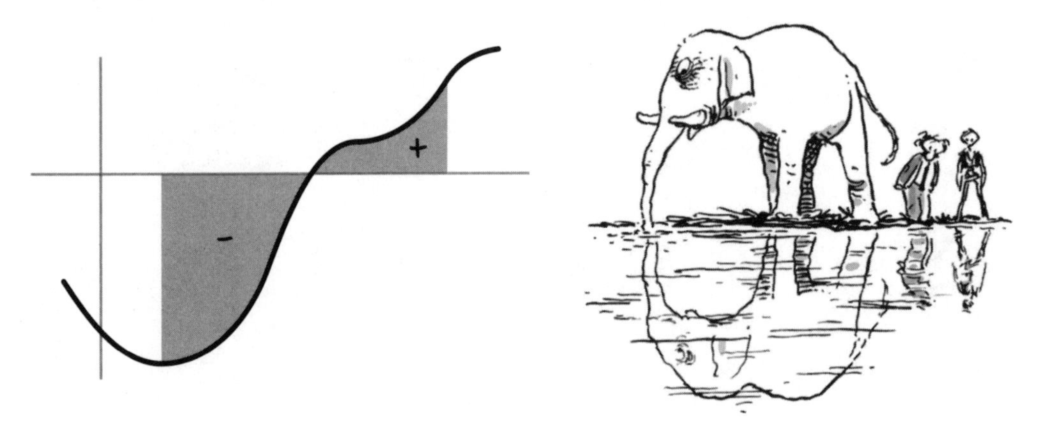
+
–

Example: EVEN THOUGH WE DON'T YET KNOW HOW TO CALCULATE DEFINITE INTEGRALS, WE CAN SEE DIRECTLY THAT

$$\int_0^{2\pi} \sin\theta \, d\theta = 0$$

BECAUSE THE REGION BETWEEN π AND 2π, WHICH LIES UNDER THE x-AXIS, IS A PERFECT MIRROR IMAGE OF THE POSITIVE REGION BETWEEN 0 AND π. THESE TWO AREAS CANCEL EACH OTHER OUT.

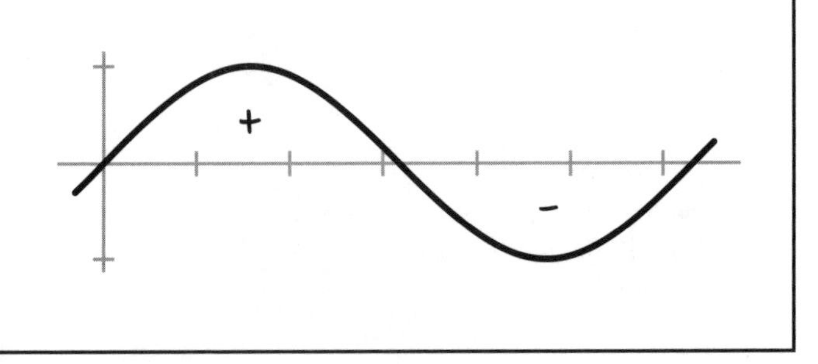
+
–

WE CAN ALSO INTEGRATE SOME FUNCTIONS THAT ARE NOT CONTINUOUS.

Example: THE WINDSHIELD WIPERS ON MOST CARS HAVE AN **INTERMITTENT WIPE** FEATURE: ELECTRIC CHARGE BUILDS UP IN A CAPACITOR IN THE CONTROL MECHANISM...

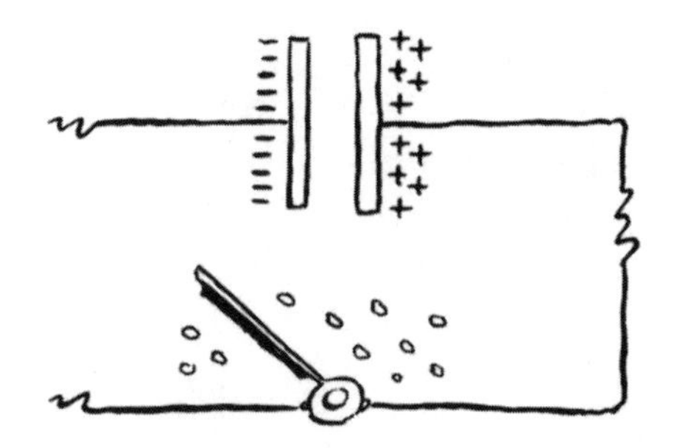

WHEN THE CHARGE REACHES A CERTAIN THRESHOLD, IT LEAPS THE GAP, COMPLETES A CIRCUIT, AND THE WIPER BLADES MAKE A SWEEP.

ZAP

THE GRAPH OF THE CHARGE IN THE CAPACITOR, AS A FUNCTION OF TIME, LOOKS LIKE THIS. IT HAS JUMPS.

EVEN SO, WE CAN INTEGRATE IT: JUST ADD UP THE AREAS WHERE THE FUNCTION IS CONTINUOUS.

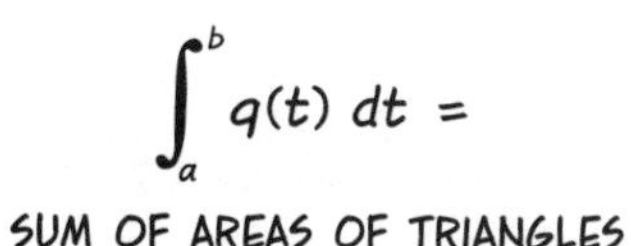

$$\int_a^b q(t)\,dt =$$

SUM OF AREAS OF TRIANGLES OR PARTS OF TRIANGLES.

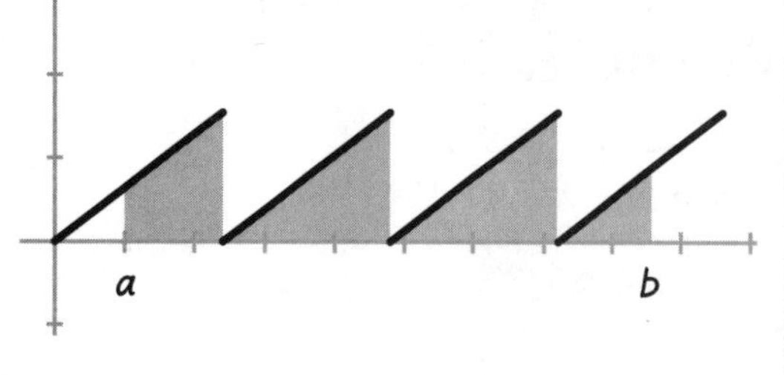
a
b

THIS ILLUSTRATES AN IMPORTANT FORMULA. IF c IS A POINT BETWEEN a AND b, THEN

$$\int_a^c f(x)\,dx + \int_c^b f(x)\,dx = \int_a^b f(x)\,dx$$

THIS IS OBVIOUS, AND WE OMIT THE PROOF. TOTAL (SIGNED) AREA IS THE SUM OF THE TWO PARTS.

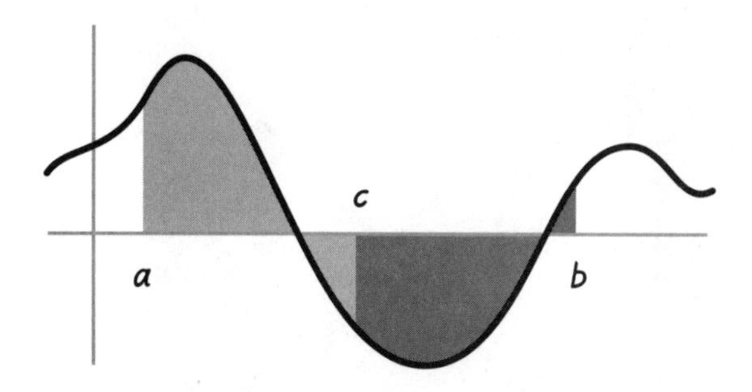
c
a
b

SO... ARE WE GOING TO GET SOME DEFINITE ANSWERS FOR THESE DEFINITE INTEGRALS?
YES!
YOU MEAN, PROVE THE AREA OF THIS TRIANGLE IS 1/2?
y = x
0
1

LET'S START BY DOING ONE THE HARD WAY—BY TAKING THE LIMIT OF RIEMANN SUMS.

Example: SHOW THAT $\int_0^1 x\,dx = \frac{1}{2}$

SUBDIVIDE THE INTERVAL [0, 1] INTO n EQUAL PARTS BY USING THE POINTS $\{0, 1/n, 2/n, \ldots, 1\}$. THEN EACH SUBINTERVAL HAS WIDTH $\Delta x = 1/n$, AND $f(x_i) = i/n$. THEN THE UPPER SUM IS

$$\sum_{i=1}^{n} f(x_i)\Delta x$$

$$= \sum_{i=1}^{n} \left(\frac{i}{n}\right)\left(\frac{1}{n}\right) = \frac{1}{n^2}\sum_{i=1}^{n} i$$

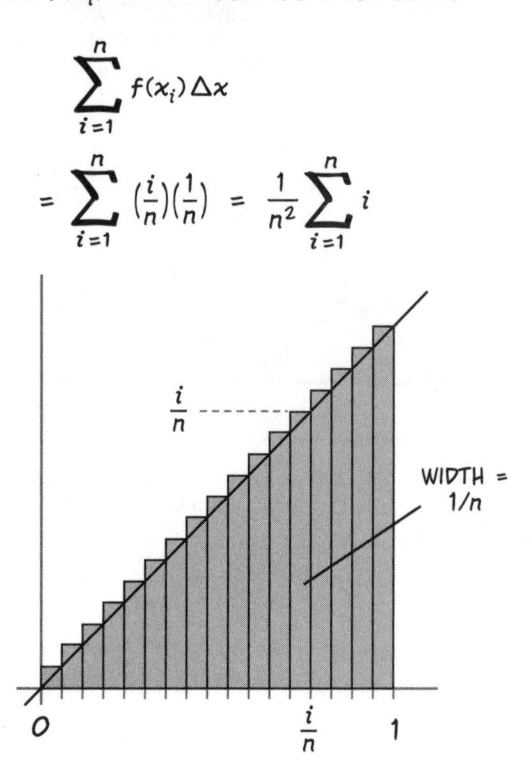
$\frac{i}{n}$
WIDTH = 1/n
0
$\frac{i}{n}$
1

NOW YOU MAY REMEMBER (OR IF NOT, LOOK IT UP!) THE FORMULA FOR THE SUM OF THE FIRST n POSITIVE INTEGERS:

$$\sum_{i=1}^{n} i = \frac{n(n+1)}{2} = \frac{n^2+n}{2}$$

THE RIEMANN SUM, THEN, IS

$$\frac{1}{n^2}\sum_{i=1}^{n} i = \frac{n^2+n}{2n^2} = \frac{1}{2} + \frac{1}{2n}$$

AS $n \to \infty$ AND THE SUBDIVISION GETS EVER FINER, THIS APPROACHES 1/2. THAT IS,

$$\int_0^1 x\,dx = \frac{1}{2}$$

O.K.... IT WAS ONLY A TRIANGLE... BUT WE WENT THROUGH ALL THAT TROUBLE TO MAKE A POINT: NEWTON AND LEIBNIZ SAVED EVERYONE A LOT OF WORK BY INVENTING CALCULUS. THEIR BIG INSIGHT INTO INTEGRALS IS SO IMPORTANT, IN FACT, THAT IT'S CALLED **THE FUNDAMENTAL THEOREM OF CALCULUS.** WE COVER IT NEXT...

OR UNCOVER IT!

AND BY THE WAY... IN CASE YOU WERE WONDERING WHY THERE WAS NO ADDED CONSTANT IN THAT LAST ANSWER, YOU MUST ALWAYS REMEMBER THAT DEFINITE INTEGRALS ARE **DEFINITE:** A DEFINITE INTEGRAL IS A SIGNED AREA, A NUMBER. THE **INDEFINITE** INTEGRAL, OR ANTIDERIVATIVE, HAS THE ADDED CONSTANT.

$$\int x\,dx = \tfrac{1}{2}x^2 + C$$

$$\int_0^1 x\,dx = \tfrac{1}{2}$$

THEN WHY IN THE **WORLD** DO THEY USE THE SAME SYMBOL?
THE ANSWER IS COMING RIGHT UP...

Problems

1. DEFINE A FUNCTION g BY

$$g(x) = 1 \quad \text{IF } 2n \le x < 2n + 1$$
$$g(x) = -1 \quad \text{IF } 2n + 1 \le x < 2n + 2$$

FOR ALL INTEGERS $n = 0, \pm 1, \pm 2, ...$ DRAW THE GRAPH OF g.

EVALUATE THE INTEGRAL

$$\int_{-4.086}^{7.358} g(x)\, dx$$

2. GIVEN THE FUNCTION $g(t) = t^2$ AND ANY NUMBER T, BUILD A RIEMANN SUM BETWEEN 0 AND T IN THE FOLLOWING WAY. SUBDIVIDE THE INTERVAL $[0, T]$ INTO n EQUAL INTERVALS BY MEANS OF THE POINTS $\{0, T/n, 2T/n, \dots iT/n, \dots , nT/n = T\}$. NOTE THAT THE LENGTH OF EACH INTERVAL IS T/n. LETTING $t_i = iT/n$, WE GET THIS RIEMANN SUM S_n:

$$S_n = \sum_{i=1}^{n} \left(\frac{iT}{n}\right)^2 \left(\frac{T}{n}\right)$$

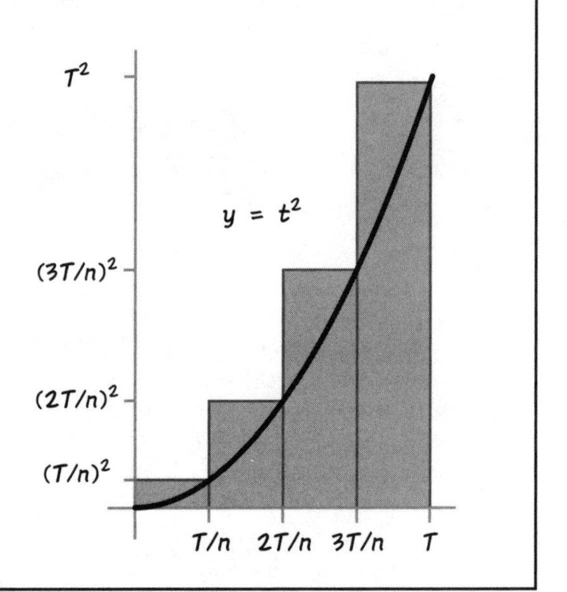

T^2
$y = t^2$
$(3T/n)^2$
$(2T/n)^2$
$(T/n)^2$
T/n $2T/n$ $3T/n$ T

SIMPLIFY THIS EXPRESSION. THEN USE THIS FORMULA (DISCOVERED BY THE ANCIENT GREEKS)...

$$\sum_{i=1}^{n} i^2 = \frac{n(n+1)(2n+1)}{6}$$

... TO DERIVE A FORMULA FOR S_n IN TERMS OF n AND T. SHOW THAT, AS $n \to \infty$,

$$S_n \to \tfrac{1}{3}T^3.$$

WHAT DO YOU MAKE OF THE FACT THAT THIS IS NEGATIVE WHEN $T < 0$?

3. USING THE FORMULA FOR THE SUM OF CUBES,

$$\sum_{i=1}^{n} i^3 = \left(\frac{n(n+1)}{2}\right)^2$$

SHOW, AS ABOVE, THAT

$$\int_0^T t^3\, dt = \tfrac{1}{4}T^4$$

4. ON P. 155, WE SHOWED A FUNCTION THAT IS NOT CONTINUOUS AT $x = 2$.

$$f(x) = \frac{1}{|x-2|} \quad \text{IF } x \ne 2$$
$$f(2) = 1$$

EXPLAIN WHY THERE IS NO UPPER SUM FOR f ON ANY INTERVAL CONTAINING $x = 2$.

Chapter 11
Fundamentally...

IN WHICH EVERYTHING COMES TOGETHER

IN CHAPTER 8, WE FOUND THAT POSITION, THE ANTIDERIVATIVE OF VELOCITY, APPEARED AS THE AREA UNDER THE GRAPH OF VELOCITY. THIS RESULT, IT TURNS OUT, WAS NO COINCIDENCE. THE INTEGRALS OF **ALL** GOOD FUNCTIONS ARE FOUND FROM THEIR ANTIDERIVATIVES! WITHOUT FURTHER ADO, THEN, HERE IS THE...

YOU'RE SURE YOU DON'T WANT SOME FURTHER ADO?
ENOUGH ADO. NO MORE ADO.

Fundamental Theorem of Calculus v.1: IF f IS A CONTINUOUS FUNCTION ON THE INTERVAL $[a, b]$, AND F IS **ANY** ANTIDERIVATIVE OF f ON $[a, b]$, THEN

$$\int_a^b f(x)\,dx = F(b) - F(a)$$

WHISK!

THIS ASTONISHING THEOREM UNITES DERIVATIVES AND INTEGRALS. IT SAYS: TO EVALUATE A DEFINITE INTEGRAL, FIRST FIND AN ANTIDERIVATIVE OF THE INTEGRAND, THEN EVALUATE THAT ANTIDERIVATIVE AT THE TWO ENDPOINTS, AND FINALLY TAKE THE DIFFERENCE! AND **THAT'S ALL!**

I AM AGAPE.

Example: FIND $\int_0^1 x\,dx$

FIRST FIND AN ANTIDERIVATIVE OF $f(x) = x$. WE KNOW THAT $F(x) = \frac{1}{2}x^2$ IS ONE. THE THEOREM THEN SAYS:

$$\int_0^1 x\,dx = F(1) - F(0)$$

$$= \tfrac{1}{2}(1)^2 - \tfrac{1}{2}(0)^2$$

$$= \mathbf{\tfrac{1}{2}}$$

AS WE SAW, WITH MUCH DIFFICULTY, THREE PAGES AGO.

STILL, IT **IS** JUST THE AREA OF A TRIANGLE...
THERE'S MORE!
y = x
0
1

Example: $\int_{-1}^5 x^3 dx$

WE KNOW THAT $F(x) = \frac{1}{4}x^4$ IS AN ANTIDERIVATIVE, SO THE INTEGRAL IS

$$F(5) - F(-1) = \tfrac{1}{4}(5)^4 - \tfrac{1}{4}(-1)^4$$

$$= \frac{625 - 1}{4} = \mathbf{156}$$

THIS DIFFERENCE IS OFTEN WRITTEN $\frac{1}{4}x^4 \Big|_{-1}^5$

Example: $\int_0^b x^n\,dx$

$G(x) = \dfrac{x^{n+1}}{n+1}$ IS AN ANTIDERIVATIVE, SO

$$\int_0^b x^n\,dx = \frac{x^{n+1}}{n+1}\Bigg|_0^b = \mathbf{\frac{b^{n+1}}{n+1}}$$

Example: $\int_0^{\pi/2} \sin\theta\,d\theta = -\cos\theta \Big|_0^{\pi/2}$

$$= -\cos(\tfrac{\pi}{2}) - (-\cos 0)$$

$$= 0 + 1 = \mathbf{1}$$

Example: $\int_0^1 \frac{1}{1+u^2}\,du = \arctan u \Big|_0^1$

$$= \arctan 1 - \arctan 0$$

$$= \frac{\pi}{4} - 0 = \mathbf{\frac{\pi}{4}}$$

(HERE WE MADE u THE VARIABLE OF INTEGRATION JUST TO REMIND YOU THAT ANY LETTER WILL DO!)

YOU'RE RIGHT! IT'S COOL! I'M **TOTALLY** SPEECHLESS...
AND YET I CAN HEAR YOU...
y = 1/(1 + u²)
π/4
0
1

THERE ARE SEVERAL WAYS TO GRASP THE FUNDAMENTAL RELATIONSHIP BETWEEN DERIVATIVES AND INTEGRALS. ONE IS TO SEE DIRECTLY WHY THE "DERIVATIVE OF THE AREA" IS THE ORIGINAL FUNCTION ITSELF. TO DO THIS, WE HAVE TO MAKE THE INTEGRAL INTO A FUNCTION.

O.K., YOU HOLD YOUR END STILL, AND I'LL SLIDE MINE AROUND...

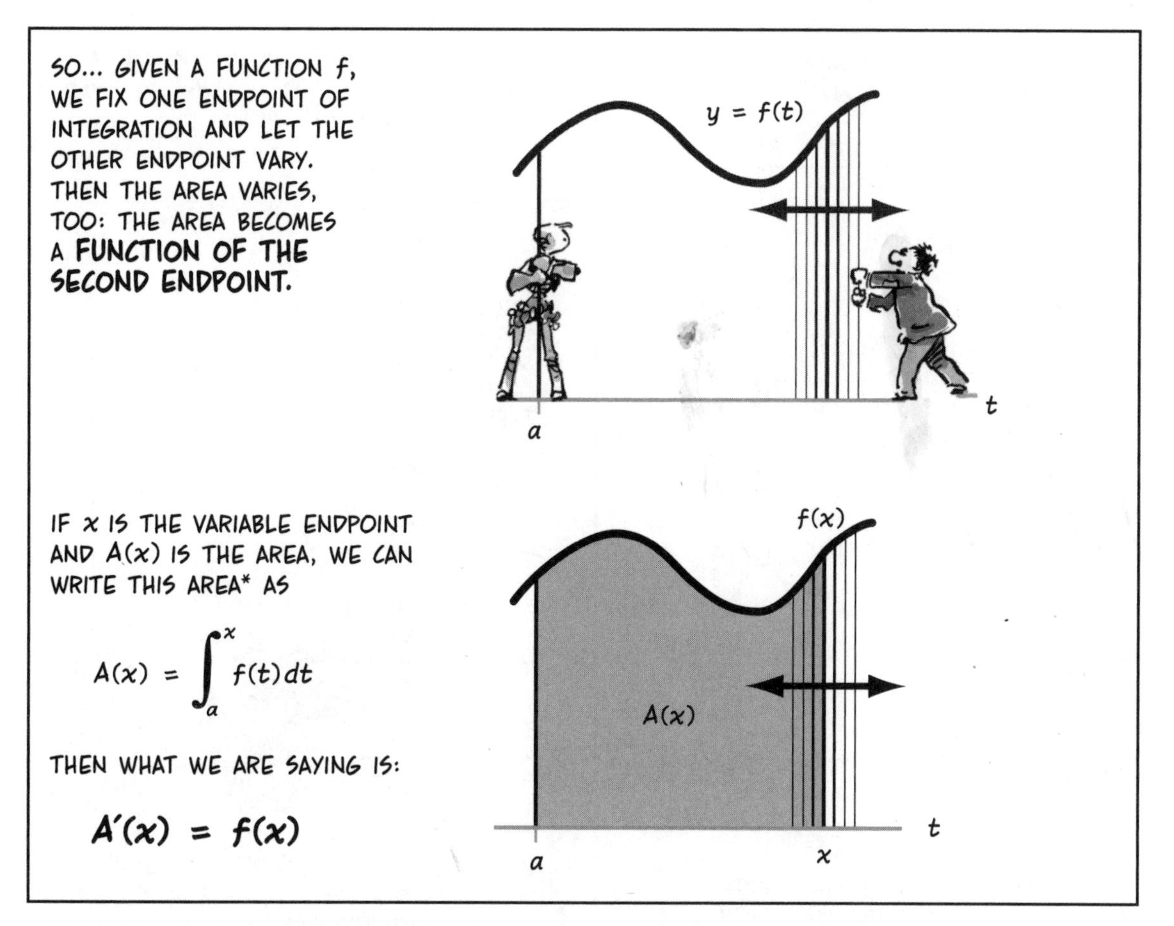
SO... GIVEN A FUNCTION f, WE FIX ONE ENDPOINT OF INTEGRATION AND LET THE OTHER ENDPOINT VARY. THEN THE AREA VARIES, TOO: THE AREA BECOMES A **FUNCTION OF THE SECOND ENDPOINT.**
$y = f(t)$
t
a
IF x IS THE VARIABLE ENDPOINT AND $A(x)$ IS THE AREA, WE CAN WRITE THIS AREA* AS
$$A(x) = \int_a^x f(t)\,dt$$
THEN WHAT WE ARE SAYING IS:
$$A'(x) = f(x)$$
$f(x)$
$A(x)$
t
a
x

*BY AREA, WE ALWAYS MEAN SIGNED AREA. WE ALSO NEED TO ALLOW THE POSSIBILITY THAT THE VARIABLE ENDPOINT FALLS TO THE **LEFT** OF a, IN WHICH CASE WE AGREE THAT

$$\int_a^x f(t)\,dt \text{ MEANS } -\int_x^a f(t)\,dt$$

HERE IS THE FORMAL STATEMENT:

Fundamental Theorem of Calculus, v.2

IF f IS CONTINUOUS, a IS IN ITS DOMAIN, AND A IS DEFINED BY

$$A(x) = \int_a^x f(t)\,dt$$

THEN A IS DIFFERENTIABLE, AND $\mathbf{A'(x) = f(x)}$.

PROOF: IF A HAS A DERIVATIVE, IT MUST BE THIS LIMIT, IF THE LIMIT EXISTS:

$$A'(x) = \lim_{h \to 0} \frac{A(x+h) - A(x)}{h}$$

BY DEFINITION OF A,

$$A(x+h) - A(x) =$$

$$\int_a^{x+h} f(t)\,dt - \int_a^x f(t)\,dt = \int_x^{x+h} f(t)\,dt$$

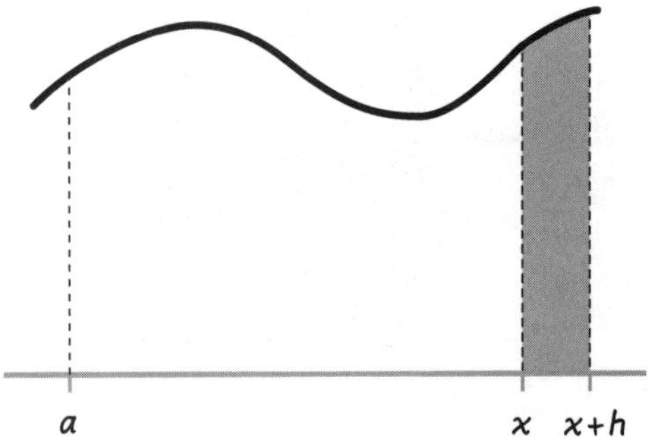

a x $x+h$

THAT STRIP HAS HEIGHT $\approx f(x)$, WIDTH $= h$, AND HENCE AREA $\approx hf(x)$, SO

$$\frac{\text{AREA}}{h} \approx \frac{hf(x)}{h} = f(x)$$

WE CAN MAKE THAT ARGUMENT PRECISE: A DEFINITE INTEGRAL IS SANDWICHED BETWEEN ITS UPPER AND LOWER SUMS:

$$mh \le \int_x^{x+h} f(t)\,dt \le Mh$$

WHERE m AND M ARE THE UPPER AND LOWER BOUNDS, RESPECTIVELY, OF f ON $[x, x+h]$. THEN

M m x $x+h$

$$m \le \frac{A(x+h) - A(x)}{h} \le M$$

AS $h \to 0$, m AND M SQUEEZE TOGETHER!

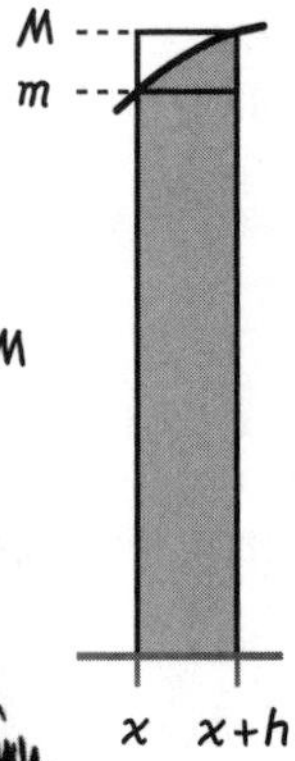

BECAUSE f IS CONTINUOUS, m AND M BOTH APPROACH $f(x)$ AS $h \to 0$, SO BY THE SANDWICH THEOREM,

$$\lim_{h \to 0} \frac{A(x+h) - A(x)}{h} = f(x)$$

LET'S RUN THROUGH THAT AGAIN, TO GET A BETTER FEEL FOR IT!
ΔA
f(x)
h
OR, AS LEIBNIZ WOULD HAVE PUT IT: dA IS THE **INFINITESIMALLY THIN** STRIP OF WIDTH dx AND HEIGHT f(x), WHICH HAS AREA f(x) dx.
dA = f(x)dx, SO
dA/dx = f(x)
dA
f(x)
dx

ΔA IS THE AREA OF THAT THIN STRIP AT THE EDGE OF THE DEFINITE INTEGRAL. THE STRIP'S WIDTH IS h, ITS HEIGHT IS APPROXIMATELY $f(x)$, SO ITS AREA IS APPROXIMATELY $hf(x)$.* THEREFORE

$$\frac{\Delta A}{h} \approx \frac{hf(x)}{h} = f(x)$$

THAT LITTLE WEDGE ON TOP IS NO MORE THAN $(M - m)h$... IN OTHER WORDS, IT'S A FLEA!

$$\Delta A = hf(x) + \text{FLEA}$$

SO $A'(x) = f(x)$.

I TOLD YOU MY NOTATION IS BETTER!

EITHER WAY, THE POINT IS THIS: **THE RATE OF CHANGE OF THE AREA AT A POINT IS GIVEN BY THE HEIGHT OF THE GRAPH THERE.**

*$f(x + h)$ IS APPROXIMATELY $f(x)$ BECAUSE f IS ASSUMED TO BE CONTINUOUS: IT CAN'T JUMP AROUND WILDLY NEAR x.

NOW WE CAN PROVE VERSION 1 OF THE FUNDAMENTAL THEOREM. IT FOLLOWS DIRECTLY FROM THE FACT THAT ANY ANTIDERIVATIVE MUST DIFFER FROM $A(x)$ BY A CONSTANT.

A(x)

Proof of Fundamental Theorem, v. 1:

WE WISH TO SHOW THAT IF G IS **ANY** ANTIDERIVATIVE OF A CONTINUOUS FUNCTION f, THEN

$$\int_a^b f(t)\,dt = G(b) - G(a)$$

PROOF: BY THE FUNDAMENTAL THEOREM, V.2, ONE ANTIDERIVATIVE A OF f IS

$$A(x) = \int_a^x f(t)\,dt$$

NOTE THAT $A(a) = 0$, SO FOR THIS ONE ANTIDERIVATIVE, ANYWAY,

$$\int_a^b f(t)\,dt = A(b) - A(a)$$

BUT G MUST DIFFER FROM A BY A CONSTANT:

$$G(x) = A(x) + C$$

SO

$$\int_a^b f(t)\,dt = A(b) - A(a)$$

$$= A(b) + C - (A(a) + C)$$

$$\mathbf{= G(b) - G(a)}$$

QUEUEW-EE-DOOODLY-GOOGLY-BOOBLY-BOBBLY-BOOP-DEE-DEE!!

Example: SHOW THAT $\int_1^x \frac{1}{t}\,dt = \ln x$ IF $x > 0$.

BY THE FUNDAMENTAL THEOREM, VERSION 1,

$$\int_1^x \frac{1}{t}\,dt = F(x) - F(1),$$

WHERE F IS ANY ANTIDERIVATIVE OF $1/t$. $F(t) = \ln t$ IS **AN** ANTIDERIVATIVE, SO

$$\int_1^x \frac{1}{t}\,dt = \ln t\,\Big|_1^x = \ln x - \ln 1 = \ln x$$

BECAUSE $\ln 1 = 0$.

y = 1/t
GRAY AREA = ln x
1 ⟶ x

NOTE THAT WHEN x < 1, THE INTEGRAL IS NEGATIVE, SINCE WE INTEGRATE FROM RIGHT TO LEFT (AND THE INTEGRAND IS POSITIVE).
y = 1/t
x ⟵ 1

Example:

$$\int_0^x \frac{1}{\sqrt{1-u^2}}\,du = \arcsin x$$

BECAUSE $\arcsin 0 = 0$.

HERE AGAIN, WE MAY HAVE TO INTEGRATE FROM RIGHT TO LEFT, AND THE ARCSINE IS NEGATIVE WHEN $-1 \le x < 0$.

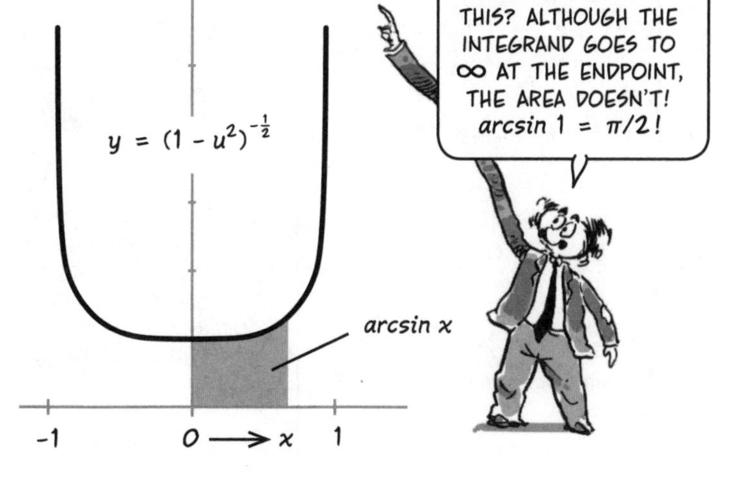

AND HOW ABOUT THIS? ALTHOUGH THE INTEGRAND GOES TO ∞ AT THE ENDPOINT, THE AREA DOESN'T! arcsin 1 = π/2!
y = (1 - u²)^(-1/2)
arcsin x
-1
0 ⟶ x
1

Problems

EVALUATE THESE INTEGRALS:

1. $\int_{-3}^{20} 6\,dx$

2. $\int_{-1}^{5} \frac{2}{3}x^4\,dx$

3. $\int_{3}^{4} (x-2)^{50}\,dx$

4. $\int_{1/2}^{2/3} (1-x)^{-2}\,dx$

5. $\int_{a}^{a+1} (a-x)^n\,dx$

WHEN IS THE INTEGRAL IN #5 NOT DEFINED?

6. $\int_{\sqrt{2}}^{2} \frac{1}{\sqrt{4-x^2}}\,dx$

7. $\int_{\pi/4}^{7\pi/2} \sin 2x\,dx$

8. $\int_{2}^{e^2+1} \frac{dx}{1-x}$

9. $\int_{4}^{25} t^{3/2} + t^{5/2} - 4t^{-7/2}\,dt$

10. $\int_{-1}^{2} \frac{3}{2}x^2 e^{(x^3+1)}\,dx$

11. $\int_{5\pi/6}^{11\pi/6} \sin^2\theta\cos\theta + \cos^2\theta\sin\theta\,d\theta$

12. SHOW THAT IF $|f(x)| \le M$ ON AN INTERVAL $[a, b]$ FOR SOME NUMBER M, THEN

$$\left|\int_a^b f(x)\,dx\right| \le M(b-a)$$

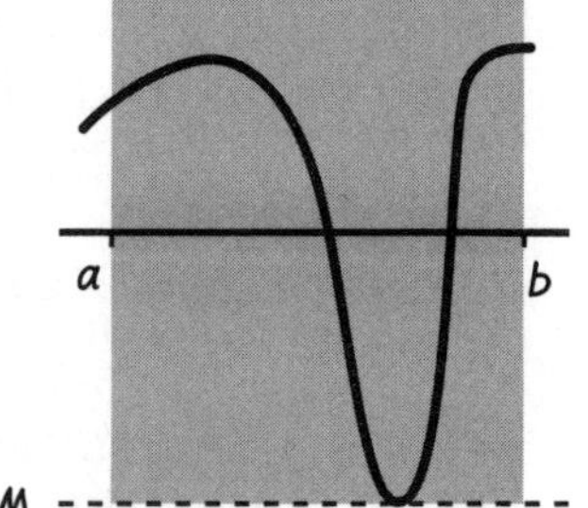

M
a
b
-M

CONCLUDE THAT IF THERE ARE TWO FUNCTIONS f AND g SUCH THAT $|f(x) - g(x)| \le \varepsilon$ ON THE WHOLE INTERVAL, THEN

$$\left|\int_a^b f(x) - g(x)\,dx\right| \le \varepsilon(b-a)$$

IN OTHER WORDS, IF TWO FUNCTIONS ARE CLOSE ON AN INTERVAL, THEIR INTEGRALS ARE CLOSE, TOO.

13. FROM ALGEBRA, RECALL THAT

$$1 - t^n = (1-t)(1 + t + t^2 + \dots + t^{n-1})$$

OR

$$\frac{1-t^n}{1-t} = 1 + t + t^2 + \dots + t^{n-1}$$

CONCLUDE THAT $1 + t + t^2 + \dots + t^{n-1}$ **IS CLOSE TO** $1/(1-t)$ **WHEN** t **IS SMALL.**

14. NOW SUBSTITUTE $t = -x^2$ TO GET

$$\frac{1}{1+x^2} \approx 1 - x^2 + x^4 - x^6 - \dots + (-1)^n x^{2n}$$

INTEGRATE FROM 0 TO 1:

$$\int_0^1 \frac{1}{1+x^2}\,dx \approx \int_0^1 1 - x^2 + x^4 - \dots + (-1)^n x^{2n}\,dx$$

EVALUATE BOTH SIDES TO FIND A FORMULA NAMED AFTER LEIBNIZ (EVEN THOUGH IT WAS DISCOVERED IN INDIA CENTURIES EARLIER!).

Chapter 12
Shape-Shifting Integrals

MORE WAYS TO FIND ANTIDERIVATIVES

TO INTEGRATE A FUNCTION, "ALL" WE HAVE TO DO IS FIND ITS ANTIDERIVATIVE. BUT THAT MAY NOT BE SO EASY... THE FUNCTION MAY NOT LOOK FAMILIAR... WE MAY NOT RECOGNIZE IT AS ANYTHING'S DERIVATIVE... IT MAY SEEM HOPELESS... SO MATHEMATICIANS HAVE DEVELOPED TOOLS FOR TINKERING WITH INTEGRALS THAT MAKE THEM EASIER TO "CRACK..."

EXCELLENT! I LOVE A GOOD TOOL!

Substitution of Variables

FROM NOW ON, WE'RE GOING TO EMBRACE LEIBNIZ NOTATION AND USE dx, dt, du, dV, dF, ETC., AS THEY WERE LITTLE QUANTITIES. DON'T WORRY ABOUT IT! IT MAKES LIFE SO MUCH EASIER, AND IT REALLY CAN'T GET YOU INTO TROUBLE...

OH, I DON'T KNOW... NEWTON SLANDERED ME ALL OVER TOWN...
#$!&##
PLAGIARIST!

BEGIN WITH THIS BASIC EQUATION, WHEN u IS A FUNCTION OF x:

$$\frac{du}{dx} = u'(x)$$

WHICH BECOMES

$$du = u'(x)\,dx$$

WHICH REALLY MEANS

$$\int du = \int u'(x)\,dx = u + C$$

WHICH WE DO KNOW TO BE TRUE, BY THE FUNDAMENTAL THEOREM!

NOW PUT ANOTHER FUNCTION v IN THE CHAIN, WHERE v IS A FUNCTION OF u. THEN AS BEFORE

$$dv = v'(u)\,du$$

PLUG IN $du = u'(x)\,dx$ TO GET

$$dv = v'(u(x))\,u'(x)\,dx$$

WHICH IS ANOTHER WAY OF WRITING THE CHAIN RULE. IT SAYS THAT

$$v + C = \int v'(u)\,du = \int v'(u(x))u'(x)\,dx$$

THERE IS SOME NOTATION HERE I DON'T QUITE REMEMBER FROM ALGEBRA 2...

WHY DOES THIS HELP? BECAUSE IT ALLOWS US TO **SIMPLIFY** OR TRANSFORM THE INTEGRAL ON THE RIGHT INTO THE ONE ON THE LEFT!!! BY **SUBSTITUTING** du FOR $u'(x)\,dx$, WE GET A MUCH SIMPLER-LOOKING INTEGRAL!!!

Example 1: FIND $\int 2t\cos(t)^2\,dt$

SET $u = t^2$, THEN $du = 2t\,dt$, AND THE INTEGRAL BECOMES

$$\int 2t\cos(t)^2\,dt = \int \cos u\,du$$

$$= \sin u + C$$

$$= \boldsymbol{\sin(t)^2 + C}$$

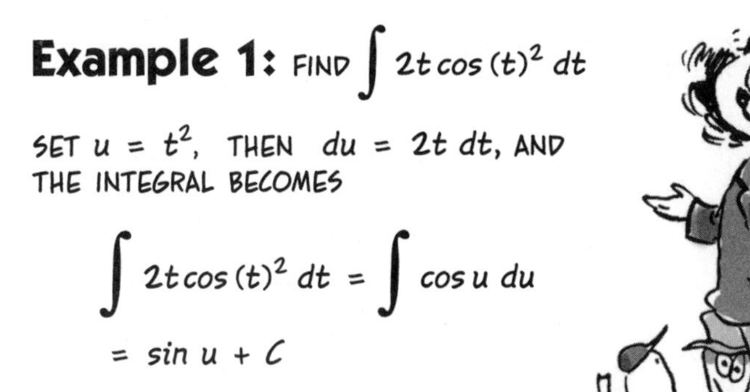
YOU MAY RECOGNIZE THIS AS A SYSTEMATIC WAY TO "GUESS AND CHECK" AS ON P. 173.

HERE IS THE PROCEDURE, STEP BY STEP:

1. LOOK FOR AN INSIDE FUNCTION u WHOSE DERIVATIVE u' ALSO APPEARS AS A FACTOR IN THE INTEGRAND.

v'
u
u'
dt

2. WRITE $du = u'(t)\,dt$ (OR $u'(x)\,dx$, OR WHATEVER THE VARIABLE IS).

v'
u
du
u'
dt

3. EXPRESS EVERYTHING IN TERMS OF u.

4. TRY THE INTEGRATION WITH RESPECT TO u. IF SUCCESSFUL, REPLACE u BY $u(t)$ IN THE ANSWER.

v'
du =
v
+C =
v
u
+C

Example 2. FIND $\int x^3 \sqrt[3]{x^4 + 9}\, dx$

HERE $u = x^4 + 9$ LOOKS LIKE A GOOD INSIDE FUNCTION, BECAUSE ITS DERIVATIVE IS $4x^3$, AND WE SEE x^3 AS A FACTOR IN THE INTEGRAND.

$$du = 4x^3 dx, \text{ SO } x^3 dx = \tfrac{1}{4} du$$

SO

$$\int x^3 \sqrt[3]{x^4 + 9}\, dx = \frac{1}{4} \int u^{1/3}\, du =$$

$$(\tfrac{1}{4})(\tfrac{3}{4}) u^{4/3} + C = \frac{3}{16} (x^4 + 9)^{4/3} + C$$

OH, YEAHHH...
X5

Example 3. FIND $\int u \sqrt{2u - 3}\, du$

SOMETIMES A SUBSTITUTION LOOKS UNPROMISING BUT WORKS ANYWAY. THIS INTEGRAND DOESN'T QUITE FIT OUR TEMPLATE, BECAUSE THE FACTOR u IS NOT THE DERIVATIVE OF THE INSIDE FUNCTION. BUT LET'S TRY ANYWAY...

$$v = 2u - 3, \quad u = \tfrac{1}{2}(v + 3), \quad du = \tfrac{1}{2} dv$$

NOW WE MUST EXPRESS EVERYTHING IN TERMS OF v:

$$\int u \sqrt{2u - 3}\, du = \int \tfrac{1}{2}(v + 3) v^{1/2} (\tfrac{1}{2})\, dv =$$

$$\frac{1}{4} \int v^{3/2} + 3v^{1/2}\, dv = \frac{1}{4} (\tfrac{2}{5}) v^{5/2} + 3(\tfrac{2}{3}) v^{3/2} + C$$

$$= \frac{(2u - 3)^{5/2}}{10} + 2(2u - 3)^{3/2} + C$$

THIS SAME SUBSTITUTION WORKS GENERALLY WITH THE INTEGRAND $u^n (au + b)^m$ FOR ANY POSITIVE INTEGER n AND ANY POWER m, AND ANY a, b, AND THEREFORE WITH $P(u)(au + b)^m$ FOR ANY POLYNOMIAL P.

Substitution and Definite Integrals

WHEN USING SUBSTITUTION IN A DEFINITE INTEGRAL, THE ENDPOINTS OF INTEGRATION MUST BE ADJUSTED TO REFLECT THE SUBSTITUTION. IF F IS AN ANTIDERIVATIVE OF f, THEN

$$\int_a^b f(u(x))\,u'(x)\,dx = F(u(b)) - F(u(a)) = \int_{u(a)}^{u(b)} f(u)\,du$$

THE ENDPOINTS a AND b ARE REPLACED BY $u(a)$ AND $u(b)$ WHEN INTEGRATING WITH RESPECT TO u.

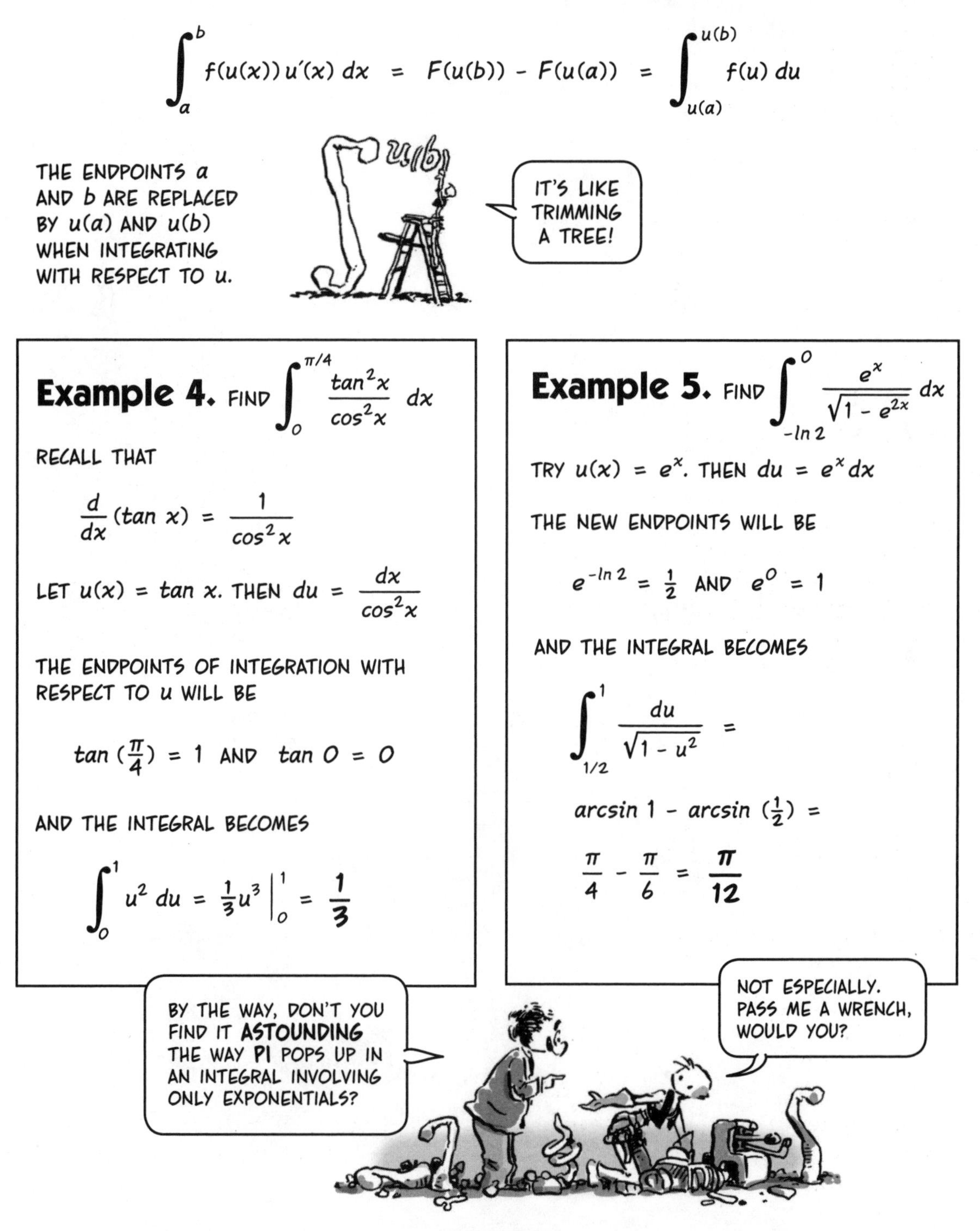

u(b)
IT'S LIKE TRIMMING A TREE!

Example 4. FIND $\displaystyle\int_0^{\pi/4} \frac{\tan^2 x}{\cos^2 x}\,dx$

RECALL THAT

$$\frac{d}{dx}(\tan x) = \frac{1}{\cos^2 x}$$

LET $u(x) = \tan x$. THEN $du = \dfrac{dx}{\cos^2 x}$

THE ENDPOINTS OF INTEGRATION WITH RESPECT TO u WILL BE

$$\tan\left(\tfrac{\pi}{4}\right) = 1 \text{ AND } \tan 0 = 0$$

AND THE INTEGRAL BECOMES

$$\int_0^1 u^2\,du = \tfrac{1}{3}u^3\Big|_0^1 = \mathbf{\frac{1}{3}}$$

Example 5. FIND $\displaystyle\int_{-\ln 2}^{0} \frac{e^x}{\sqrt{1 - e^{2x}}}\,dx$

TRY $u(x) = e^x$. THEN $du = e^x dx$

THE NEW ENDPOINTS WILL BE

$$e^{-\ln 2} = \tfrac{1}{2} \text{ AND } e^0 = 1$$

AND THE INTEGRAL BECOMES

$$\int_{1/2}^{1} \frac{du}{\sqrt{1 - u^2}} =$$

$$\arcsin 1 - \arcsin\left(\tfrac{1}{2}\right) =$$

$$\frac{\pi}{4} - \frac{\pi}{6} = \mathbf{\frac{\pi}{12}}$$

BY THE WAY, DON'T YOU FIND IT **ASTOUNDING** THE WAY **PI** POPS UP IN AN INTEGRAL INVOLVING ONLY EXPONENTIALS?
NOT ESPECIALLY. PASS ME A WRENCH, WOULD YOU?

MAKING A VARIABLE SUBSTITUTION WORKS A SORT OF **SHAPE-SHIFTING OPERATION** ON INTEGRALS. IT'S AMAZING, REALLY... A HORRIBLE-LOOKING INTEGRAL MAY TURN INTO SOMETHING COMPLETELY DIFFERENT AND EVEN SIMPLE AND FAMILIAR!

$$\int \frac{\tan^2 x}{\cos^2 x}\, dx \quad \text{BECOMES} \quad \int u^2\, du \qquad (u = \tan x,\ du = dx/(\cos^2 x))$$

$$\int \frac{2x}{1 + x^2}\, dx \quad \text{BECOMES} \quad \int \frac{dy}{y} \qquad (y = 1 + x^2,\ dy = 2x\,dx)$$

$$\int x^2 \sqrt{1 + x}\, dx \quad \text{BECOMES} \quad \int t^{5/2} - t^{3/2} + t^{1/2}\, dt \qquad (t = 1 + x,\ dt = dx)$$

$$\int \frac{e^t}{1 + e^{2t}}\, dt \quad \text{BECOMES} \quad \int \frac{dv}{1 + v^2} \qquad (v = e^t,\ dv = e^t\, dt)$$

TALK ABOUT POWER...

THIS IS, IN FACT, THE MAIN IDEA BEHIND SUCCESSFUL INTEGRATION: GIVEN AN UNFAMILIAR INTEGRAL, **TINKER WITH IT UNTIL IT LOOKS LIKE ONE YOU RECOGNIZE.**

HM... I WONDER WHAT ELSE IS IN THAT TOOL BOX...

Integration by Parts

IS BASED ON THE PRODUCT RULE FOR DIFFERENTIATION:
$(uv)' = uv' + vu'$ OR
$d(uv) = udv + vdu$
INTEGRATING GIVES
$uv = \int u\,dv + \int v\,du$
WHICH SOME INTEGRAL MECHANIC SAW WAS MORE PRODUCTIVE IF REARRANGED LIKE SO:
$\int u\,dv = uv - \int v\,du$
ONE INTEGRAL...
IN TERMS OF ANOTHER!
Example 5. FIND $\int 3x^2 \ln x\,dx$
SUBSTITUTION DOESN'T HELP HERE... BUT WE DO SEE A CANDIDATE FOR dv:
$3x^2 dx = d(x^3)$
ACCORDINGLY, TRY
$v(x) = x^3,\ dv = 3x^2\,dx$
$u(x) = \ln x,\ du = \frac{1}{x}\,dx$
SO
$\int 3x^2 \ln x\,dx = uv - \int v\,du$
$= x^3 \ln x - \int (x^3)(\frac{1}{x})\,dx$
$= x^3 \ln x - \int x^2\,dx$
$= x^3 \ln x - \frac{1}{3}x^3 + C$
WE CAN CHECK THE ANSWER BY DIFFERENTIATING:
$\frac{d}{dx}(x^3 \ln x - \frac{1}{3}x^3) =$
$3x^2 \ln x + \frac{x^3}{x} - x^2 =$
$3x^2 \ln x + x^2 - x^2 =$
$3x^2 \ln x$
THIS IS THE ORIGINAL INTEGRAND.
I CAN'T WAIT TO TRY THIS ONE OUT...

Example 6. FIND $\int \ln x \, dx$

YOU MAY WONDER WHERE v IS, BUT IN FACT, THIS IS VERY MUCH LIKE THE PREVIOUS EXAMPLE. JUST SET $dv = dx$!

$$u = \ln x, \quad du = \frac{1}{x}, \quad v = x$$

AND

$$\int \ln x \, dx = x \ln x - \int x\left(\frac{1}{x}\right) dx =$$

$$x \ln x - \int dx = \mathbf{x \ln x - x + C}$$

I STRONGLY RECOMMEND THAT YOU CHECK ALL ANSWERS BY DIFFERENTIATING!
BUT I'M HAVING SO MUCH FUN DOING THIS...

Example 7. FIND $\int x \cos x \, dx$

HERE WE HAVE A CHOICE OF dv: EITHER $\cos x \, dx = d(\sin x)$ OR $x \, dx = d(\frac{1}{2}x^2)$.

YOU SHOULD CONVINCE YOURSELF THAT THE LATTER OPTION ONLY MAKES THINGS WORSE... SO INSTEAD WE GO WITH THE FIRST ONE:

$u = x, \quad du = dx, \quad dv = d(\sin x), \quad v = \sin x,$ AND THEN

$$\int x \cos x \, dx = x \sin x - \int \sin x \, dx = \mathbf{x \sin x + \cos x + C}$$

Example 8. FIND $\int x^2 \sin x \, dx$

PROCEED AS IN EXAMPLE 7:

$$u = x^2, \quad du = 2x \, dx,$$

$$dv = \sin x \, dx, \quad v = -\cos x$$

$$\int x^2 \sin x \, dx = -x^2 \cos x - \int 2x(-\cos x) \, dx =$$

$$-x^2 \cos x + 2\int x \cos x \, dx =$$

$$\mathbf{-x^2 \cos x + 2x \sin x + 2\cos x + C}$$

THIS IS THE INTEGRAL FROM EXAMPLE 7...

EXAMPLES 7 AND 8 SHOW HOW TO HANDLE THESE INTEGRALS (n BEING A POSITIVE INTEGER):

$$\int x^n \sin x \, dx \quad \text{OR} \quad \int x^n \cos x \, dx$$

WE "BOOTSTRAP" OUR WAY UP: INTEGRATION BY PARTS PRODUCES A SIMILAR INTEGRAL, BUT WITH THE FACTOR x^{n-1} IN PLACE OF x^n... WE AGAIN INTEGRATE BY PARTS... AND SO ON, UNTIL THE INTEGRAND IS $\sin x$ OR $\cos x$ ALONE.

WOW! IT REALLY WORKS!

Example 9. FIND $\int \sin^2 x \, dx$

OUR ONLY HOPE IS TO TRY

$$u = \sin x, \quad du = \cos x \, dx,$$

$$dv = \sin x \, dx, \quad v = -\cos x$$

IN WHICH CASE

$$\int \sin^2 x \, dx = -\sin x \cos x + \int \cos^2 x \, dx$$

THE SECOND INTEGRAL, WITH $\cos^2 x$, LOOKS JUST AS BAD AS THE FIRST ONE... BUT $\cos^2 x = 1 - \sin^2 x$... SO WE TRY PLUGGING THIS INTO THE RIGHT-HAND INTEGRAL AND REARRANGING:

$$2\int \sin^2 x \, dx = -\sin x \cos x + \int dx$$

$$= -\sin x \cos x + x + C$$

SO

$$\int \sin^2 x \, dx = \tfrac{1}{2}(-\sin x \cos x + x) + C$$

LET'S CHECK THE ANSWER:

$$\frac{d}{dx}\left(-\tfrac{1}{2}\sin x \cos x + \tfrac{1}{2}x\right)$$

$$= -\tfrac{1}{2}(\cos^2 x - \sin^2 x) + \tfrac{1}{2}$$

$$= -\tfrac{1}{2}(1 - 2\sin^2 x) + \tfrac{1}{2}$$

$$= \sin^2 x - \tfrac{1}{2} + \tfrac{1}{2} = \sin^2 x$$

THE SAME TRIG IDENTITY ALLOWS US TO BOOTSTRAP OUR WAY TO ALL INTEGRALS OF THE FORM

$$\int \sin^m x \cos^n x \, dx.$$

Problems

READY, SET... INTEGRATE!

1. $\int \frac{x}{1+x^2}\,dx$

2. $\int x(1+x^2)^{-2}\,dx$

3. $\int \sin t\, e^{n\cos t}\,dt$

4. $\int \tan u\,du$

HINT: EXPRESS THE TANGENT IN TERMS OF SINE AND COSINE.

5. $\int x^2(3x-1)^{-1/2}\,dx$

6. $\int \sqrt{1-x^2}\,dx$

HINT: SUBSTITUTE $x = \cos\theta$, USE A TRIG IDENTITY, AND REFER TO EXAMPLE 9. DON'T FORGET TO CONVERT THE ANSWER BACK INTO AN EXPRESSION INVOLVING x.

7. $\int_0^1 (x^3+x+1)(\sqrt{2x+5})\,dx$

8. $\int e^x \sin x\,dx$

9. $\int te^{-t}\,dt$

10. $\int_1^5 (\ln x)^2\,dx$

11. $\int (\ln x)^3\,dx$

12. $\int_0^x \arctan v\,dv$

HERE IS A GRAPH OF THE NATURAL LOGARITHM, $y = \ln t$. REMEMBER, THIS IS ALSO THE GRAPH $t = e^y$, BECAUSE THE LOG AND THE EXPONENTIAL ARE INVERSE FUNCTIONS. THIS IMPLIES THAT THE SHADED REGION HAS AREA

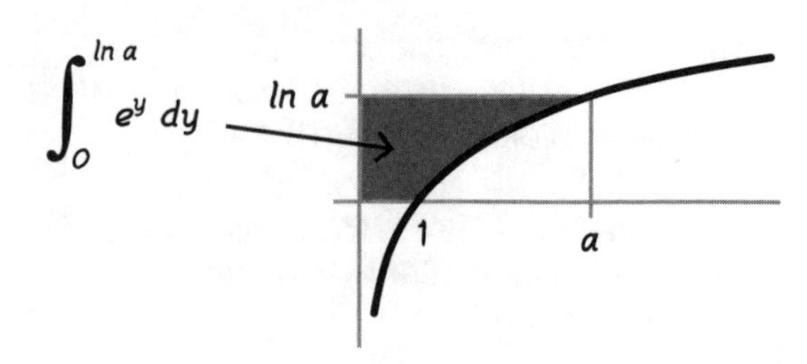

$\int_0^{\ln a} e^y\,dy$
ln a
1
a

SEE THAT? THE AREA UNDER THE LOGARITHM GRAPH IS THE AREA OF A RECTANGLE MINUS THE SHADED AREA... OR:

$$\int_1^a \ln t\,dt = a\ln a - \int_0^{\ln a} e^y\,dy$$

$$= \mathbf{a\ln a - a + 1}$$

THIS AGREES WITH WHAT WE FOUND BY INTEGRATION BY PARTS.

13. APPLY THE SAME IDEA TO $\int_0^x \arctan v\,dv$

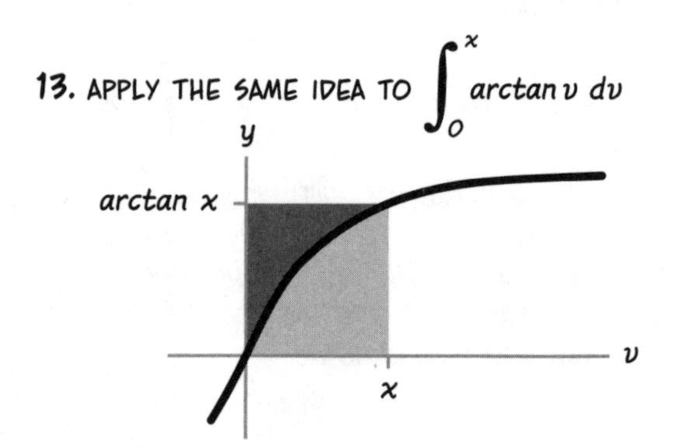

y
arctan x
v
x

YOUR ANSWER MAY LOOK DIFFERENT FROM WHAT YOU FOUND IN PROBLEM 12. IF SO, REFER TO THIS TRIANGLE TO WRESTLE IT INTO SHAPE...

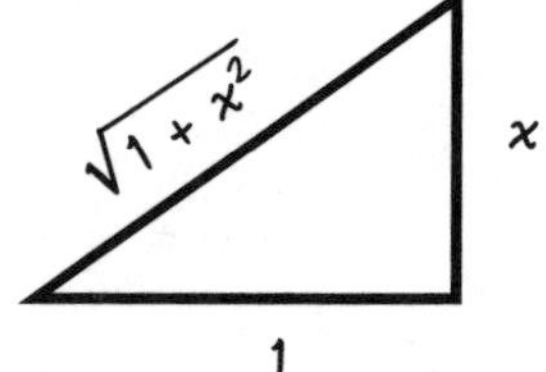

$\sqrt{1+x^2}$
x
1

Chapter 13
Using Integrals

THIS STUFF REALLY IS GOOD FOR SOMETHING, YOU KNOW?

INTEGRALS ARE EVERYWHERE... YOU ONLY NEED EYES TO SEE THEM.

IN THIS CHAPTER, WE'LL FIND INTEGRALS AT WORK (AND AT PLAY?) IN GEOMETRY, PHYSICS, ECONOMICS, STATISTICS, BUSINESS... JUST ABOUT ANYWHERE THAT THINGS PILE UP.

DID I MENTION THAT INTEGRALS UNLOCK THE SECRETS OF THE UNIVERSE?

Areas and Volumes

WE CAN FIND THE **AREA BETWEEN TWO GRAPHS** BY INTEGRATING THE **DIFFERENCE** BETWEEN THE TWO FUNCTIONS.

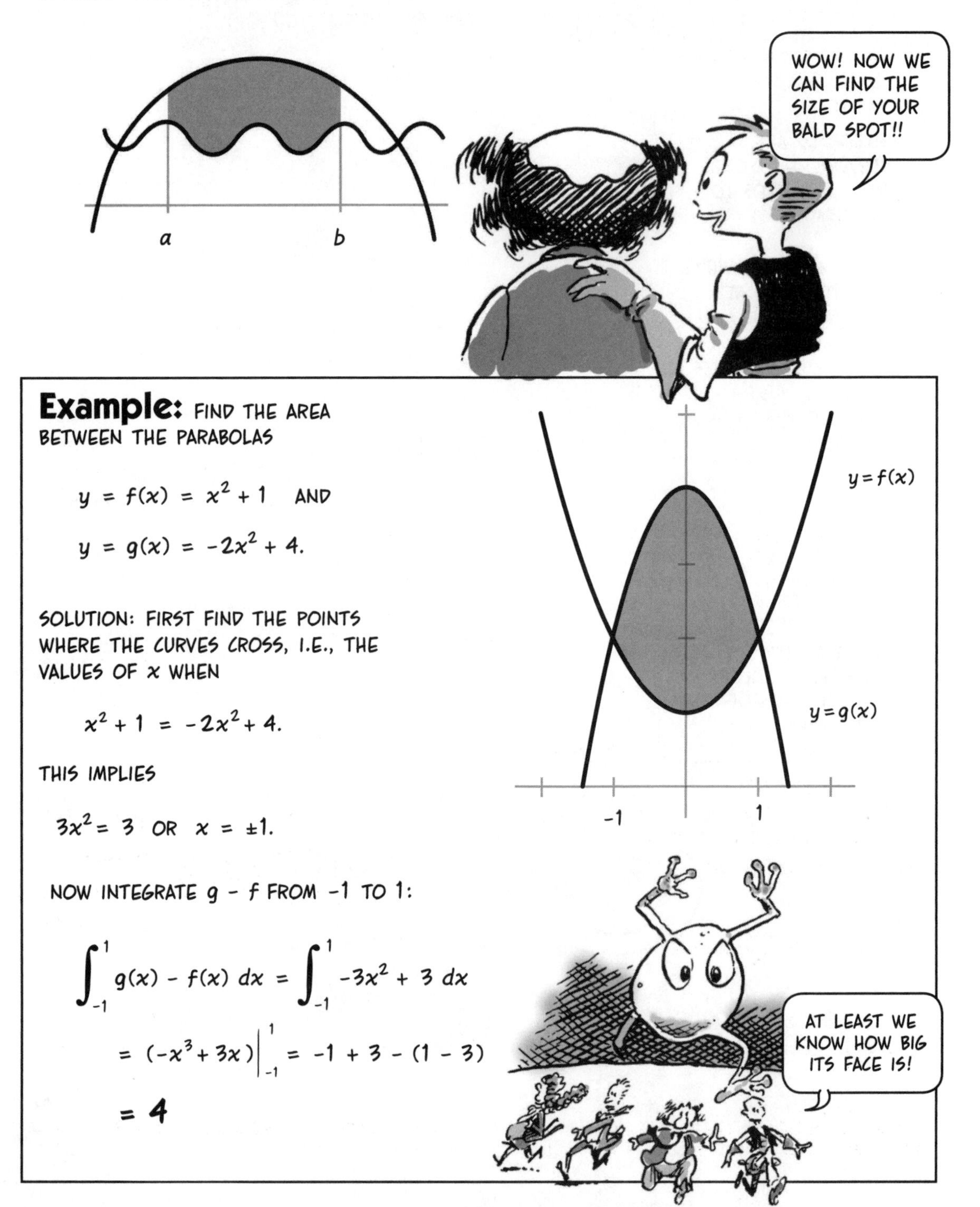

a
b
WOW! NOW WE CAN FIND THE SIZE OF YOUR BALD SPOT!!
y = f(x)
y = g(x)
-1
1
AT LEAST WE KNOW HOW BIG ITS FACE IS!

Example: FIND THE AREA BETWEEN THE PARABOLAS

$$y = f(x) = x^2 + 1 \quad \text{AND}$$

$$y = g(x) = -2x^2 + 4.$$

SOLUTION: FIRST FIND THE POINTS WHERE THE CURVES CROSS, I.E., THE VALUES OF x WHEN

$$x^2 + 1 = -2x^2 + 4.$$

THIS IMPLIES

$$3x^2 = 3 \quad \text{OR} \quad x = \pm 1.$$

NOW INTEGRATE $g - f$ FROM -1 TO 1:

$$\int_{-1}^{1} g(x) - f(x)\, dx = \int_{-1}^{1} -3x^2 + 3\, dx$$

$$= (-x^3 + 3x)\Big|_{-1}^{1} = -1 + 3 - (1 - 3)$$

$$= \mathbf{4}$$

IN THE REAL WORLD, WE MIGHT SEE SOMETHING LIKE THIS: HERE IS A VELOCITY FUNCTION $v = v(t)$ THAT DESCRIBES A CAR ACCELERATING FROM A STOP, BEGINNING AT THE ZERO-POINT OF THE ROAD. THE AREA UNDER THE CURVE BETWEEN 0 AND T,

$$\int_0^T v(t)\,dt$$

IS THE CAR'S POSITION AT TIME T.

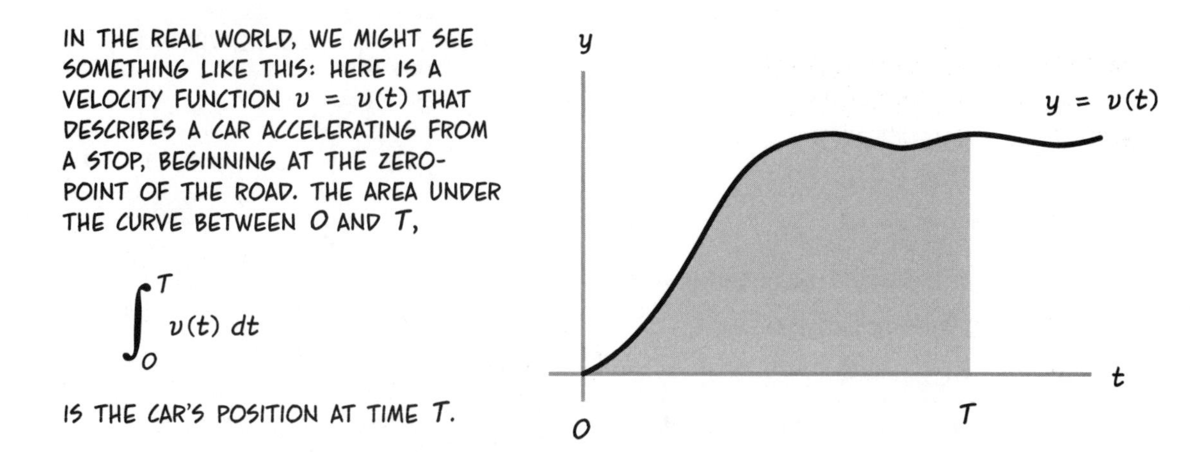

y
y = v(t)
t
0
T

IF AN AUDI (A) AND A BMW (B) BOTH PULL AWAY FROM THE SAME STOP SIGN SIMULTANEOUSLY, THE GRAPHS OF THEIR VELOCITIES MIGHT LOOK LIKE THIS:*

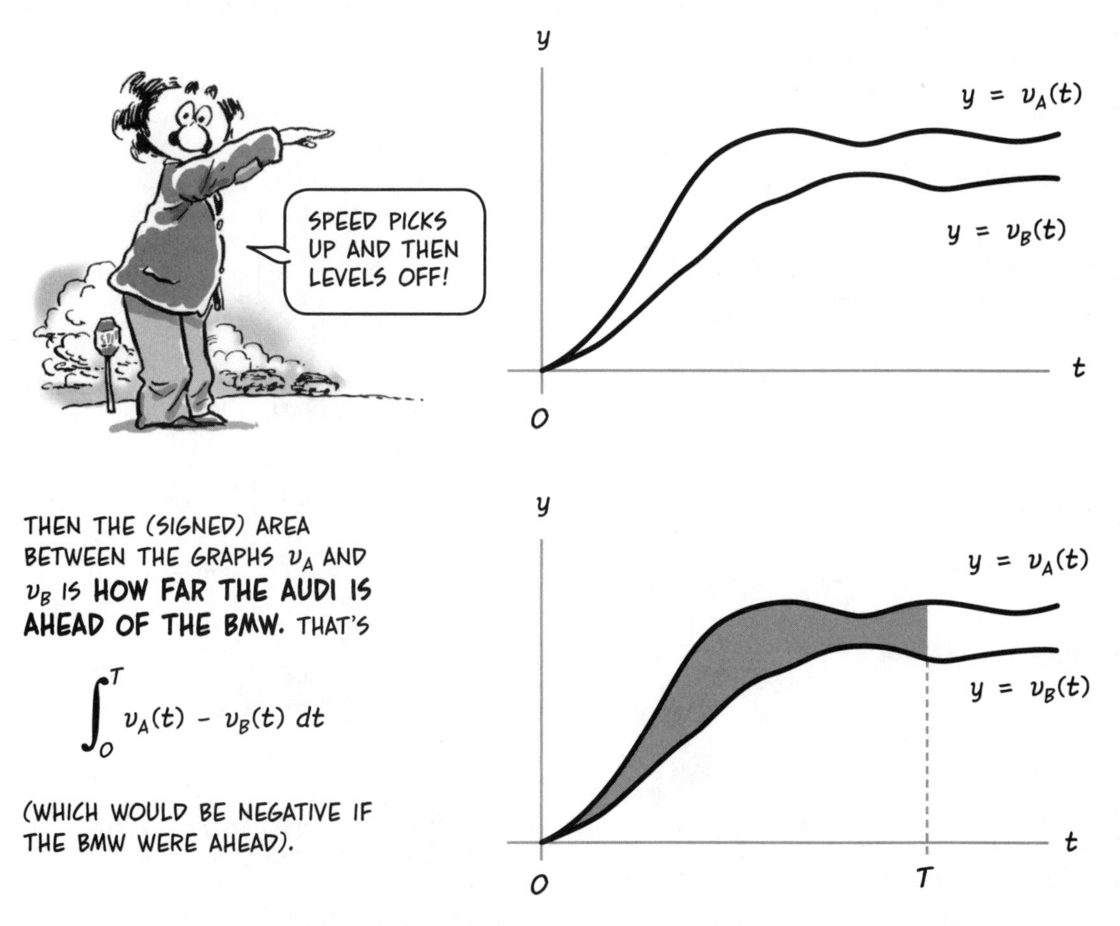

SPEED PICKS UP AND THEN LEVELS OFF!
y
y = v_A(t)
y = v_B(t)
t
0
y
y = v_A(t)
y = v_B(t)
t
0
T

THEN THE (SIGNED) AREA BETWEEN THE GRAPHS v_A AND v_B IS **HOW FAR THE AUDI IS AHEAD OF THE BMW.** THAT'S

$$\int_0^T v_A(t) - v_B(t)\,dt$$

(WHICH WOULD BE NEGATIVE IF THE BMW WERE AHEAD).

*THIS ASSUMES THAT THE BMW ACTUALLY CAME TO A COMPLETE STOP. I HAVE NEVER WITNESSED THIS HAPPENING MYSELF, BUT I REMAIN HOPEFUL THAT IT MAY HAPPEN SOMEDAY.

IN A SIMPLE CASE, THE AUDI'S VELOCITY MIGHT BE

$$v_A(t) = 3t \text{ M/SEC FOR THE FIRST 10 SECONDS}$$
$$= 30 \text{ M/SEC AFTER } t = 10 \text{ SEC.}$$

AND SUPPOSE THE BMW'S VELOCITY IS:

$$v_B(t) = 5t \text{ M/SEC FOR THE FIRST 4 SECONDS}$$
$$= 20 \text{ M/SEC AFTER } t = 4$$

IN THE EARLY GOING, THE BMW OUTPACES THE AUDI...

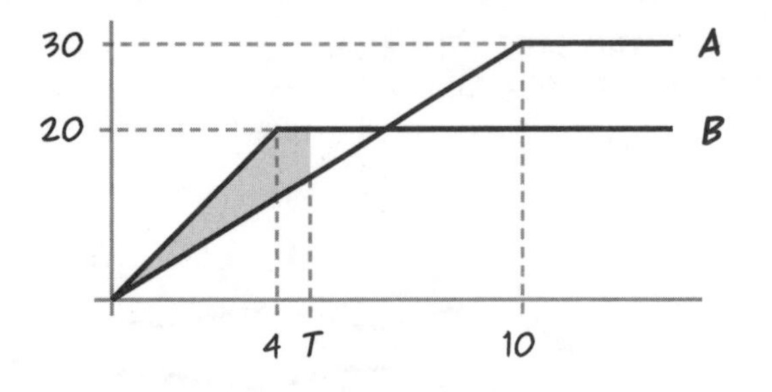
30
20
A
B
4
T
10

BUT AS T GETS LARGE, THE AUDI PULLS AHEAD. THE DARK AREA ON TOP WILL EVENTUALLY EXCEED THE LIGHTER GRAY AREA.

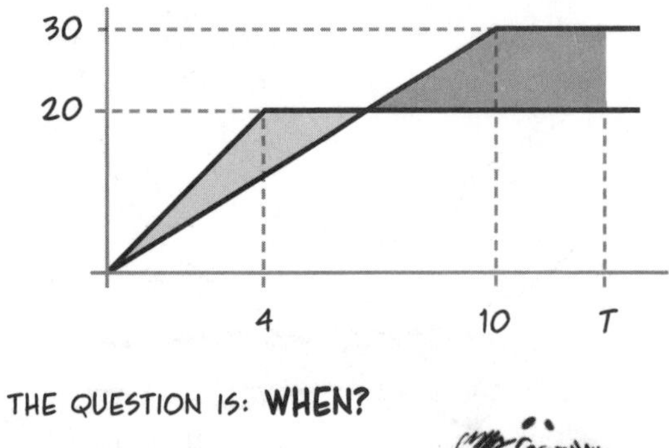
30
20
4
10
T

THE QUESTION IS: **WHEN?**

HM. I THOUGHT THE QUESTION WAS, WHY DO I DRIVE A RUSTED-OUT SUZUKI?

WHEN $T \geq 10$ SECONDS, THE CARS' POSITIONS ARE

$$s_A(T) = \int_0^{10} 3t\,dt + 30(T - 10)$$

$$s_B(T) = \int_0^{4} 5t\,dt + 20(T - 4)$$

THE INTEGRAL FOR THE "TRIANGULAR" PART WHEN SPEED IS PICKING UP...

THE REST FOR THE "RECTANGULAR" PART!

EVALUATING THE INTEGRALS GIVES:

$$s_A(T) = \frac{3}{2}t^2 \Big|_0^{10} + 30(T - 10)$$
$$= 150 + 30T - 300$$
$$= 30T - 150$$

$$s_B(T) = \frac{5}{2}t^2 \Big|_0^{4} + 20(T - 4)$$
$$= 20T - 40$$

THE AUDI PASSES THE BMW WHEN THEIR POSITIONS ARE EQUAL:

$$s_A(T) = s_B(T)$$
$$30T - 150 = 20T - 40$$
$$10T = 110$$
$$T = \mathbf{11} \text{ SECONDS}$$

An Area Using Polar Coordinates

POLAR COORDINATES, WRITTEN (r, θ), ARE AN ALTERNATIVE TO ORDINARY "RECTANGULAR" COORDINATES x AND y. ANY POINT P IN THE PLANE IS UNIQUELY SPECIFIED BY ITS DISTANCE r FROM THE ORIGIN AND THE ANGLE θ BETWEEN THE HORIZONTAL AXIS AND THE LINE SEGMENT OP.

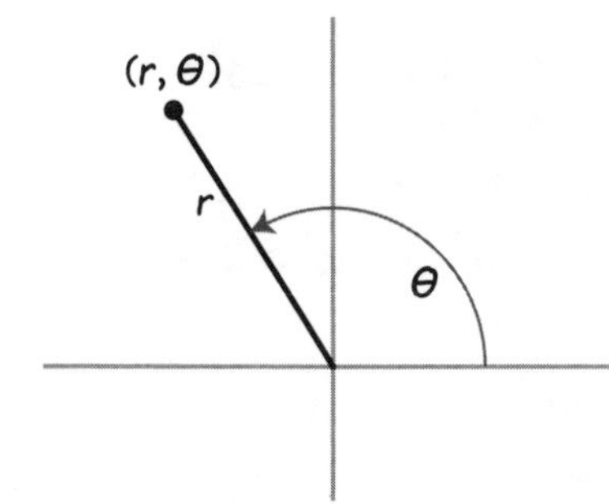
(r, θ)
r
θ

THE RELATIONSHIP BETWEEN THE COORDINATES IS

$$r^2 = x^2 + y^2 \qquad \tan\theta = \frac{y}{x} \quad (0 \le \theta < 2\pi)$$

WE CAN USE THE VARIABLE r TO DERIVE THE **AREA OF A CIRCLE** BY INTEGRATION.

GIVEN A CIRCLE OF RADIUS R, SUBDIVIDE THE RADIUS INTO MANY SHORT INTERVALS OF LENGTH Δr. THESE DIVIDE THE CIRCLE INTO MANY NARROW RINGS OF THICKNESS Δr.

IF r_i IS THE RADIUS OF A RING, THEN THE RING HAS AREA $\approx 2\pi r_i \Delta r$. (IMAGINE THE RING AS A THIN RIBBON THAT YOU COULD STRAIGHTEN OUT INTO A LONG, NARROW RECTANGLE, WITH LENGTH ABOUT $2\pi r$ AND HEIGHT Δr.)

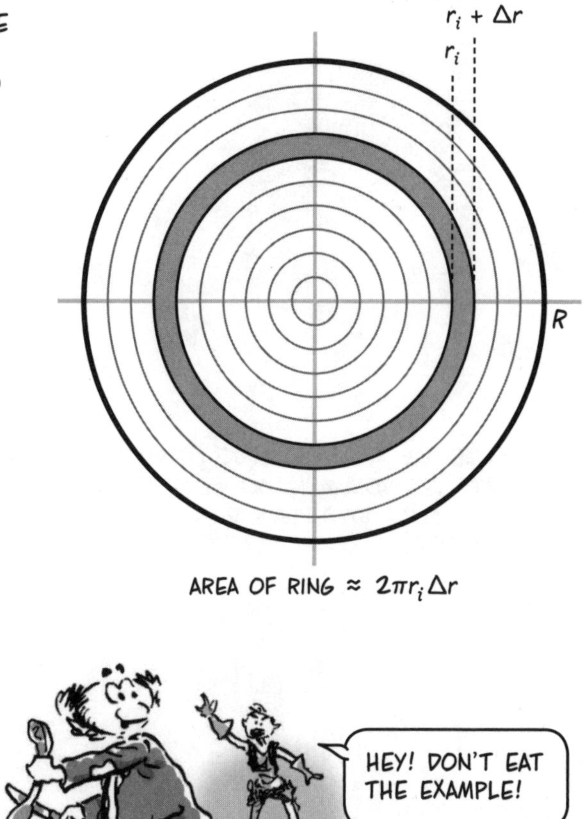
r_i + Δr
r_i
R
Δr
2πr
HEY! DON'T EAT THE EXAMPLE!

AREA OF RING $\approx 2\pi r_i \Delta r$

THE WHOLE CIRCLE THEN HAS AN APPROXIMATE AREA OF $\sum 2\pi r_i \Delta r$, AND AS $\Delta r \to 0$, THIS BECOMES

$$\int_0^R 2\pi r\, dr = \pi r^2 \Big|_0^R = \pi R^2$$

MOST OF US HAVE BEEN HEARING THAT A CIRCLE'S AREA IS πr^2 EVER SINCE GRADE SCHOOL. BUT WE HAD TO WAIT FOR CALCULUS TO PROVE IT! ROUND THINGS ARE THAT MUCH MORE DIFFICULT THAN SQUARE THINGS.

I CAN'T WAIT TO TAKE CALCULUS...

HERE'S ANOTHER ROUND THING WE CAN CALCULATE NOW:

Volume of a Sphere: A SPHERE IS ROUND EVERY WHICH WAY! HOW DO WE DEAL WITH THAT?

WELL, THE WAY OF THE INTEGRAL IS TO CUT IT INTO SLICES. LET'S TRY THAT...

EACH SLICE HAS A CURVED EDGE (HARD TO CALCULATE THE VOLUME!), SO WE APPROXIMATE EACH SLICE BY A PLAIN DISK WITH A STRAIGHT SIDE.

NOW ADD THE VOLUMES OF ALL THE DISKS, LET THEIR THICKNESS GO TO ZERO...

UM... WHAT'S THE VOLUME OF A DISK?

SAY THE SPHERE HAS RADIUS R AND ITS CENTER AT THE ORIGIN. ALONG THE x-AXIS, SUBDIVIDE THE INTERVAL $[-R, R]$ BY POINTS $\{x_0, x_1, \ldots, x_i, \ldots, x_n\}$ INTO MANY SHORT INTERVALS OF LENGTH Δx. THEN A CROSS-SECTION THROUGH THE POINT x_i HAS RADIUS $\sqrt{R^2 - x_i^2}$, BY THE PYTHAGOREAN THEOREM.

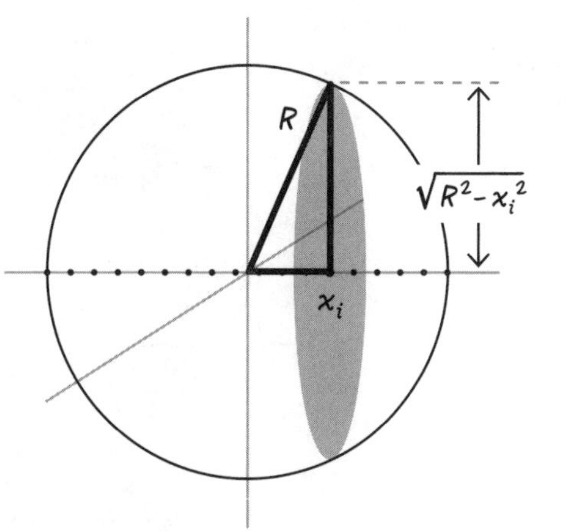
R
$\sqrt{R^2 - x_i^2}$
x_i

A DISK'S VOLUME IS THE PRODUCT OF ITS HEIGHT TIMES THE AREA OF ITS BASE. HERE THE BASE HAS AREA

$$\pi(\sqrt{R^2 - x_i^2})^2 = \pi(R^2 - x_i^2)$$

ITS HEIGHT IS Δx, SO THE VOLUME IS

$$\text{BASE} \cdot \text{HEIGHT} = (\pi R^2 - \pi x_i^2)\Delta x$$

ADDING TOGETHER THE VOLUMES OF **ALL** DISKS GIVES

$$\sum_{i=1}^{n} (\pi R^2 - \pi x_i^2)\Delta x$$

LETTING $\Delta x \to 0$ PRODUCES AN INTEGRAL!

$$V = \int_{-R}^{R} \pi R^2 - \pi x^2 \, dx$$

$$= \pi R^2 x \Big|_{-R}^{R} - \frac{1}{3}\pi x^3 \Big|_{-R}^{R}$$

$$= 2\pi R^3 - \frac{2}{3}\pi R^3 = \frac{4}{3}\pi R^3$$

SOMETHING ELSE YOU "KNEW" ALREADY!
O.K., I ADMIT IT... IT'S JUST SOMETHING I HEARD FROM TEACHERS IN SCHOOL!

WHAT WORKS FOR THE SPHERE ALSO WORKS FOR MANY OTHER VOLUMES THAT CAN BE APPROXIMATED BY A STACK OF DISKS, ESPECIALLY **SOLIDS OF REVOLUTION** MADE BY SPINNING A CURVE AROUND AN AXIS.

Cone: A CONE IS MADE BY ROTATING THE LINE $y = ax$ AROUND THE y-AXIS. IF THE HEIGHT OF THE CONE IS H, THEN THE RADIUS OF THE BASE IS H/a. WE MAKE SLICES PERPENDICULAR TO THE y-AXIS AND INTEGRATE WITH RESPECT TO y. AT A POINT y_i, THE CROSS-SECTION HAS RADIUS y_i/a.

H/a

$y = ax$

H y_i y_i/a

THEN THE CIRCLE'S AREA IS $\pi(y_i/a)^2$ AND A THIN CYLINDER OF HEIGHT dy HAS VOLUME

$$\pi \frac{{y_i}^2}{a^2}\, dy$$

INTEGRATING THE SLICES GIVES THE CONE'S VOLUME:

$$V = \int_0^H \pi \frac{y^2}{a^2}\, dy = \frac{1}{3} \frac{\pi}{a^2} y^3 \Big|_0^H$$

$$= \frac{1}{3} \pi \frac{H^3}{a^2}$$

ANOTHER FORMULA I THOUGHT I KNEW...

THE CONE'S BASE HAS RADIUS H/a, SO ITS AREA IS $\frac{1}{3}\pi(H/a)^2$. THE VOLUME IS THEREFORE ONE-THIRD THE AREA OF THE BASE TIMES THE HEIGHT.

Paraboloid: THIS SOLID IS GENERATED BY ROTATING THE PARABOLA $y = ax^2$ AROUND THE y-AXIS. WHAT IS ITS VOLUME UP TO A HEIGHT H?

LET'S WHIZ THROUGH THIS ONE: A CROSS-SECTION THROUGH y HAS RADIUS $\sqrt{(y/a)}$ AND AN AREA OF $(\pi y/a)$. SO A THIN SLICE OF DEPTH Δy HAS VOLUME $(\pi y \Delta y/a)$, AND THE VOLUME OF THE PARABOLOID IS

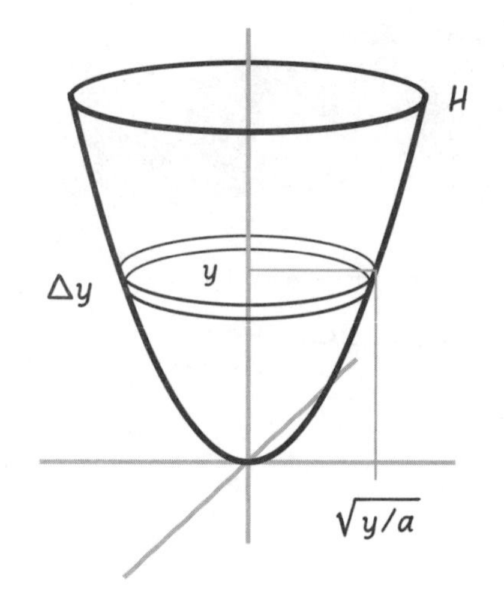

H
y
Δy
√y/a

$$V = \int_0^H \frac{\pi y}{a}\,dy = \frac{1}{2}\frac{\pi y^2}{a}\bigg|_0^H$$

$$= \frac{1}{2}\frac{\pi H^2}{a}$$

CAN YOU SHOW THAT THIS IS HALF THE AREA OF THE BASE TIMES THE HEIGHT? WHAT IS THE RADIUS OF THE BASE?

SOMETIMES IT IS MORE CONVENIENT TO FIND THESE ROTATIONALLY SYMMETRICAL VOLUMES BY INTEGRATING THE VOLUMES OF THIN CYLINDRICAL SHELLS INSTEAD OF DISKS. FOR INSTANCE, IN THE PREVIOUS EXAMPLE, WE COULD HAVE—BUT WAIT... WHAT'S THIS—?

BA-BOOF!
THE GLUE FACTORY HAS EXPLODED!!

ALL RIGHT, LET'S DO A DIFFERENT EXAMPLE...

HOW DEEP IS IT?
DUNNO... I CAN'T READ THE DIPSTICK...

Example

AN EXPLOSION AT A GLUE FACTORY BURIES THE SURROUNDING COUNTRYSIDE IN A LAYER OF VISCOUS YELLOW GLOP IN A SYMMETRICAL, CIRCULAR MOUND. MEASUREMENTS REVEAL THAT THE DEPTH OF THE GLUE DIMINISHES WITH DISTANCE FROM THE CENTER. IN FACT, $D(r)$, THE DEPTH IN METERS AT A DISTANCE OF r KILOMETERS, TURNS OUT TO FOLLOW A FORMULA:

$$D(r) = 2e^{-3r^2} \text{ METERS}$$

WHAT IS THE TOTAL VOLUME OF GLUE, IN CUBIC METERS, WITHIN A RADIUS OF 5 KILOMETERS?

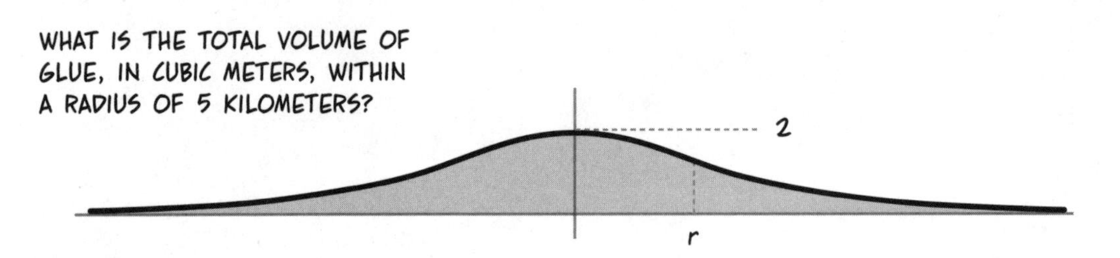

2
r

THE GLUE FORMS A VOLUME OF REVOLUTION, BUT INSTEAD OF INTEGRATING OVER y, FROM TOP TO BOTTOM, LET'S INTEGRATE **OUTWARD,** WITH RESPECT TO r.

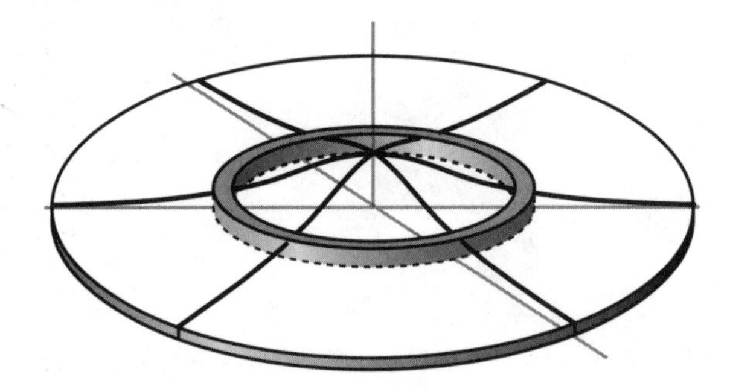

BETWEEN TWO NEARBY DISTANCES r AND $r + dr$, THE DEPTH OF GLUE IS NEARLY CONSTANT, NAMELY $2e^{-3r^2}$ METERS. THUS THE NARROW RING OF GLUE BETWEEN THESE TWO DISTANCES HAS APPROXIMATE VOLUME

$$dV \approx 2\pi r \cdot (2e^{-3r^2}) \cdot 10^6 \, dr \text{ cubic meters.*}$$

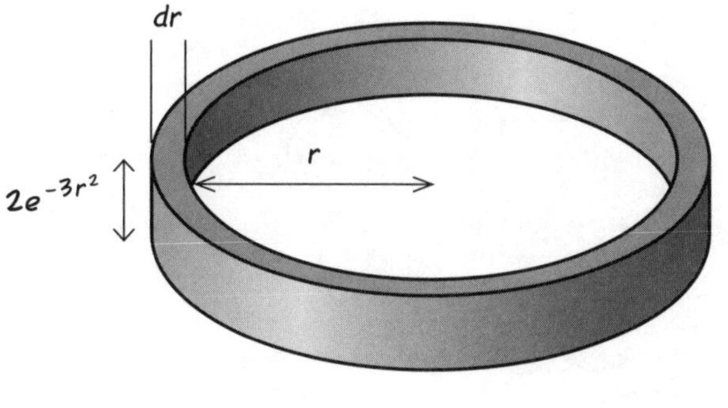

dr

r

$2e^{-3r^2}$

(AS ON P. 209, THINK OF THE RING AS A THIN, FLAT NOODLE THAT CAN BE UNCURLED TO FORM A RECTANGULAR BLOCK.)

THE VOLUME OUT TO 5 KM IS THIS INTEGRAL:

$$V(5) = 10^6 \int_0^5 4\pi r e^{-3r^2} \, dr$$

$$= (4\pi)10^6 \int_0^5 r e^{-3r^2} \, dr$$

WE FIND THIS BY A STRAIGHTFORWARD SUBSTITUTION

$$u = -3r^2, \quad du = -6r\,dr$$

$$u(0) = 0, \quad u(5) = -75$$

THEN

$$4\pi 10^6 \int_0^5 r e^{-3r^2} \, dr = 4\pi 10^6 \int_0^{-75} -(1/6)e^u \, du$$

$$= -(2/3)10^6 \pi e^u \Big|_0^{-75}$$

$$= (2/3)10^6 \pi (e^0 - e^{-75})$$

= APPROXIMATELY **2.1 MILLION** CUBIC METERS OF GLUE.

ALL RIGHT, **"EXCALIBUR,"** ONWARD!

*10^6 IS A CONVERSION FACTOR, NECESSARY BECAUSE WE MEASURED BOTH r AND Δr IN KILOMETERS, AND THE DEPTH IN METERS. 1 KM = 10^3 M.

Improper Integrals

WE JUST CALCULATED HOW MUCH GLUE LANDED WITHIN A RADIUS OF 5 KM OF GROUND ZERO... BUT WHAT IF WE WANTED TO KNOW THE **TOTAL** VOLUME OF GLUE OUT THERE?

THE GLUE TO INFINITY?

WE'D LIKE TO WRITE THAT AS AN INTEGRAL WITH AN **INFINITE LIMIT:**

$$10^6 \int_0^{\infty} 4\pi r e^{-3r^2}\, dr$$

A GUY CAN DREAM, CAN'T HE?

(WE IMAGINE THAT THIS PARTICULAR GLUE FACTORY SITS ON AN INFINITE, FLAT PLANE, NOT THE CURVED SURFACE OF THE EARTH.)

AN INTEGRAL INVOLVING INFINITY IS CALLED AN **IMPROPER** INTEGRAL, AN UNFORTUNATE NAME, SINCE IT'S JUST AS GOOD AS ANY OTHER INTEGRAL, REALLY.

I FEEL SORRY FOR IT...
TSK
TSK
TSK
TSK

AFTER THE GLUE BLAST, THE VOLUME OF GLUE (IN CUBIC METERS) WITHIN A RADIUS OF R KM WAS

$$V(R) = 10^6 \int_0^R 4\pi r e^{-3r^2}\, dr$$

$$= -(2/3)\pi 10^6\, e^{-3r^2} \Big|_0^R$$

$$= (2/3)\pi 10^6 (1 - e^{-3R^2})$$

AS $R \to \infty$, THE SECOND TERM GOES TO ZERO, SO

$$\lim_{R\to\infty} V(R) = (2/3)\pi 10^6$$

A FINITE AMOUNT OF STUFF SPREAD OVER AN INFINITE REGION!
SOLVENT

WE SAY AN IMPROPER INTEGRAL **CONVERGES** WHEN THIS LIMIT IS FINITE:

$$\lim_{x\to\infty} \int_a^x f(t)\, dt$$

IN THAT CASE WE **DEFINE** THE IMPROPER INTEGRAL TO BE THIS LIMIT:

$$\int_a^\infty f(t)\, dt = \lim_{x\to\infty} \int_a^x f(t)\, dt$$

AS WE JUST SAW, THE INTEGRAL FROM THE GLUE FACTORY EXAMPLE CONVERGES.

$$10^6 \int_0^\infty 4\pi r e^{-3r^2}\, dr =$$

$(\frac{2}{3}\pi)\, 10^6$ **CUBIC METERS.**

AT LEAST SOME GOOD CAME OF THIS HORRIBLE TRAGEDY: BETTER UNDERSTANDING...
LET'S HOPE IT STICKS...
SQUINCH
SQUINCH
SQUINCH

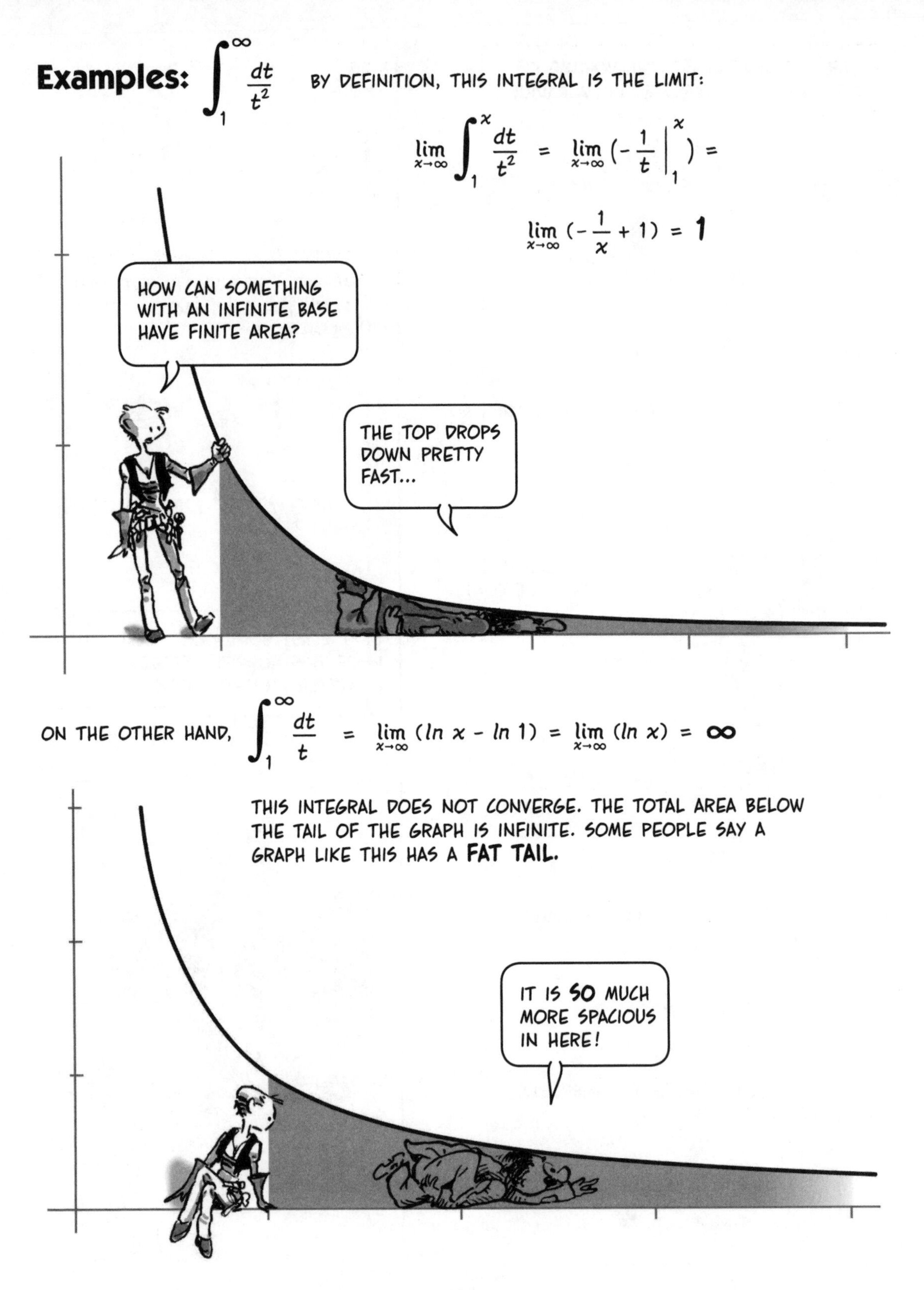
Examples: $\int_1^{\infty} \frac{dt}{t^2}$ BY DEFINITION, THIS INTEGRAL IS THE LIMIT:
$\lim_{x \to \infty} \int_1^{x} \frac{dt}{t^2} = \lim_{x \to \infty} \left(-\frac{1}{t}\Big|_1^x\right) =$
$\lim_{x \to \infty} \left(-\frac{1}{x} + 1\right) = 1$
HOW CAN SOMETHING WITH AN INFINITE BASE HAVE FINITE AREA?
THE TOP DROPS DOWN PRETTY FAST...
ON THE OTHER HAND, $\int_1^{\infty} \frac{dt}{t} = \lim_{x \to \infty} (\ln x - \ln 1) = \lim_{x \to \infty} (\ln x) = \infty$
THIS INTEGRAL DOES NOT CONVERGE. THE TOTAL AREA BELOW THE TAIL OF THE GRAPH IS INFINITE. SOME PEOPLE SAY A GRAPH LIKE THIS HAS A **FAT TAIL.**
IT IS **SO** MUCH MORE SPACIOUS IN HERE!

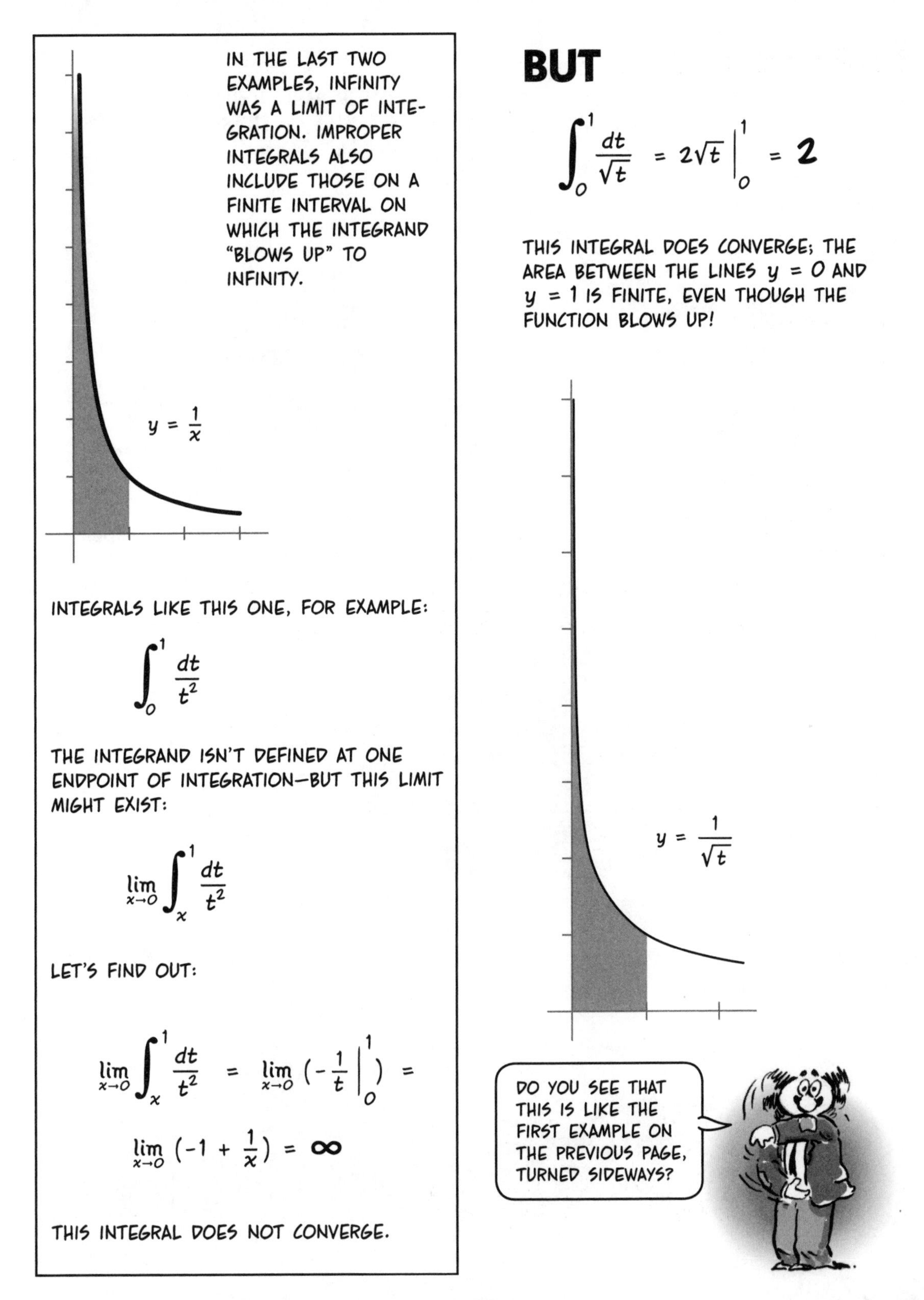
IN THE LAST TWO EXAMPLES, INFINITY WAS A LIMIT OF INTEGRATION. IMPROPER INTEGRALS ALSO INCLUDE THOSE ON A FINITE INTERVAL ON WHICH THE INTEGRAND "BLOWS UP" TO INFINITY.
$y = \frac{1}{x}$
INTEGRALS LIKE THIS ONE, FOR EXAMPLE:
$\int_0^1 \frac{dt}{t^2}$
THE INTEGRAND ISN'T DEFINED AT ONE ENDPOINT OF INTEGRATION—BUT THIS LIMIT MIGHT EXIST:
$\lim_{x \to 0} \int_x^1 \frac{dt}{t^2}$
LET'S FIND OUT:
$\lim_{x \to 0} \int_x^1 \frac{dt}{t^2} = \lim_{x \to 0} \left(-\frac{1}{t} \Big|_0^1 \right) =$
$\lim_{x \to 0} \left(-1 + \frac{1}{x} \right) = \infty$
THIS INTEGRAL DOES NOT CONVERGE.
BUT
$\int_0^1 \frac{dt}{\sqrt{t}} = 2\sqrt{t} \Big|_0^1 = 2$
THIS INTEGRAL DOES CONVERGE; THE AREA BETWEEN THE LINES $y = 0$ AND $y = 1$ IS FINITE, EVEN THOUGH THE FUNCTION BLOWS UP!
$y = \frac{1}{\sqrt{t}}$
DO YOU SEE THAT THIS IS LIKE THE FIRST EXAMPLE ON THE PREVIOUS PAGE, TURNED SIDEWAYS?

Density

AS WE ALL KNOW, A FEATHER-FILLED PILLOW, EVEN A BIG ONE, DOESN'T WEIGH MUCH.

ON THE OTHER HAND, A CUBIC METER OF LEAD HAS A MASS OF **11,340 KILOGRAMS,** MORE THAN **TEN TONS** (!).

LEAD AND FEATHERS HAVE DIFFERENT **DENSITIES.** A GIVEN VOLUME OF LEAD HAS A LARGER MASS THAN A GIVEN VOLUME OF FEATHERS (OR WATER, OR COPPER, BUT NOT GOLD! GOLD IS EVEN DENSER THAN LEAD).

AREN'T YOU GRATEFUL IT ISN'T GOLD?
PAINFULLY SO.

B.C. (BEFORE CALCULUS), WE WOULD DEFINE THE DENSITY OF AN OBJECT AS ITS MASS DIVIDED BY ITS VOLUME.

DENSITY =

MASS
VOLUME

BUT NOW WE'RE MORE SOPHISTICATED THAN THAT! NOW WE CAN IMAGINE MATERIALS WITH **VARIABLE DENSITY:** STUFF WHERE THE MATERIAL IS MORE OR LESS DENSE, DEPENDING ON WHERE YOU SAMPLE IT...
LIKE A LEAD-COATED FEATHER?
SURE, WHY NOT?

THE **ATMOSPHERE,** FOR EXAMPLE... AIR THINS OUT AS ALTITUDE INCREASES... THE DENSITY AT SEA LEVEL IS FAR GREATER THAN AT 5,000 METERS ABOVE...

GASP GASP GASP
AH!
AND THIS MEANS WHAT, EXACTLY?

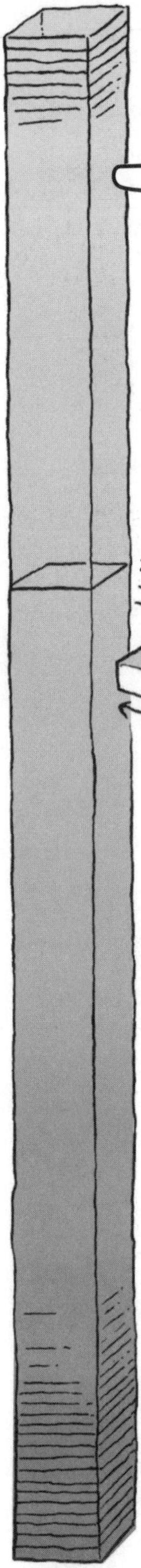

HERE IS A SQUARE COLUMN OF AIR, ONE METER ON A SIDE.

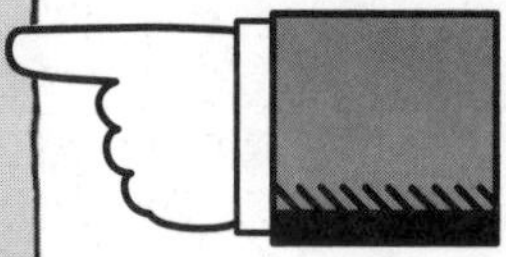

CALL $M(x)$ THE TOTAL MASS OF AIR FROM THE GROUND UP TO x. THEN A SLICE OF DEPTH dx HAS MASS dM AND VOLUME $(1)\cdot(1)\cdot dx = dx\ \text{M}^3$.

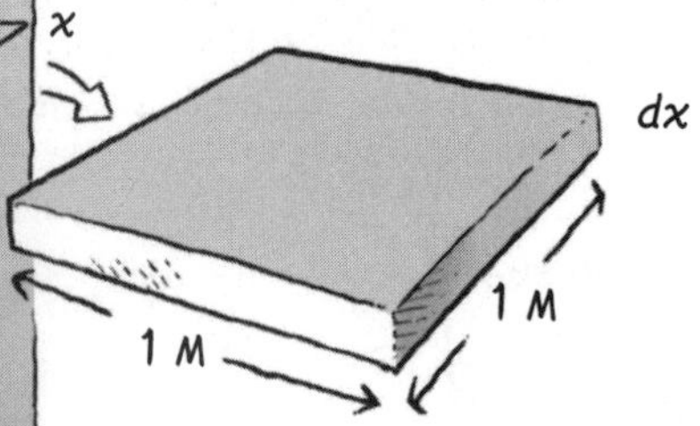

x

dx

1 M

1 M

IF THE SLICE IS THIN, THE AIR IN IT HAS UNIFORM DENSITY, AND

$$D(x) = \frac{dM}{dx}$$

SO

$$M = \int D(x)\,dx$$

THE TOTAL MASS IS THE **INTEGRAL OF THE DENSITY.** THIS AMOUNTS TO ADDING UP THE MASSES OF ALL THESE "PIZZA BOXES" OF AIR.

MEASUREMENTS OF AIR SAMPLES SHOW THAT ATMOSPHERIC DENSITY $D(x)$ AT HEIGHT x METERS IS

$$D(x) = 1.28e^{-0.000124x}\ \text{KG/M}^3$$

SO THE TOTAL MASS OF A 1-METER SQUARE COLUMN OF AIR 10,000 METERS TALL IS

$$M = \int_0^{10{,}000} 1.28e^{-0.000124x}\,dx =$$

$$(1.28)\left(\frac{-1}{0.000124}\right)e^{-0.000124x}\Big|_0^{10{,}000}$$

$$\approx -2980 + 10{,}320$$

$= \mathbf{7{,}340}$ KILOGRAMS OF AIR

WHERE'S THE PIZZA?

Other Dense Things

THE SAME APPROACH WORKS WITH **POPULATION DENSITY.** IT VARIES FROM PLACE TO PLACE.

SUPPOSE **EASY STREET** RUNS FROM ONE SIDE OF TOWN TO THE OTHER. WE CAN COUNT THE NUMBER OF RESIDENTS IN EACH BLOCK TO GET A POPULATION DENSITY IN TERMS OF **PEOPLE PER BLOCK.** BECAUSE OF THE HIGH-RISES IN THE CENTER AND THE CROWDED SLUMS AT THE OUTSKIRTS, THIS DENSITY VARIES. (FOR SIMPLICITY, LET'S ASSUME THERE ARE NO CROSS STREETS WHERE THE DENSITY WOULD BE ZERO.)

WE COULD MEASURE THE DENSITY ALONG A SHORT SLICE OF EASY STREET... AND A SHORTER ONE... AND SHORTER... UNTIL WE'RE THINKING OF POPULATION DENSITY AS VARYING **CONTINUOUSLY** ALONG THE STREET.

HEY, IT'S MY OLD FRIEND SLICEMAN! SHAKE, BUDDY!
WITH WHAT?

THE POPULATION DENSITY FUNCTION OPERATES IN THE SAME WAY AS MASS DENSITY. IF $P(x)$ IS THE NUMBER OF PEOPLE LIVING BETWEEN $-\infty$ AND x (I.E., ANYWHERE WEST OF x), THEN A SLICE AT POINT x OF WIDTH dx CONTAINS dP PEOPLE, AND

$$D(x) = \frac{dP}{dx}$$

SO

$$P = \int D(x)\,dx$$

YOU CENSUS TAKERS HAVE GONE TOO FAR THIS TIME...

IF a AND b ARE TWO STREET ADDRESSES, THEN $\int_a^b D(x)\,dx = P(b) - P(a)$ IS THE NUMBER OF PEOPLE LIVING BETWEEN POINTS a AND b.

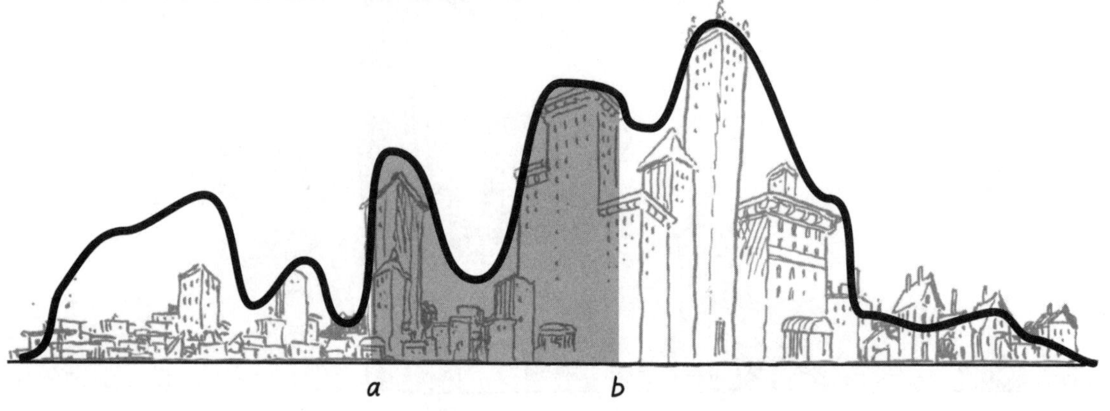
a
b

IN PARTICULAR, INTEGRATING FROM (BEYOND) ONE END OF THE STREET TO THE OTHER,

$$\int_{-\infty}^{\infty} D(x)\,dx = \text{THE TOTAL POPULATION OF EASY STREET.}$$

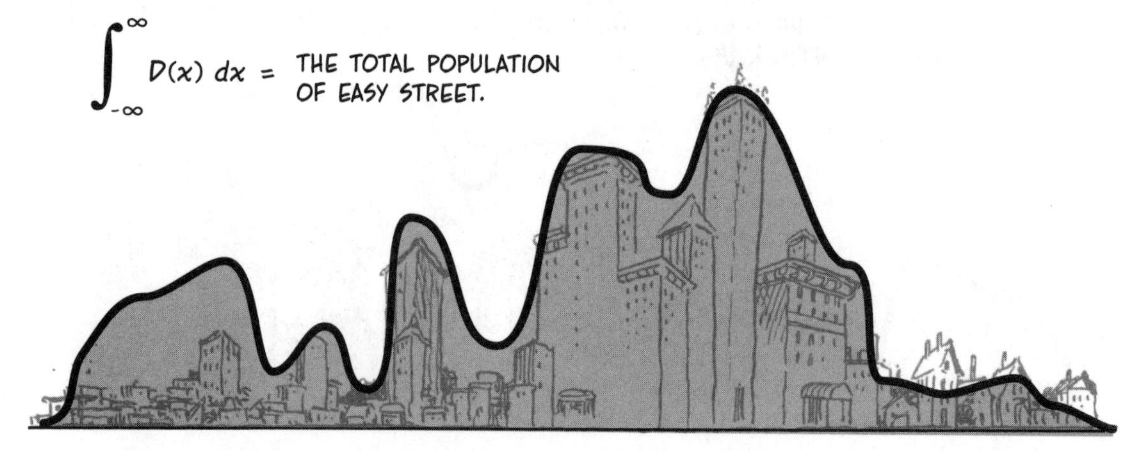

IF n PEOPLE LIVE ON A PART OF EASY STREET, THEN n/N IS THE **FRACTION** THEY MAKE UP OF THE TOTAL POPULATION N. THIS MEANS THAT THE FUNCTION $p(x) = D(x)/N$ HAS THESE PROPERTIES:

$$\int_{-\infty}^{\infty} p(x)\,dx = 1$$

$$\int_{a}^{b} p(x)\,dx = \left\{ \begin{array}{l} \text{FRACTION OF THE POPU-} \\ \text{LATION WITH ADDRESSES} \\ \text{BETWEEN } a \text{ AND } b. \end{array} \right.$$

THAT LATTER NUMBER IS ALSO INTERPRETED AS THE **PROBABILITY** THAT A RANDOMLY CHOSEN PERSON LIVES BETWEEN a AND b.

HI THERE! WHAT'S YOUR ADDRESS?

A **PROBABILITY DENSITY** (OR **PROBABILITY DISTRIBUTION**) IS ANY NON-NEGATIVE FUNCTION p WITH

$$\int_{-\infty}^{\infty} p(x)\,dx = 1$$

EVERY "RANDOM VARIABLE"—MEANING A RANDOM SYSTEM WITH NUMERICAL OUTCOMES, SUCH AS BLINDLY CHOOSING A RESIDENT OF EASY STREET AND ASKING FOR AN ADDRESS—HAS A PROBABILITY DENSITY p. THE ENTIRE FIELD OF STATISTICS IS BASED ON PROBABILITY DENSITIES.

HOW EXCITING! WILL YOU EXPLAIN THIS MORE?
PROBABLY NOT...

More Uses for Integrals (QUICKIE VERSION):
IN PHYSICS, WHEN A CONSTANT FORCE F PUSHES A BODY A DISTANCE d, THE WORK DONE IS THE PRODUCT
WORK = FORCE × DISTANCE
0
d
BUT WHAT IF THE FORCE VARIES WITH POSITION?
BE RIGHT BACK. I'M—ER—FORCED TO TAKE A SHORT BREAK...
0
d

YOU GUESSED IT: IF $F(x)$ IS THE FORCE EXERTED AT POINT x, THEN $\int_a^b F(x)\,dx$ IS THE WORK DONE BETWEEN a AND b.

OVER A SHORT INTERVAL dx, THE FORCE IS NEARLY CONSTANT, THE WORK ON THAT INTERVAL IS F(x)dx, ETC., ETC., ETC....
BY THE WAY, WHY ARE THESE "BODIES" IN PHYSICS ALWAYS BLOCKS OR BALLS?
DUNNO... BUT THIS ONE SEEMS TO BE A BLOCK OF CHEESE...

SPEAKING OF FORCE, WATER EXERTS ONE. AT ANY DEPTH, THE WEIGHT OF WATER ABOVE PUSHES IN ALL DIRECTIONS... THE DEEPER YOU GO, THE HARDER IT PUSHES BECAUSE OF THE INCREASED WEIGHT ABOVE.

WATER PRESSURE IS THE **FORCE PER UNIT AREA,** MEASURED IN UNITS CALLED KILOPASCALS (*kPa*). (ONE KILOPASCAL IS 1000 NEWTONS PER SQUARE METER.) AT DEPTH x, THE PRESSURE IS GIVEN BY

$$P(x) = 9.8x \text{ kPa}$$

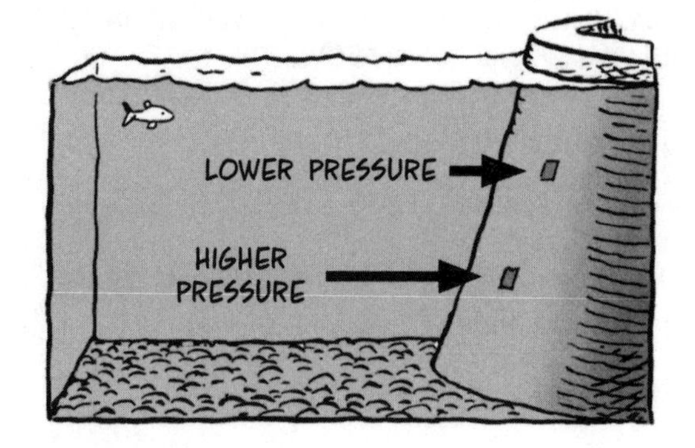
LOWER PRESSURE
HIGHER PRESSURE

SUPPOSE A DAM HOLDS BACK A MASS OF WATER. AT ANY DEPTH x, THE PRESSURE IS CONSTANT ALONG A THIN HORIZONTAL STRIP OF THICKNESS dx. THE FORCE ON THE SLICE IS THE PRESSURE TIMES THE AREA OF THE SLICE. THAT AREA IS $W(x)\,dx$, WHERE $W(x)$ IS THE LENGTH OF CURVE OF THE DAM AT THAT DEPTH. IF $F(x)$ IS THE TOTAL FORCE FROM 0 TO x, THEN

$$dF = 9.8x\,W(x)\,dx$$

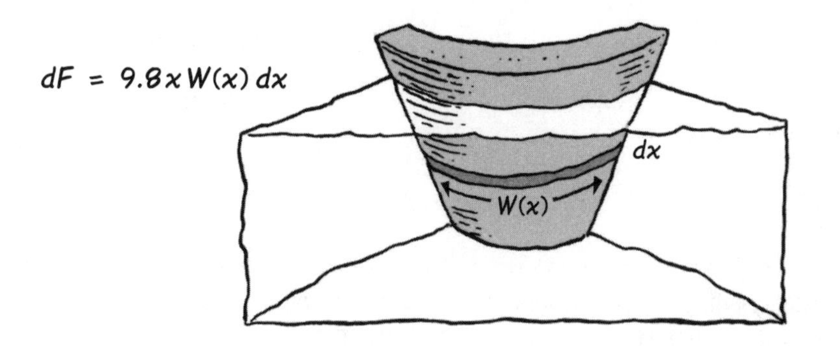
dx
W(x)

IF A DAM HOLDS BACK WATER TO A DEPTH OF D METERS, THEN THE TOTAL FORCE ON THE DAM IS

$$\int_0^D 9.8x\,W(x)\,dx \text{ KILONEWTONS}$$

INTEGRATION ENABLES ENGINEERS TO ASSESS STRESSES ON DAMS, BRIDGES, AND OTHER STRUCTURES.

AND THEY'RE USUALLY PRETTY GOOD AT IT, TOO!

…EM ON PAGE 124, WE GAVE A FORMULA FOR THE … WATER IN A HEMISPHERICAL BOWL. DERIVE THAT …. IF THE WATER IS D UNITS DEEP, BEGIN BY FINDING …OLUME OF THE BOWL **ABOVE** THE WATER, OR

$$\int_0^{R-D} \pi(R^2 - y^2)\, dy$$

SUBTRACT THIS FROM $\frac{2}{3}\pi R^3$, THE VOLUME OF THE HEMISPHERE, TO FIND THE VOLUME OF THE WATER.

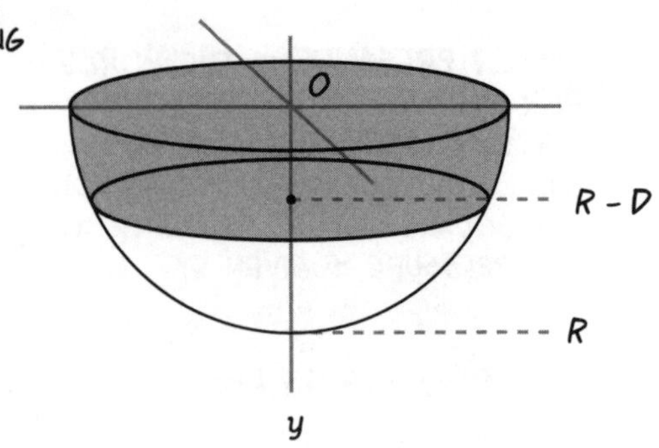
O
R - D
R
y

2. FIND $\int_0^1 \ln x \, dx$

HINT: TO FIND $\lim_{x \to 0} x \ln x$, LET $y = 1/x$ AND USE L'HÔPITAL'S RULE TO FIND

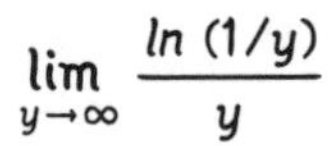

$$\lim_{y \to \infty} \frac{\ln (1/y)}{y}$$

3. CALCULATE THE VOLUME OF THE PARABOLOID ON P. 213 BY USING CONCENTRIC CYLINDERS INSTEAD OF DISKS.

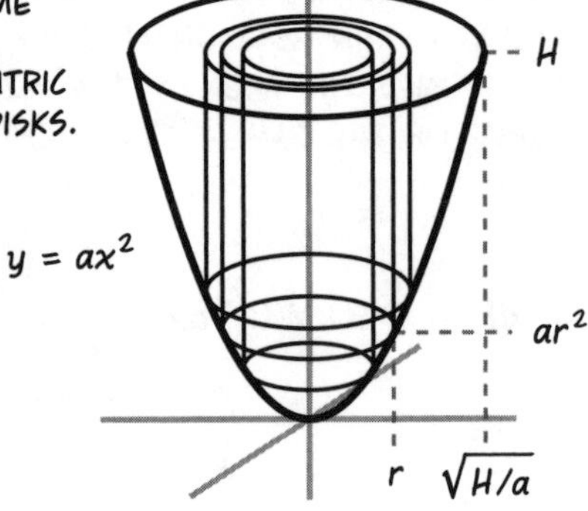
H
y = ax²
ar²
r
√H/a

4. ROTATE THE CURVE $y = 1/x$ AROUND THE x-AXIS TO MAKE A SORT OF "INFINITE TRUMPET." WHAT IS ITS VOLUME TO THE RIGHT OF $x = 1$?

5. AN IDIOT ENGINEER DESIGNS A PERFECTLY FLAT, VERTICAL, TRAPEZOIDAL DAM (CURVED IS MUCH STRONGER!) 300 METERS ACROSS AT THE TOP, 200 METERS AT THE BOTTOM, AND 200 METERS HIGH. IF IT HOLDS BACK A BODY OF WATER 175 METERS DEEP, WHAT IS THE TOTAL FORCE OF THE WATER ON THE DAM?

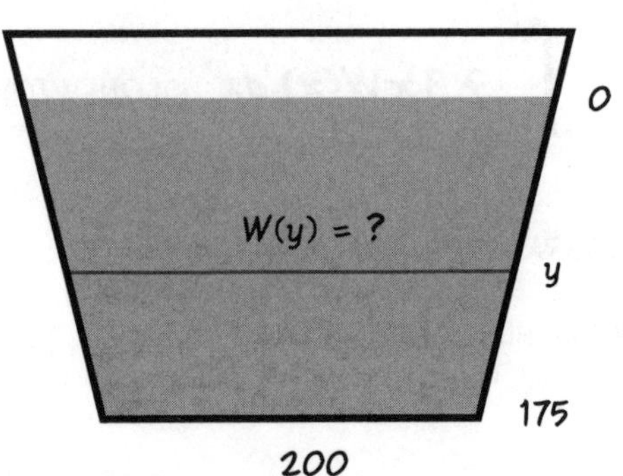
300
O
W(y) = ?
y
175
200

Chapter 14
What's Next?

READER, THIS BOOK IS ONLY THE BEGINNING... THERE'S SO MUCH MORE YOU CAN DO WITH CALCULUS. IT'S A POWERFUL TOOL, USED IN ALL THE SOCIAL, BIOLOGICAL, AND PHYSICAL SCIENCES, ENGINEERING, ECONOMICS, AND STATISTICS, AND ITS IDEAS HAVE BEEN EXTENDED BY MANY GENERATIONS OF MATHEMATICIANS SINCE NEWTON AND LEIBNIZ.

ENLIGHTENMENT
of a certain kind
Caution: Road
Work Ahead

HERE ARE A FEW MORE ADVANCED TOPICS YOU MAY EXPECT TO ENCOUNTER ALONG THE WAY:

Differential Equations

BESIDES DISCOVERING CALCULUS, NEWTON ALSO LAID DOWN A FAMOUS PHYSICAL LAW RELATING FORCE AND VELOCITY:

$$F = \frac{d}{dt}(mv)$$

ANY EQUATION THAT CONTAINS DERIVATIVES, AS THIS ONE DOES, IS CALLED A **DIFFERENTIAL EQUATION.**

ANOTHER DIFFERENTIAL EQUATION IS HOOKE'S LAW, OR THE SPRING EQUATION. IF A MASS m IS DISPLACED x UNITS FROM THE SPRING'S NEUTRAL POSITION AND RELEASED, THEN AT ANY TIME ITS ACCELERATION IS PROPORTIONAL TO ITS DISPLACEMENT:

$$x''(t) = \frac{k}{m}x(t) \quad \text{OR, GIVEN NEWTON'S FIRST LAW, } F = kx$$

(k IS A CONSTANT DEPENDING ON THE STIFFNESS OF THE SPRING.)

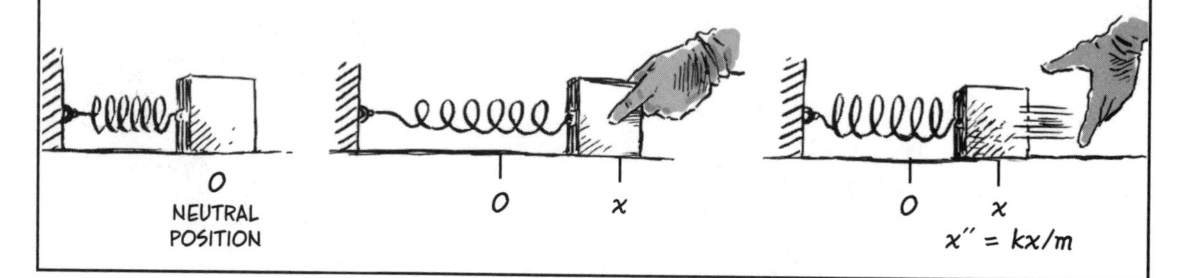
O
NEUTRAL POSITION
O
x
O
x
x'' = kx/m

THE UNIVERSE IS DESCRIBED BY DIFFERENTIAL EQUATIONS, AND SOLVING THEM IS JOB #1 IN SCIENCE.

O.K., I THINK I'VE OPTIMIZED MY CLASS SCHEDULE...

Many Variables

THIS DESCRIBES FUNCTIONS THAT VARY OVER REGIONS OF SPACE, INSTEAD OF JUST ALONG THE x-AXIS. SINCE THE SPACE WE LIVE IN HAS AT LEAST THREE DIMENSIONS, THIS IS OBVIOUSLY AN IMPORTANT SUBJECT!

WELL, AT LEAST **THEY** LIVE IN 3 DIMENSIONS...
OH, YEAH... **THEM...**

Sequences and Series

HOW DOES YOUR POCKET CALCULATOR DO SINES AND COSINES? WOULD IT SURPRISE YOU TO KNOW THAT

$$\sin x \approx x - \frac{x^3}{6} + \frac{x^5}{120} - \frac{x^7}{5040} + \ldots$$

OH? IT WOULDN'T? WELL, NEVER MIND THEN...

Path & Surface Integrals

THESE ARE WAYS TO INTEGRATE ALONG CURVES AND ACROSS SURFACES, RATHER THAN BORING OLD STRAIGHT LINES.

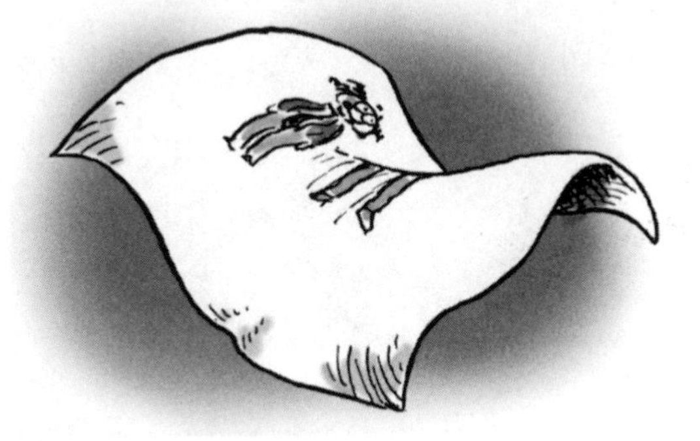

Complex Variables

WHEN WE DO CALCULUS WITH THE MISLEADINGLY NAMED "IMAGINARY" NUMBER $i = \sqrt{-1}$, MIND-BENDING THINGS HAPPEN!

NOT ONLY ARE COMPLEX VARIABLES THE "RIGHT" WAY TO DESCRIBE ELECTRICITY, QUANTUM MECHANICS, AND OTHER BRANCHES OF PHYSICS, BUT THEY REVEAL DEEP MATHEMATICAL RELATIONSHIPS, SUCH AS THE ASTONISHING EQUATION:

$e^{i\pi} + 1 = 0$

POSSIBLY THE MOST IMPRESSIVE THING ABOUT ADVANCED CALCULUS, THOUGH, IS THAT ALL OF IT STILL DEPENDS ON TWO BASIC IDEAS, THE DERIVATIVE AND THE INTEGRAL, INVENTED BY TWO GUYS MORE THAN 300 YEARS AGO. HERE'S TO 'EM, SAY I!

PLAGIARIST!

INDEX

HMMM... DOES $\frac{d}{dx}$ COME BEFORE $\int$?

DON'T STOP!
KEEP GOING...

KEEP GOING...

!

About the Author

LARRY GONICK HAS DONE A WHOLE LOT OF BOOKS THAT USE CARTOONS TO EXPLAIN BIG SUBJECTS. HE HAS TWO DEGREES IN MATH FROM HARVARD AND TAUGHT CALCULUS THERE BEFORE DROPPING OUT TO PROTEST THE SHORTAGE OF CARTOONS IN THE CURRICULUM.

∞
COMICS!